An Integrated Approach to Mechanical Engineering
Volume I

An Integrated Approach to Mechanical Engineering Volume I

Edited by **Bilroy Lockhart**

NY RESEARCH PRESS

New York

Published by NY Research Press,
23 West, 55th Street, Suite 816,
New York, NY 10019, USA
www.nyresearchpress.com

An Integrated Approach to Mechanical Engineering: Volume I
Edited by Bilroy Lockhart

International Standard Book Number: 978-1-63238-046-3 (Hardback)

Printed in China.

Contents

Preface

A widely used stream of engineering; mechanical engineering is primarily based upon the principles of design, study, development and construction of all tangible devices and systems. Even though this discipline of engineering can be traced back several thousand years; it emerged as a dominant field during the industrial revolution. Since then, mechanical engineering has undergone dynamic advancements in technology. We do know mechanical as a distinct stream of engineering; yet it overlaps and facilitates other streams of engineering such as aerospace engineering, civil engineering, electrical engineering, manufacturing engineering and many others.

Mechanical engineering is one of the most important engineering disciplines that present tangible solutions for all the problems, technical work minimization, prosperity and the quality of life. Rapidly increasing demands have been increasing its importance more and more. The many areas within the scope of mechanical engineering include transportation, power generation, energy conversion, machine design, manufacturing and automation, the control of system. The purpose of the book is to present to the engineers in industrial areas and to the academicians, the state-of-the-art information on the most important topics of modern mechanical engineering. The aim of contributors of this book was to spread out new methods and techniques developed by scientists to be used in all spheres of mechanical engineering. Although Mechanical engineering is one of the oldest divisions of engineering, the research in this field has been constantly growing.

I would like to pay my regards to all the researchers and contributors of this book. I would also like to thank the publisher for the continuous support and trust in me. Lastly, I wish to thank my friends and family for supporting me throughout my life.

Editor

1

Theoretical Analysis of the Effect of a Bottom Reflector on a Vertical Multiple-Effect Diffusion Solar Still Coupled with a Basin-Type Still

Hiroshi Tanaka

Department of Mechanical Engineering, Kurume National College of Technology, Komorino, Kurume, Fukuoka 830-8555, Japan

Correspondence should be addressed to Hiroshi Tanaka; tanakad@kurume-nct.ac.jp

Academic Editor: Jamel Orfi

The effect of a flat plate bottom reflector on a vertical multiple-effect diffusion solar still coupled with a basin-type still is analyzed theoretically at 30°N latitude. The still has a right-angled triangular cross-section consisting of a horizontal basin liner, a sloping double glass cover, and the vertical multiple-effect diffusion still (multiple-effect section) at the vertical rear wall. A geometrical model was constructed to calculate the amount of solar radiation reflected by the reflector and absorbed on both the basin liner and the first partition of the multiple-effect section. The reflected radiation from the reflector can be a thermal input to the multiple-effect section both in cases where the reflected radiation is absorbed on the basin liner and on the first partition. It is predicted that the overall daily productivity of the still can be increased about 16%, 13%, and 29% on the spring and autumn equinoxes, the winter solstice, and the summer solstice, respectively, by adding the external reflector and setting the reflector inclination to the proper angle according to seasons.

1. Introduction

A solar still can be used to get sanitary fresh water from saline water using only solar energy. Various types of solar stills have been presented and studied as reviewed by Tiwari et al. [1], Kaushal and Varun [2], Sampathkumar et al. [3], and Kabeel and El-Agouz [4]. Among them, multiple-effect diffusion solar stills, which consist of a series of closely spaced parallel partitions in contact with saline-soaked wicks, have great potential because of their high productivity and simplicity [5–23].

In the multiple-effect diffusion still, solar thermal energy absorbed on the first partition is used to evaporate the saline water in the wick of the first partition. Water vapor diffuses through the humid air layer between partitions and condenses on the second partition. The latent heat of condensation causes further evaporation from the wick of the second partition. In this manner, the evaporation and condensation process is repeated, and the solar energy is recycled on all partitions in the still to increase distillate productivity.

In a multiple-effect diffusion still, the size of diffusion gaps between partitions is a very important parameter. Smaller diffusion gaps can increase distillate productivity considerably, mainly because the smaller gaps reduce thermal resistance through each single gap as well as throughout the whole still. Accordingly, vertical multiple-effect diffusion stills have an advantage over inclined ones, since the deformation of partition plates due to gravity can be reduced, and then the diffusion gaps can be smaller for the vertical stills than for the inclined ones [10, 17–20, 22, 23]. If the partition plates are deformed at small diffusion gaps, contact between the evaporating wick and facing condensing surface is liable to be occurred, which leads to the contamination of distilled water with saline water and/or the reabsorption of distilled water by the wick.

Tanaka et al. [17–19] theoretically and experimentally studied a vertical multiple-effect diffusion solar still coupled with a basin-type still shown in Figure 1. The still is a single slope basin-type still with a double glass cover and a vertical multiple-effect diffusion still (or multiple-effect section) which is attached to the vertical rear wall of the basin-type

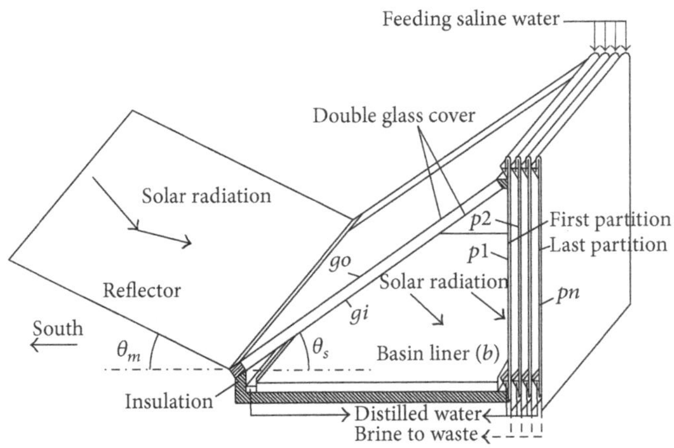

FIGURE 1: Schematic diagram of a vertical multiple-effect diffusion still coupled with a basin-type still using a flat plate bottom reflector.

still. Water in the basin liner is heated by solar energy and evaporated. Water vapor from basin liner condenses on the inner glass cover and the vertical rear wall, which is the first partition of the multiple-effect section. Latent heat of condensation on the inner glass cover is just released to the surroundings through the double glass cover, while the latent heat of condensation on the rear wall can be used as a heat source for the multiple-effect section. Therefore, the latent heat of condensation as well as solar energy directly absorbed on the rear wall is utilized repeatedly in the multiple-effect section as mentioned above.

The vertical multiple-effect diffusion still coupled with the basin-type still has very simple structure; that is, the still can be constructed by adding the outer glass cover and some partitions to the conventional single slope basin-type still, which is the simplest solar still. However, the still is highly productive, and Tanaka et al. [19] reported that a still with 5 mm diffusion gaps between 11 partitions in the multiple-effect section produces 19.0–24.0 kg/m^2day of distillate per unit base area at 20.7–22.4 MJ/m^2day solar radiation on the glass cover in outdoor experiments, which is in good agreement with theoretical predictions calculated from the heat and mass transfer model developed by Tanaka et al. [17].

Recently, many studies on the effect of flat plate external reflectors on basin-type stills have been reported [24–37], although few reports had been presented before the year 2000 [38–40]. The results showed that an external reflector can increase the solar radiation absorbed on the basin liner as well as the distillate productivity of the still if the reflector angle is set properly. Among them, Tanaka [34] theoretically studied a basin-type still with a flat plate external bottom reflector extending from the front wall of the still and found that the increase in the still's daily productivity obtained by adding an external bottom reflector and setting its inclination to the optimum angle is predicted to be about 21%, 23%, and 12% on the spring equinox and summer and winter solstice days.

A flat plate bottom reflector can be applied to the vertical multiple-effect diffusion still coupled with a basin-type still, but in this case, the reflected radiation not only on the basin liner but also on the rear wall (the first partition of the multiple-effect section) should be taken into consideration, since the reflected radiation on the rear wall can be used directly as thermal input to the multiple-effect section. To analyze the external bottom reflector on the basin-type still, Tanaka [34] presented a geometrical model by introducing the mirror-symmetric plane of a basin liner to the external reflector and determined the amount of radiations reflected from the external reflector and then absorbed on the basin liner by calculating the direct radiation that goes through the external reflector and is then absorbed on the mirror-symmetric plane. However, the reflected radiation on the rear wall cannot be calculated with this model.

Therefore, in this paper, a new geometrical model is presented to calculate the amount of solar radiation reflected from the external bottom reflector and then absorbed on the basin liner as well as on the rear wall and determine the effect of the external reflector on the distillate productivity of a vertical multiple-effect diffusion still coupled with basin-type still at 30°N latitude.

2. Vertical Multiple-Effect Diffusion Solar Still Coupled with Basin-Type Still Using Flat Plate External Bottom Reflector

The proposed still is shown in Figure 1. The still has a right-angled triangular cross-section consisting of a horizontal basin liner, a sloping double glass cover facing due south, and a vertical multiple-effect diffusion still (multiple-effect section) at the vertical rear wall. The basin liner and the vertical rear wall (the first partition of the multiple-effect section) are assumed to be painted black. The bottom and side walls are insulated. The two inner side walls are assumed to be covered with highly reflective materials. The multiple-effect section consists of a series of closely spaced vertical parallel partitions in contact with saline-soaked wicks. Saline water is fed to the wicks constantly. A flat plate external bottom reflector is attached to the front wall of the still.

Direct and diffuse solar radiation as well as reflected solar radiation from the external reflector transmits through the double glass cover and is absorbed on the basin liner (b)

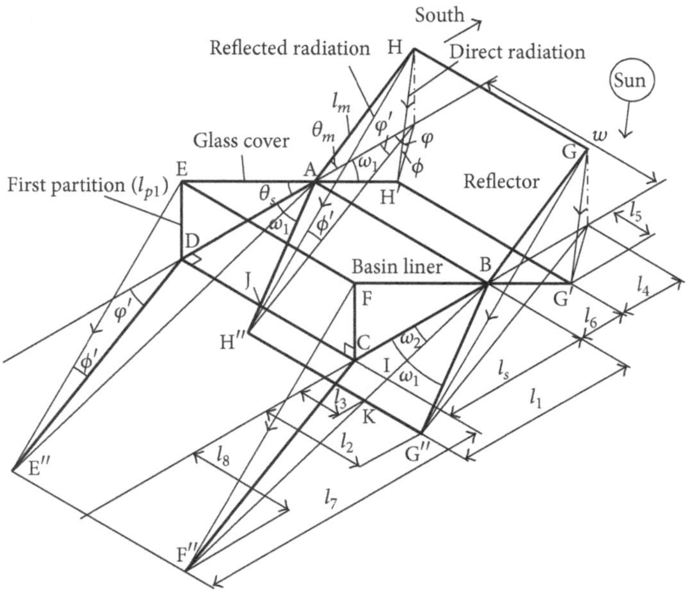

FIGURE 2: The shadow of the still and reflector and the reflected projection of the reflector on a horizontal surface caused by direct radiation. $l_1 = l_m\{\sin\theta_m \tan(\pi - \omega_3 - 2\theta_m) - \cos\theta_m\}$, $l_2 = l_1 \tan\omega_1$, $l_3 = l_1 \tan\omega_2$, $l_4 = l_m \sin\theta_m \cos\varphi / \tan\phi$, $l_5 = l_m \sin\theta_m \sin|\varphi| / \tan\phi$, $l_6 = l_m \cos\theta_m - l_4$, $l_7 = l_s \tan\theta_s \cos\varphi' / \tan\phi'$, $l_8 = l_s \tan\theta_s \sin|\varphi'| / \tan\phi'$, $\tan\omega_1 = l_5/l_6$, $\tan\omega_2 = l_8/(l_7 + l_s)$, $\tan\varphi' = l_2/(l_1 + l_m\cos\theta_m)$, and $\tan\phi' = l_m \sin\theta_m \cos\varphi'/(l_1 + l_m \cos\theta_m)$.

and the first partition ($p1$). Water vapor from the basin liner condenses on the inner glass cover (gi) and the first partition ($p1$) to become distilled water. The latent heat of condensation and the solar radiation directly absorbed on the first partition is utilized repeatedly in the multiple-effect section.

The double glass cover maintains a high temperature on the inner glass cover. This causes an increase in the ratio of condensate on the first partition $p1$ to that on the inner glass cover gi and a reduction in radiative and convective heat loss through the glass cover.

Tanaka et al. [18] showed that the average daily productivity of the still for four typical days (the spring and autumn equinoxes and the summer and winter solstices) was predicted to be maximum when the glass cover inclination is about 40° to 45°. Therefore, the inclination of the glass cover θ_s is determined as 40° in this paper. The inclination angle of the reflector θ_m is changed in the calculation to find the optimum angle of θ_m.

The basin liner and the external reflector are determined to be the same size at 1 m^2 (1 m × 1 m). So the area of each glass plate of the double glass cover and each partition plate in the multiple-effect section would be about 1.31 m^2 and 0.84 m^2.

3. Theoretical Analysis

3.1. Reflected Radiation from External Reflector Absorbed on the Basin Liner and the First Partition. To simplify the calculations, it is assumed that the same amount of solar radiation obstructed by one side wall hits the other inner side wall since both inner side walls are assumed to be covered with highly reflective materials. This assumption would be valid if the still's width is sufficient in relation to the length of the basin liner. The height of the front wall and thickness of the double glass cover are assumed to be negligible in calculating the absorption of solar radiation.

The reflected radiation from the external reflector cannot all be absorbed on the basin liner and the first partition, and part of or all of the reflected radiations would escape to the surroundings without hitting the basin liner and the first partition. Figure 2 shows a schematic diagram of the shadow of the reflector as well as a projection of the reflected sunrays from the reflector on a horizontal surface caused by direct solar radiation. ABCD is the basin liner, ABFE is the double glass cover, CDEF is the first partition, and ABGH is the external reflector. l_s and l_m are the length of the basin liner and reflector, and θ_s and θ_m are the angle of the glass cover and reflector from the horizontal. w is the width of both the still and the reflector. φ and ϕ are the azimuth and altitude angles of the sun. The glass cover is facing due south.

The shadow of the reflector and the projection of the reflected sunrays from the reflector caused by direct radiation are shown as ABG′H′ and ABG″H″, respectively. Assuming that radiation with the same azimuth (φ') and altitude angles (ϕ') as the reflected sunrays from the reflector hits the glass cover, the shadow of the glass cover would be an area shown as ABF″E″. The amount of reflected radiation from the reflector absorbed on the basin liner and the first partition can be determined as the overlapping area of the reflected projection of the reflector (ABG″H″) and the shadow of the glass cover (ABF″E″) shown as trapezoid ABKH″. The residue shown as

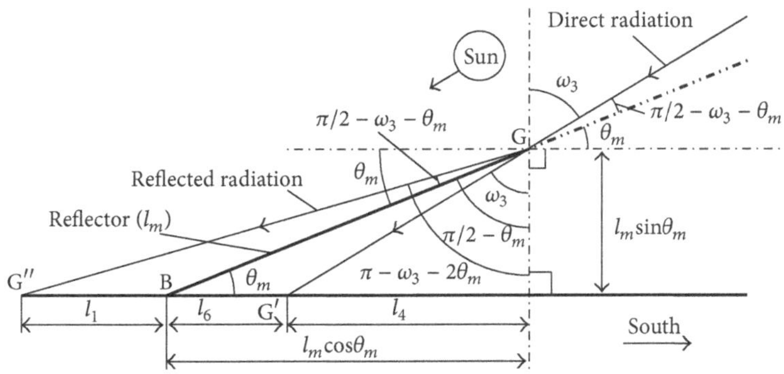

FIGURE 3: Side view of the external reflector. $l_1 = l_m\{\sin\theta_m \tan(\pi-\omega_3-2\theta_m) - \cos\theta_m\}$ and $\tan\omega_3 = l_4/l_m\sin\theta_m = (l_m\sin\theta_m\cos\varphi/\tan\phi)/(l_m\sin\theta_m) = \cos\varphi/\tan\phi$.

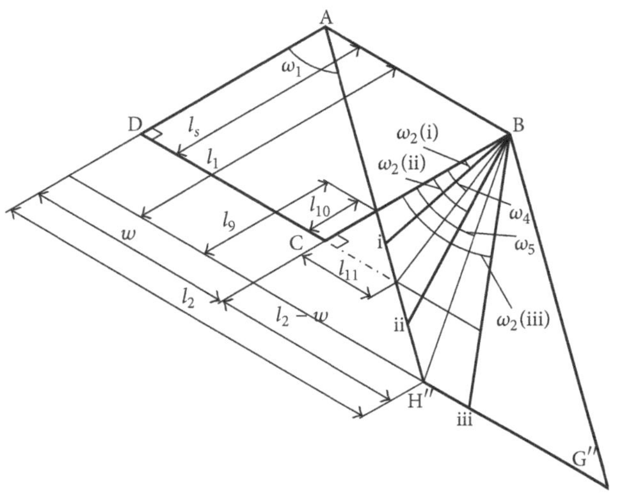

FIGURE 4: Reflected projection and shadow of the glass cover when $l_2 > w$. $l_9 = (l_2-w)/\tan\omega_1$, $l_{10} = l_9 - (l_1 - l_s)$, $l_{11} = l_{10}\tan\omega_1$, $\tan\omega_4 = l_{11}/l_s$, and $\tan\omega_5 = (l_2-w)/l_1$.

triangle BG''K would escape to the ground. It follows that the portion of the reflected radiation absorbed on either the basin liner or the first partition can be determined as area ABIJ or JIKH'', respectively. Since the reflected radiation from the reflector is condensed or diluted (i.e., the length of the shadow (l_6) and reflected projection (l_1) of the reflector are not equal), the intensity of the reflected radiation from the reflector on a horizontal surface should be determined as $G_{dr}\times l_6/l_1$. Here, G_{dr} is direct solar radiation on a horizontal surface. Therefore, the solar radiation reflected from the reflector and absorbed on the basin liner, $Q_{re,b}$, or the first partition, $Q_{re,p1}$, can be expressed as

$$Q_{re,b} = G_{dr}\frac{l_6}{l_1}\{\tau_g(\beta')\}^2\rho_m\alpha_b\times s_b, \quad (1)$$

$$s_b = l_s\{w-0.5l_s(\tan\omega_1-\tan\omega_2)\}, \quad (2)$$

$$Q_{re,p1} = G_{dr}\frac{l_6}{l_1}\{\tau_g(\beta')\}^2\rho_m\alpha_{p1}\times s_{p1}, \quad (3)$$

$$s_{p1} = (l_1-l_s)[w-0.5\{l_2-l_3+l_s(\tan\omega_1-\tan\omega_2)\}], \quad (4)$$

where τ_g is the transmittance of each glass cover, β' is the incident angle of the reflected sunlight from the reflector to the glass cover, ρ_m is the reflectance of the external reflector, α_b and α_{p1} are the absorptance of the basin liner and the first partition, and s_b and s_{p1} are the area of ABIJ and JIKH''.

To calculate (1) to (4), the length of the reflected projection of the reflector, l_1, has to be determined. The side view of the external reflector is shown in Figure 3. Since the incident angle and reflection angle of the sunrays for the reflector can be expressed as $\pi/2-\omega_3-\theta_m$, the length of the reflected projection of the reflector, l_1, can be determined with the angle $\omega_3(=\tan^{-1}(\cos\varphi/\tan\phi))$ as

$$l_1 = l_m\{\sin\theta_m\tan(\pi-\omega_3-2\theta_m)-\cos\theta_m\}. \quad (5)$$

Therefore, $Q_{re,b}$ and $Q_{re,p1}$ can be calculated with the lengths l_1 to l_8 and angles ω_1 to ω_3 shown in Figures 2 and 3.

When $Q_{re,b}$ and $Q_{re,p1}$ are calculated, there are some exceptions as follows.

(1) When length l_2 is longer than the still's width w as shown in Figure 4, s_b and s_{p1} should be determined as follows:

case i ($\omega_2 < \omega_4$):

$$s_b = \frac{0.5w^2\sin(\pi/2-\omega_1)\sin(\pi/2+\omega_2)}{\sin(\omega_1-\omega_2)},$$
$$s_{p1} = 0 \quad (6)$$

case ii ($\omega_4 \le \omega_2 < \omega_5$):

$$s_b = 0.5\{l_s w + l_s^2(\tan\omega_2-\tan\omega_4)\},$$

$$s_{p1} = \left(0.5\{l_s(\tan\omega_2 - \tan\omega_4)\}^2 \sin\left(\frac{\pi}{2} - \omega_1\right) \times \sin\left(\frac{\pi}{2} + \omega_2\right)\right) \times (\sin(\omega_1 - \omega_2))^{-1}, \quad (7)$$

case iii ($\omega_2 \geq \omega_5$):

$$s_b = 0.5\left\{l_s w + l_s^2(\tan\omega_2 - \tan\omega_4)\right\},$$

$$s_{p1} = \frac{1}{2}\{l_s(\tan\omega_5 - \tan\omega_4)\}^2 \times \frac{\sin(\pi/2 - \omega_1)\sin(\pi/2 + \omega_5)}{\sin(\omega_1 - \omega_5)} + \frac{1}{2}\left(l_1^2 - l_s^2\right)(\tan\omega_2 - \tan\omega_5). \quad (8)$$

(2) When length l_1 is shorter than l_s, the reflected radiation from the reflector does not hit the first partition. So s_{p1} should be zero, and s_b should be determined as follows:

(a) $l_2 \leq w$:

$$s_b = l_1\{w - 0.5(l_2 - l_3)\},$$

(b) $l_2 > w$ and $\omega_2 < \omega_5$:

$$s_b = \frac{0.5w^2\sin(\pi/2 - \omega_1)\sin(\pi/2 + \omega_2)}{\sin(\omega_1 - \omega_2)}, \quad (9)$$

(c) $l_2 > w$ and $\omega_2 \geq \omega_5$:

$$s_b = 0.5\left\{l_1 w + l_1^2(\tan\omega_2 - \tan\omega_5)\right\}.$$

(3) When the reflection angle $\pi/2 - \omega_3 - \theta_m$ is larger than the reflector inclination θ_m ($\pi/2 - \omega_3 - \theta_m > \theta_m$), all of the reflected radiations will escape to the surroundings. So both $Q_{re,b}$ and $Q_{re,p1}$ will be zero.

(4) During the months of April to August, when the sun moves north, that is, the absolute value of the azimuth angle of the sun, $|\varphi|$, is larger than $\pi/2$ in the morning and evening, $Q_{re,b}$ and $Q_{re,p1}$ should be calculated by determining lengths l_1 and l_6 as

$$l_1 = l_m\{\sin\theta_m\tan(\pi + \omega_3 - 2\theta_m) - \cos\theta_m\}$$
$$l_6 = l_m\cos\theta_m + l_4. \quad (10)$$

(5) When the altitude angle of the sun is very small and the reflector would shade a part of the basin liner in the early morning and late evening, $Q_{re,b}$ and $Q_{re,p1}$ should be zero.

3.2. Absorption of Direct and Diffuse Radiation on the Basin Liner and the First Partition. Direct solar radiation absorbed on the basin liner, $Q_{dr,b}$, can be determined as

$$Q_{dr,b} = G_{dr}\left\{\tau_g(\beta)\right\}^2\alpha_b \times l_s w, \quad (11)$$

where β is the incident angle of the sunrays to the glass cover. Direct solar radiation absorbed on the first partition, $Q_{dr,p1}$, can be determined by assuming that the azimuth and altitude angles of the reflected sunrays, φ' and ϕ', in Figure 2 would be those of the direct radiation, φ and ϕ, and considering the shadow area of the first partition (DCF″E″). This can be expressed as

$$Q_{dr,p1} = G_{dr}\left\{\tau_g(\beta)\right\}^2\alpha_{p1} w l_{p1}\frac{\cos\varphi}{\tan\phi}, \quad (12)$$

where l_{p1} is the height of the first partition.

When the sun moves north during the months of April to August, $Q_{dr,p1}$ would be zero, and $Q_{dr,b}$ should be determined taking into consideration the shadow of the first partition on the basin liner as

$$Q_{dr,b} = G_{dr}\left\{\tau_g(\beta)\right\}^2\alpha_b \times \left\{l_s w - l_{p1}\frac{\cos\varphi}{\tan\phi}\left(w - 0.5l_{p1}\frac{\sin|\varphi|}{\tan\phi}\right)\right\}. \quad (13)$$

Diffuse solar radiation absorbed on the basin liner and the first partition, $Q_{df,b}$ and $Q_{df,p1}$, can be determined assuming that diffuse radiation comes uniformly from all directions in the sky dome and can be expressed as

$$Q_{df,b} = G_{df}\left(\tau_g\right)_{df,b}\alpha_b \times l_s w,$$
$$Q_{df,p1} = G_{df}\left(\tau_g\right)_{df,p1}\alpha_{p1} \times l_{p1} w, \quad (14)$$

where G_{df} is the diffuse solar radiation on a horizontal surface and $(\tau_g)_{df,b}$ and $(\tau_g)_{df,p1}$ are functions of the inclination of the glass cover, θ_s, and is calculated by integrating the transmittance of the double glass cover for diffuse radiation from all directions in the sky dome. This can be expressed as

$$\left(\tau_g\right)_{df,b} = 0.484 - 2.82 \times 10^{-3} \times \theta_s + 1.22 \times 10^{-5} \times \theta_s^2, \ \theta_s\,[°],$$
$$\left(\tau_g\right)_{df,p1} = 0.253 + 3.88 \times 10^{-3} \times \theta_s - 4.47 \times 10^{-5} \times \theta_s^2, \ \theta_s\,[°]. \quad (15)$$

3.3. Absorption of Solar Radiation on the Glass Cover. All of the reflected or direct solar radiations which go through the double glass cover should be absorbed on the basin liner and the first partition. So the amount of the solar radiation absorbed on each glass cover can be determined by adding the area of (1) and (3) for reflected radiation and (11) and (12) for direct radiation. Diffuse solar radiation absorbed on each glass cover can be determined with the same assumption for $Q_{df,b}$ and $Q_{df,p1}$.

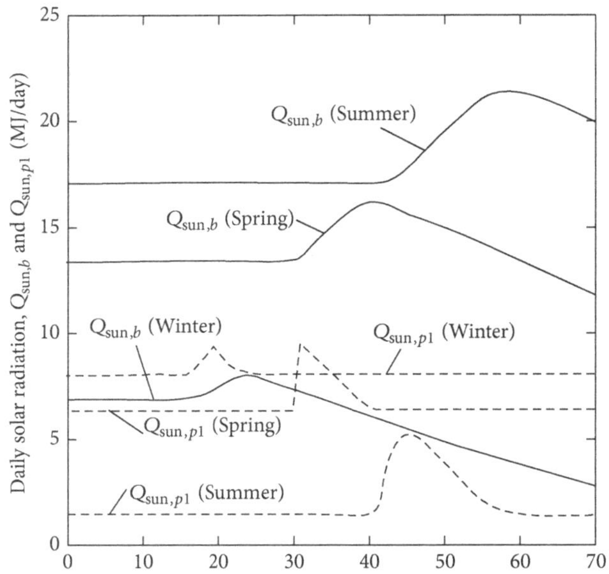

FIGURE 5: Daily solar radiation absorbed on the basin liner, $Q_{sun,b}$, and the first partition, $Q_{sun,p1}$, varying with reflector inclination θ_m on the spring equinox and the summer and winter solstice days at 30°N.

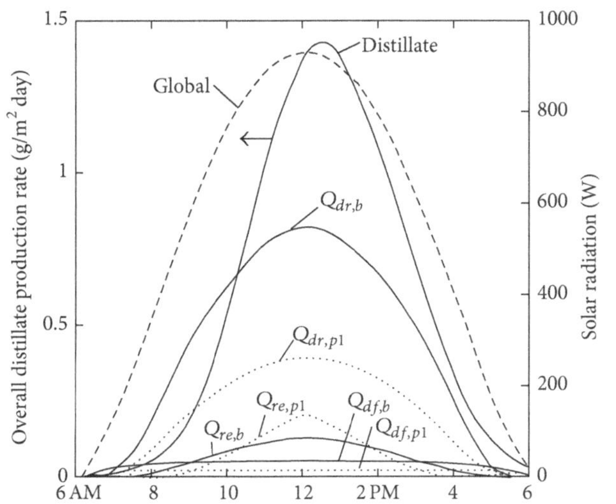

FIGURE 6: Time variation of overall distillate production rate (Distillate), global solar radiation on a horizontal surface (Global), and absorption of direct (Q_{dr}), diffuse (Q_{df}), and reflected (Q_{re}) solar radiation on the basin liner and the first partition on a spring equinox day at $\theta_m = 35°$.

3.4. Effect of Shadow of External Reflector. When the altitude angle of the sun ϕ is very small, the external reflector will shade a part of the basin liner. In the calculations, this effect is taken into account and was described in a previous paper in detail [34].

A longer reflector length l_m and/or a greater reflector inclination θ_m would cast a shadow on even the first partition $p1$. But for our calculations, a shadow on the first partition was not observed with the numerical conditions employed in this paper.

3.5. Heat and Mass Transfer in the Still. Heat and mass transfer in the still was described in a previous paper in detail [17]. The equations for the solar radiation absorbed on the basin liner, the first partition and each glass cover, the energy balance equations for all components of the still, and the equations of properties were solved together to find the temperatures, heat and mass transfer rates, and the distillate production rates on the inner glass cover and each partition throughout the day. Temperatures of each component of the still were set to be equal to the ambient air temperature at just before sunrise for the initial values. The weather and design conditions are listed in Table 1.

4. Results and Discussions

Figure 5 shows the variation of the daily amount of solar radiation absorbed on the basin liner, $Q_{sun,b}$, and on the first partition, $Q_{sun,p1}$, with reflector inclination θ_m on three typical days (spring equinox and summer and winter solstice days) at 30°N latitude. $Q_{sun,b}$ and $Q_{sun,p1}$ are the sum of the direct, diffuse, and reflected radiation absorbed on the basin liner and the first partition, respectively. $Q_{sun,b}$ is greatest in summer and smallest in winter, while $Q_{sun,p1}$ is greatest in winter and smallest in summer, and the reflector inclination which can increase $Q_{sun,b}$ and $Q_{sun,p1}$ is smallest in winter and greatest in summer. This is because the altitude angle of the sun becomes nearly vertical in summer and is lowest in winter. Solar radiation absorbed on both the basin liner and the first partition can be increased by setting the reflector inclination to the proper angle according to the seasons on these three days.

On each day, both $Q_{sun,b}$ and $Q_{sun,p1}$ start to increase at a certain reflector inclination, and $Q_{sun,p1}$ has a sharp peak, while $Q_{sun,b}$ has a gentle peak. The reason for this is as follows: all of the reflected sunrays from the external reflector escape to the surroundings during $\theta_m < \pi/2 - \omega_3 - \theta_m$ (see Figure 3), and after θ_m becomes more than the angle $\pi/2 - \omega_3 - \theta_m$ by increasing the reflector inclination, reflected sunrays start to hit the first partition and the basin liner. A greater reflector inclination θ_m results in a shorter length of l_1 (shown in Figure 2), so $Q_{re,p1}$ rapidly decreases to zero, but $Q_{re,b}$ increases for a while and decreases to zero with an increase in reflector inclination θ_m. Further, when θ_m is larger than about 35° or 60° on the winter solstice or spring equinox, the external reflector shades the basin liner. So $Q_{sun,b}$ is less than that $\theta_m = 0°$ which is equivalent to the still without a reflector.

The time variation of the overall distillate production rate (Distillate), that is, the sum of distillate production rates on all condensing surfaces (gi and $p1$ to $p10$), global solar radiation on 1 m^2 horizontal surface (Global), and absorption of direct, diffuse, and reflected solar radiation on the basin liner ($Q_{dr,b}$, $Q_{df,b}$, and $Q_{re,b}$) and the first partition ($Q_{dr,p1}$, $Q_{df,p1}$, and $Q_{re,p1}$) on a spring equinox day at $\theta_m = 35°$ is shown in

TABLE 1: Design and weather conditions.

$l_s = l_m = w = 1$ m, $\theta_s = 40°$
Number of partitions = 10
Diffusion gaps between partitions = 5 mm
Base area = 1.05 m^2
Thickness of glass plate and partition plate = 3 mm and 0.5 mm
Air gap between double glass cover = 10 mm
Depth of basin water = 10 mm
Thickness and thermal conductivity of insulation = 50 mm and 0.04 W/mK
$\alpha_b = \alpha_{p1} = 0.9$, $\alpha_{g_i} = \alpha_{g_o} = 0.08$, and $\rho_m = 0.8$
$\tau_g(\beta) = 2.642\cos\beta - 2.163\cos^2\beta - 0.320\cos^3\beta + 0.719\cos^4\beta$ [17]
Feed rate of saline water to wicks: twice as large as the steady-state evaporation rate from wicks which have been calculated on the assumption that solar radiation will stay constant at the maximum of the day.
G_{dr}, G_{df}: Bouguer's and Berlage's equations [41] with solar constant of 1370 W/m^2 and transmittance of atmosphere of 0.7 at 30°N latitude
T_a = 25°C, 33°C, 30°C, and 20°C on spring equinox, summer solstice, autumn equinox, and winter solstice
Ambient wind velocity = 1 m/s

Figure 6. At $\theta_m = 35°$, the reflected sunrays can be absorbed on both the basin liner and the first partition (Figure 5). In this paper, the distillate production rate as well as the daily amount of distillate is defined as those per unit base area of the still. The overall distillate production rate peaks about 30 min later than that of the global solar radiation because of the heat capacity of the still. Absorption of direct solar radiation ($Q_{dr,b}$ and $Q_{dr,p1}$) is dominant, and the amounts of diffuse solar radiation ($Q_{df,b}$ and $Q_{df,p1}$) are very low on both the basin liner and the first partition.

Figure 7 shows the temperature distribution in the still at the peak of the overall distillate production rate and the daily amount of distillate on each condensing surface on a spring equinox day at $\theta_m = 35°$. Overall daily productivity (Total) is shown in 1/5 scale. The temperature is highest at the basin liner (b) and decreases as it passes through the double glass cover and the multiple-effect section to the ambient air (a). The daily amount of distillate from gi is greater than that on $p1$ since the temperature differences between the basin liner and the inner glass cover gi ($T_b - T_{gi}$) and the first partition $p1$ ($T_b - T_{p1}$) are almost the same, but the area of the inner glass cover is larger than that of the first partition. The daily amount of distillate on the second partition $p2$ is greater than that on $p1$ since the latent heat of condensation in addition to the solar radiation absorbed on the first partition is utilized to evaporate the saline water from the wick of the first partition. The daily amount of distillate decreases progressively from $p2$ to the last partition $p10$, since the thermal energy is consumed to heat up saline water fed to each wick, and the mean temperature of evaporating and condensing surfaces of each distilling cell decreases progressively to the outer partition.

To evaluate whether it is better to absorb the reflected sunrays on the basin liner or on the first partition, the variation in the daily amount of solar radiation absorbed on the basin liner $Q_{\text{sun},b}$ and the first partition $Q_{\text{sun},p1}$, the daily amount of distillate on the inner glass cover, $m_{d,gi}$, and the first partition, $m_{d,p1}$, with reflector inclination θ_m on a spring equinox day is shown in Figure 8. Both the absorption of solar radiation, $Q_{\text{sun},b}$ and $Q_{\text{sun},p1}$, are shown in 1/10 scale. The temperature difference between the basin liner and the inner glass cover or the first partition ($T_b - T_{gi}$ or $T_b - T_{p1}$) at the peak of overall distillate production rate is also shown in Figure 8. Reflected sunrays from the external reflector start to hit both the basin liner and the first partition at about $\theta_m = 30°$, and $Q_{\text{sun},p1}$ (or $Q_{re,p1}$) has a sharp peak at about $\theta_m = 31°$, and $Q_{\text{sun},b}$ (or $Q_{re,b}$) has a gentle peat at around $\theta_m = 40°$.

When $Q_{\text{sun},p1}$ peaks (θ_m = about 31°), the daily amount of distillate on the inner glass cover $m_{d,gi}$ increases, while that of the first partition $m_{d,p1}$ slightly decreases. This is because the greater $Q_{\text{sun},p1}$ causes a higher temperature on $p1$ (or less temperature difference $T_b - T_{p1}$) and a higher ratio of condensation of water vapor from the basin water on the inner glass cover to that on the first partition. In a range of θ_m = 31° to 40°, $Q_{\text{sun},p1}$ decreases but $Q_{\text{sun},b}$ increases, and this causes an increase in difference of the temperature difference between $T_b - T_{p1}$ and $T_b - T_{gi}$, since the absorption of solar energy on the first partition decreases, while that on the inner glass stays almost the same. Therefore, the daily amount of distillate on $p1$, $m_{d,p1}$, greatly increases, while that on gi, $m_{d,gi}$, slightly decreases with an increase in reflector inclination θ_m in this range. When θ_m is larger than 40°, $Q_{\text{sun},p1}$ remains constant ($Q_{re,p1}$ is zero), and $Q_{\text{sun},b}$ decreases with an increase in θ_m. This causes a decrease in difference of the temperature differences between $T_b - T_{p1}$ and $T_b - T_{gi}$, and $m_{d,p1}$ decreases faster than $m_{d,gi}$ with an increase in θ_m.

Figure 9 shows the daily amount of distillate on condensing surfaces when $Q_{re,p1}$ is max ($\theta_m = 31°$) and $Q_{re,b}$ is max ($\theta_m = 40°$) on a spring equinox day. The results for a still without a reflector (NRS) are also shown. When $Q_{re,p1}$ is max, the daily amount of distillate on gi and $p1$ is almost the

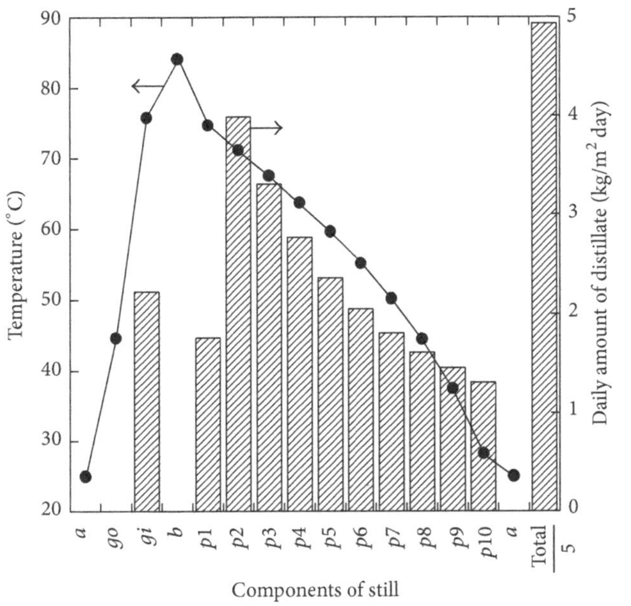

FIGURE 7: Temperature distribution at the peak of the overall distillate production rate and the daily amount of distillate in the still on a spring equinox day at $\theta_m = 35°$.

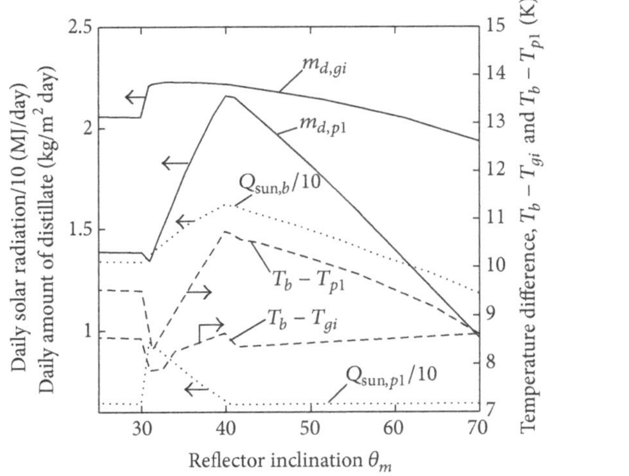

FIGURE 8: Variation of daily solar radiation absorbed on the basin liner $Q_{sun,b}$ and the first partition $Q_{sun,p1}$, daily amount of distillate on the inner glass cover $m_{d,gi}$ and the first partition $m_{d,p1}$, and temperature difference between basin water and inner glass cover, $T_b - T_{gi}$, and the first partition, $T_b - T_{p1}$, with reflector inclination θ_m on a spring equinox day.

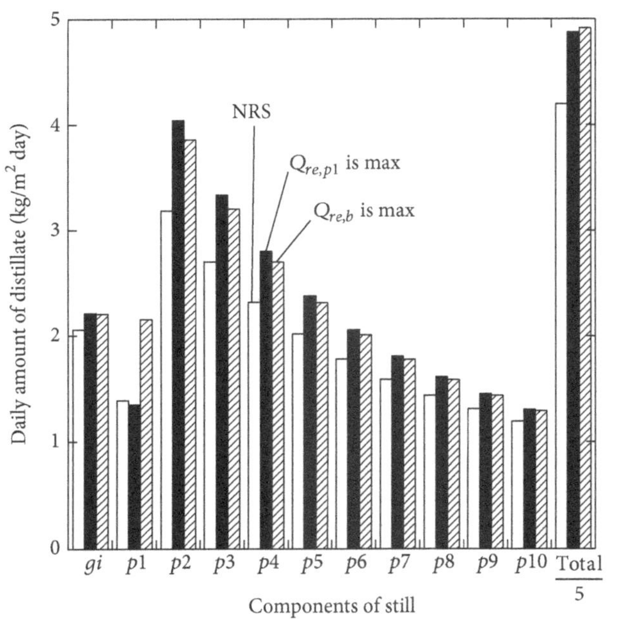

FIGURE 9: The daily amount of distillate on condensing surfaces when $Q_{re,p1}$ is max ($\theta_m = 31°$) and $Q_{re,b}$ is max ($\theta_m = 40°$) on a spring equinox day. NRS shows the still without the external reflector.

same with those of NRS, but the daily amount of distillate from $p2$ to $p10$ are greater than for NRS. This shows that the reflected radiation directly absorbed on the first partition $p1$ can be used adequately as a heat source for the multiple-effect section without escaping to the surroundings through the double glass cover. When $Q_{re,b}$ is max, the daily amount of distillate on gi is almost the same, while that on the first partition $p1$ is considerably greater than for NRS, and the daily amount of distillate from $p2$ to $p10$ would also be greater than for NRS. This indicates that the reflected radiation absorbed on the basin liner can be transferred to the multiple-effect section as a form of latent heat of condensation. As a result, the reflected radiation from the external reflector can be used in the multiple-effect section adequately both in cases where the reflected radiation is absorbed on the basin liner or on the first partition.

Figure 10 shows variations of (a) overall daily productivity of the still and (b) the increase ratio with reflector inclination θ_m on four typical days (spring and autumn equinoxes and summer and winter solstice days) at 30°N. The increase ratio is the value comparing a still without an external reflector and defined as

$$\text{increase ratio} = (\text{overall daily productivity of a still with a reflector}) / (\text{overall daily productivity of a still without a reflector}). \quad (16)$$

Daily global solar radiation on a horizontal surface is about 23.5, 22.6, 30.4, and 12.5 MJ/m²day on spring and autumn equinoxes and summer and winter solstice days, respectively. As mentioned above, reflected radiation from the external reflector can be used adequately as a heat source for the multiple-effect section both in cases where the reflected radiation is absorbed on the basin liner or on the first partition. Therefore, the overall daily productivity and increase ratio have gentle peaks around $Q_{sun,b}$, and $Q_{sun,p1}$ (or $Q_{re,b}$ and $Q_{re,p1}$) have peaks, except for on the winter solstice.

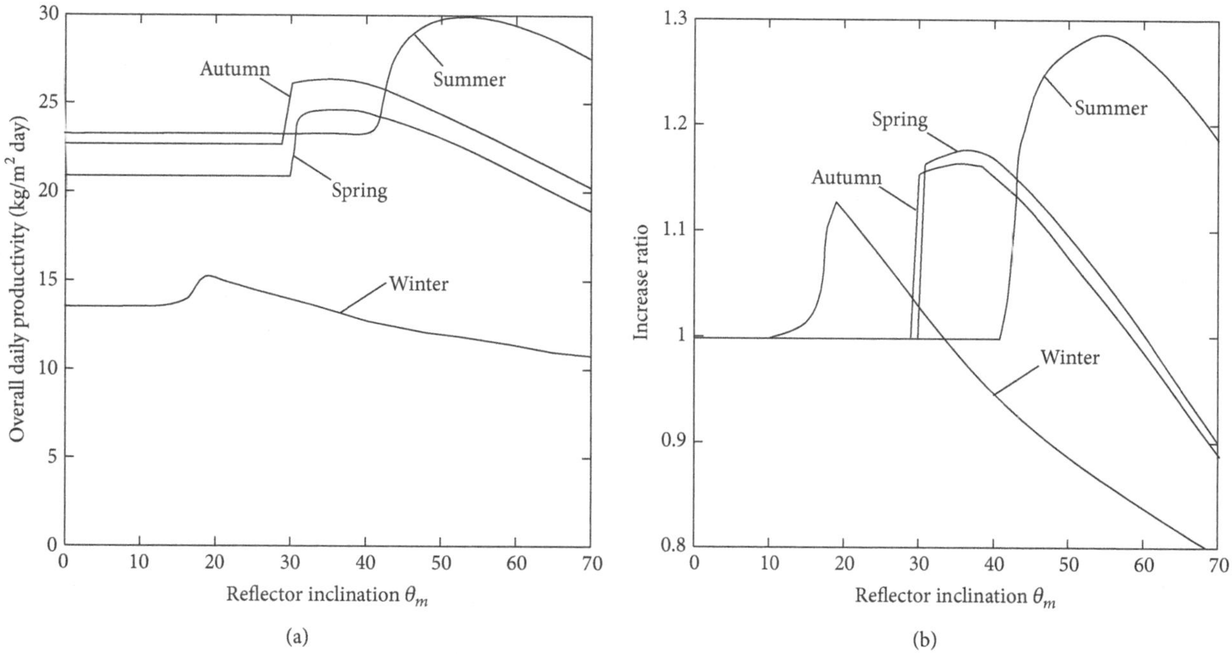

FIGURE 10: Variation of (a) overall daily productivity of the still and (b) increase ratio with reflector inclination on four days (spring and autumn equinoxes and summer and winter solstice days) at 30°N.

Only on the winter solstice, the altitude angle of the sun is very low, and $Q_{\text{sun},p1}$ is greater than $Q_{\text{sun},b}$ (Figure 5), so the overall daily productivity and increase ratio have a sharp peak when $Q_{\text{sun},p1}$ has maximum value ($\theta_m = 20°$).

The overall daily productivity of the still can be increased by using a flat plate bottom reflector and setting the inclination of the reflector to the proper angle according to the seasons. The reflector inclination should be set to be about 30° to 40° on the spring and autumn equinoxes, 50° to 60° on the summer solstice, and around 20° on the winter solstice. Increase ratio would be 1.16, 1.13, and 1.29 on the spring and autumn equinoxes, the winter solstice, and summer solstice, respectively, when the glass cover inclination is set at 40° at 30°N.

5. Conclusions

The effect of an external flat plate bottom reflector on a vertical multiple-effect diffusion solar still coupled with a basin-type still was analyzed theoretically when the glass cover inclination is 40° from a horizontal at 30°N latitude. The results of this work are summarized as follows.

(1) A geometrical model to calculate the amount of solar radiation reflected by an external reflector and then absorbed on the basin liner as well as the vertical rear wall (the first partition of the multiple-effect section) was constructed.

(2) Solar radiation absorbed on both the basin liner and the first partition can be increased by adding the external reflector and setting its inclination to the proper angle according to seasons.

(3) The reflected radiation from the reflector can be utilized adequately both in cases where the reflected radiation is absorbed on the basin liner and on the first partition.

(4) The reflector inclination should be set to be around 30° to 40° on the spring and autumn equinoxes, 50° to 60° on the summer solstice, and 20° on the winter solstice, respectively.

(5) The external reflector can increase the overall daily productivity of the still about 16%, 13%, and 29% on the spring and autumn equinoxes, the winter solstice, and the summer solstice, respectively.

Symbols

G_{df}, G_{dr}:	Diffuse and direct solar radiation on a horizontal surface, W/m^2
l:	Length, m
m_d:	Daily amount of distillate, kg/m^{2}day
pi:	ith partition
Q_{df}, Q_{dr}, Q_{re}:	Absorption of diffuse, direct, and reflected solar radiation, W
Q_{sun}:	Absorption of solar radiation, W
s:	Area, m^2
T:	Temperature, K
w:	Width, m.

Greek

α: Absorptance
β: Incident angle of sunrays to glass cover
β': Incident angle of reflected sunrays to glass cover
ϕ, φ: Altitude and azimuth angles of the sun
ϕ', φ': Altitude and azimuth angles of reflected sunrays
θ: Inclination angle
ρ: Reflectance
τ_g: Transmittance of glass cover.

Subscripts

a: Ambient air
b: Basin
gi, go: Inner and outer glass cover
m: Reflector
pi: ith partition
s: Still.

References

[1] G. N. Tiwari, H. N. Singh, and R. Tripathi, "Present status of solar distillation," *Solar Energy*, vol. 75, no. 5, pp. 367–373, 2003.

[2] A. Kaushal and Varun, "Solar stills: a review," *Renewable and Sustainable Energy Reviews*, vol. 14, no. 1, pp. 446–453, 2010.

[3] K. Sampathkumar, T. V. Arjunan, P. Pitchandi, and P. Senthilkumar, "Active solar distillation—a detailed review," *Renewable and Sustainable Energy Reviews*, vol. 14, no. 6, pp. 1503–1526, 2010.

[4] A. E. Kabeel and S. A. El-Agouz, "Review of researches and developments on solar stills," *Desalination*, vol. 276, no. 1–3, pp. 1–12, 2011.

[5] R. V. Dunkle, "Solar water distillation: the roof type still and a multiple effect diffusion still," in *Proceedings of the International Heat Transfer Conference*, part 5, pp. 895–902, International Developments in Heat Transfer, University of Colorado, 1961.

[6] P. I. Cooper and J. A. Appleyard, "The construction and performance of a three-effect, wick-type, tilted solar still," *Sun at Work*, vol. 12, pp. 4–8, 1967.

[7] M. M. Elsayed, K. Fathalah, J. Shams, and J. Sabbagh, "Performance of multiple effect diffusion stills," *Desalination*, vol. 51, no. 2, pp. 183–199, 1984.

[8] K. Tsumura, M. Yamashita, H. Watanabe, and H. Inaba, "Relation between temperature of solar radiation receiving plate and volume of distilled water by direct solar multiple effect distillers," *Bulletin of the Society of Sea Water Science*, vol. 39, no. 3, pp. 129–135, 1985.

[9] K. Tanaka, H. Okamura, M. Kan, and K. Watanabe, "A fundamental study of the performance of a multiple effect solar still (2nd report, experimental analysis on the upward heating type still under the actual outdoor insolation)," *Transactions of the Japan Society of Mechanical Engineers B*, vol. 53, no. 496, pp. 3771–3779, 1987.

[10] T. Kiatsiriroat, S. C. Bhattacharya, and P. Wibulswas, "Performance analysis of multiple effect vertical still with a flat plate solar collector," *Solar and Wind Technology*, vol. 4, no. 4, pp. 451–457, 1987.

[11] S. Toyama, T. Aragaki, H. M. Salah, and K. Murase, "Dynamic characteristics of a multistage thermal diffusion type solar distillator," *Desalination*, vol. 67, pp. 21–32, 1987.

[12] R. Ouahes, C. Ouahes, P. le Goff, and J. le Goff, "A hardy, high-yield solar distiller of brackish water," *Desalination*, vol. 67, pp. 43–52, 1987.

[13] H. Okamura, K. Tanaka, and K. Watanabe, "Study on the performance of multiple effect solar still (experimental results and simulation analysis under outdoor insolation for downward heating type stills)," *Japan Solar Energy Society*, vol. 14, no. 6, pp. 27–34, 1988.

[14] K. Ohshiro, T. Nosoko, and T. Nagata, "A compact solar still utilizing hydrophobic poly(tetrafluoroethylene) nets for separating neighboring wicks," *Desalination*, vol. 105, no. 3, pp. 207–217, 1996.

[15] B. Bouchekima, B. Gros, R. Ouahes, and M. Diboun, "Performance study of the capillary film solar distiller," *Desalination*, vol. 116, no. 2-3, pp. 185–192, 1998.

[16] H. M. Yeh and C. D. Ho, "Energy and mass balances in multiple-effect upward solar distillers with air flow through the last-effect unit," *Energy*, vol. 25, no. 4, pp. 325–337, 2000.

[17] H. Tanaka, T. Nosoko, and T. Nagata, "A highly productive basin-type-multiple-effect coupled solar still," *Desalination*, vol. 130, no. 3, pp. 279–293, 2000.

[18] H. Tanaka, T. Nosoko, and T. Nagata, "Parametric investigation of a basin-type-multiple-effect coupled solar still," *Desalination*, vol. 130, no. 3, pp. 295–304, 2000.

[19] H. Tanaka, T. Nosoko, and T. Nagata, "Experimental study of basin-type, multiple-effect, diffusion-coupled solar still," *Desalination*, vol. 150, no. 2, pp. 131–144, 2002.

[20] H. Tanaka and Y. Nakatake, "A vertical multiple-effect diffusion-type solar still coupled with a heat-pipe solar collector," *Desalination*, vol. 160, no. 2, pp. 195–205, 2004.

[21] K. Fukui, T. Nosoko, H. Tanaka, and T. Nagata, "A new maritime lifesaving multiple-effect solar still design," *Desalination*, vol. 160, no. 3, pp. 271–283, 2004.

[22] H. Tanaka and Y. Nakatake, "A simple and highly productive solar still: a vertical multiple-effect diffusion-type solar still coupled with a flat-plate mirror," *Desalination*, vol. 173, no. 3, pp. 287–300, 2005.

[23] T. Nosoko, T. Kinjo, and C. D. Park, "Theoretical analysis of a multiple-effect diffusion still producing highly concentrated seawater," *Desalination*, vol. 180, no. 1–3, pp. 33–45, 2005.

[24] H. Tanaka and Y. Nakatake, "Theoretical analysis of a basin type solar still with internal and external reflectors," *Desalination*, vol. 197, no. 1–3, pp. 205–216, 2006.

[25] H. Tanaka and Y. Nakatake, "Effect of inclination of external flat plate reflector of basin type still in winter," *Solar Energy*, vol. 81, no. 8, pp. 1035–1042, 2007.

[26] P. I. Ayav and G. Atagündüz, "Theoretical and experimental investigations on solar distillation of IZTECH campus area seawater," *Desalination*, vol. 208, no. 1–3, pp. 169–180, 2007.

[27] B. S. Kumar, S. Kumar, and R. Jayaprakash, "Performance analysis of a "V" type solar still using a charcoal absorber and a boosting mirror," *Desalination*, vol. 229, no. 1–3, pp. 217–230, 2008.

[28] S. Shanmugan, P. Rajamohan, and D. Mutharasu, "Performance study on an acrylic mirror boosted solar distillation unit utilizing seawater," *Desalination*, vol. 230, no. 1–3, pp. 281–287, 2008.

[29] H. Tanaka, "Effect of inclination of external reflector of basin type still in summer," *Desalination*, vol. 242, no. 1–3, pp. 205–214, 2009.

[30] H. Tanaka, "Experimental study of a basin type solar still with internal and external reflectors in winter," *Desalination*, vol. 249, no. 1, pp. 130–134, 2009.

[31] A. J. N. Khalifa and H. A. Ibrahim, "Effect of inclination of the external reflector on the performance of a basin type solar still at various seasons," *Energy for Sustainable Development*, vol. 13, no. 4, pp. 244–249, 2009.

[32] H. Tanaka, "Monthly optimum inclination of glass cover and external reflector of a basin type solar still with internal and external reflector," *Solar Energy*, vol. 84, no. 11, pp. 1959–1966, 2010.

[33] A. J. N. Khalifa and H. A. Ibrahim, "Effect of inclination of the external reflector of simple solar still in winter: an experimental investigation for different cover angles," *Desalination*, vol. 264, no. 1-2, pp. 129–133, 2010.

[34] H. Tanaka, "A theoretical analysis of basin type solar still with flat plate external bottom reflector," *Desalination*, vol. 279, pp. 243–251, 2011.

[35] A. Madhlopa and C. M. Johnstone, "Computation of solar radiation distribution in a solar still with internal and external reflectors," *Solar Energy*, vol. 85, no. 2, pp. 217–233, 2011.

[36] P. Monowe, M. Masale, N. Nijegorodov, and V. Vasilenko, "A portable single-basin solar still with an external reflecting booster and an outside condenser," *Desalination*, vol. 280, pp. 332–338, 2011.

[37] M. Boubekri and A. Chaker, "Yield of an improved solar still: numerical approach," *Energy Procedia*, vol. 6, pp. 610–617, 2011.

[38] M. A. S. Malik, G.N. Tiwari, A. Kumar, and M. S. Sodha, *Solar Distillation*, Pergamon Press, London, UK, 1982.

[39] A. A. Al-Karaghouli and A. N. Minasian, "A floating-wick type solar still," *Renewable Energy*, vol. 6, no. 1, pp. 77–79, 1995.

[40] A. El-Bahi and D. Inan, "Analysis of a parallel double glass solar still with separate condenser," *Renewable Energy*, vol. 17, no. 4, pp. 509–521, 1999.

[41] Japan Solar Energy Society, *Solar Energy Utilization Handbook*, Onkodo Press, Tokyo, Japan, 1985.

2

Design and Application of a Fiber Bragg Grating Tension Sensor for Anchor Rope

Jun Huang,[1] Zude Zhou,[1] Dongsheng Zhang,[2] Juntao Chen,[1] Litong Li,[2] and Xiwang Deng[2]

[1] *School of Mechanical and Electronic Engineering, Wuhan University of Technology, Wuhan, Hubei 430070, China*
[2] *National Engineering Laboratory for Optical Fiber Sensors, Wuhan University of Technology, Wuhan, Hubei 430070, China*

Correspondence should be addressed to Jun Huang; huang_jun_0218@126.com

Academic Editor: Shane Xie

A fiber Bragg grating (FBG) tension sensor for anchor rope has been proposed and implemented in the full-scale impact test of rockfall protection barriers in this paper. According to the measurement principle of the tension sensor, the wavelength shift difference of the two FBGs which, respectively, measure the positive and negative strain in the spring element is utilized as sensing signal. Therefore, the force sensitivity is increased and the temperature cross-sensitivity is compensated. Experimental results indicate that the measurement sensitivity is 14.85 pm/kN in a range from 0 to 150 kN. The dynamic forces on anchor ropes in barriers have been clearly recorded by the designed sensors during the impact test, and the variation of the impact forces could be mainly divided into three stages. This kind of tension sensor is expected to be widely used for force's quasi-distributed measurement or remote monitoring in industrial and mechanical fields.

1. Introduction

Anchor ropes have been widely applied in industry and manufacture fields. For the traditional force transducers or tension sensors mainly based on electrical strain gauges or piezoelectric materials, there are many disadvantages which restrict their utilization in harsh environment, since they are influenced by electromagnetic interference (EMI) and difficult to realize force's remote or multipoint measurement as discussed elsewhere [1–3]. Therefore, there is a lack of an efficient and effective method to realize multipoint measurement or long-term online monitoring of the tensile forces on anchor ropes at a remote distance.

In recent years, fiber-Bragg-grating- (FBG-) based sensors have been intensively investigated and widely implemented, due to their attractive characteristics such as small size, immunity to EMI, remote sensing, environmental ruggedness, and massive multiplexing capabilities, as discussed elsewhere [4–6]. Thanks to these excellent transmission capabilities, the distance to the measuring point can be many kilometers, and a lot of measuring points can be multiplexed in a single optical fiber as discussed by López-Higuera et al. [7]. Given their intrinsic capability to measure a multitude of parameters such as strain, temperature, pressure, acceleration, chemical and biological agents, and many others, FBG-based sensors have been developed for a widespread variety of sensing and monitoring applications in civil engineering, aerospace, marine, oil and gas, composites, smart structures, and biomedical devices as studied elsewhere [8–11]. A load cell based on fiber Bragg grating for icing monitoring on high-voltage overhead transmission lines was presented by Ma et al. [12]. FBG sensors were encapsulated into steel strand for tension monitoring of a tendon as proposed by Kim et al. [13]. An FBG pull sensor based on loop thin-wall section beam was presented by Liu et al. [14].

In this paper, an FBG tension sensor for anchor rope has been designed and studied. Measurement principle of the tension sensor is introduced. Using the two FBGs' wavelength shift difference as sensing signal, the measurement sensitivity is improved and the temperature cross-sensitivity is avoided. The dynamic impact forces acting on the anchor ropes in the rockfall protection barrier during the full-scale impact test

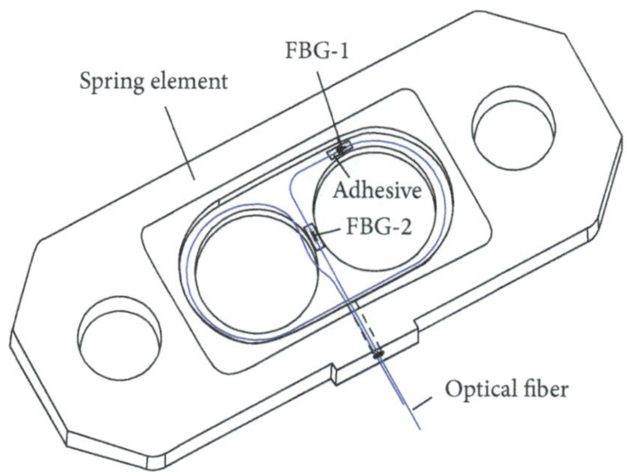

FIGURE 1: The structure of the FBG tension sensor.

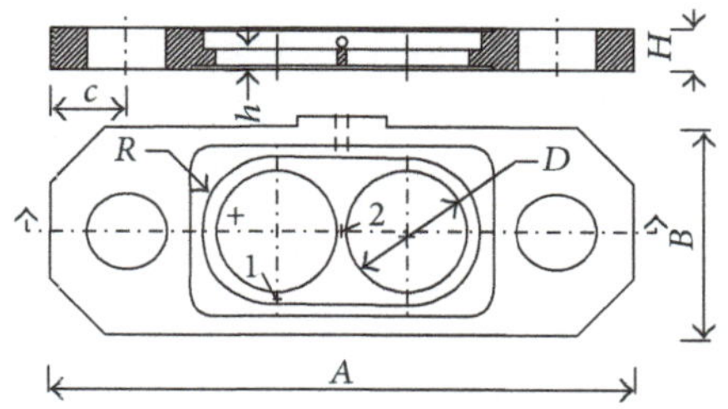

FIGURE 2: The schematic diagram of the spring element.

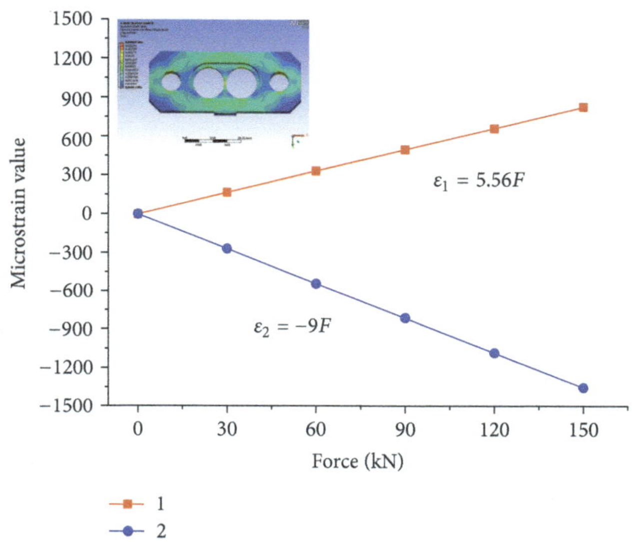

FIGURE 3: Strain values of ε_1 and ε_2 versus force.

have been accurately monitored by the FBG tension sensors, and the variation of the impact forces could be mainly divided into three stages. This kind of FBG tension sensor could be widely used for force's quasi-distributed measurement or real-time remote monitoring in industrial and mechanical fields.

2. The FBG Tension Sensor

The structure schematic diagram of the FBG tension sensor is depicted in Figure 1. The tension sensor mainly consists of two FBGs and a spring element. The two FBGs (FBG-1, FBG-2) in one optical fiber are bonded on the surface of the spring element with epoxy adhesive as shown in Figure 1. The spring element converts linearly the force into strain, and the two FBGs are used to detect the strain distributed on the surface of the spring element.

2.1. Fiber Bragg Grating. Fiber Bragg grating is a spectral filter realized by the principle of Bragg reflection. It is written by exposing the optical fiber to a periodic pattern of ultraviolet light. The Bragg wavelength shift of FBG $\Delta\lambda$ is simultaneously modulated by strain and temperature. The wavelength variation response to axial strain change $\Delta\varepsilon$ and temperature change ΔT could be given by [15]

$$\frac{\Delta\lambda}{\lambda} = (1 - p_e)\,\Delta\varepsilon + (\alpha_f + \zeta)\,\Delta T, \tag{1}$$

where λ is the initial wavelength of FBG p_e, α_f, and ζ are, respectively, the effective photoelastic coefficient, the thermal expansion coefficient, and the thermal-optic coefficient of fused silica fiber. According to (1), the influence of temperature fluctuation should be eliminated in order to obtain pure strain variation.

2.2. Measurement Principle. The spring element of the sensor is a metal plate with two circular holes in a flat oval cavity as shown in Figure 2. When an external tension F is applied, the spring element deforms and strain is produced on its surface. FBG-1 and FBG-2 bonded on the surface of the spring element sense the strain ($\varepsilon_1, \varepsilon_2$) at their placed points (point 1 and 2), respectively. The wavelength shifts of FBG-1 and FBG-2 are obtained by (1)

$$\Delta\lambda_{\text{FBG-1}} = (1 - p_e)\,\lambda_{\text{FBG-1}}\Delta\varepsilon_1 + (\alpha_f + \zeta)\,\lambda_{\text{FBG-1}}\Delta T_1, \tag{2}$$

$$\Delta\lambda_{\text{FBG-2}} = (1 - p_e)\,\lambda_{\text{FBG-2}}\Delta\varepsilon_2 + (\alpha_f + \zeta)\,\lambda_{\text{FBG-2}}\Delta T_2. \tag{3}$$

As the two FBGs in one sensor are fabricated with approximate initial wavelengths ($\lambda_{\text{FBG-1}} = 1556.101\,\text{nm}$, $\lambda_{\text{FBG-2}} = 1561.895\,\text{nm}$) and the initial wavelengths are much larger than the wavelength shifts ($\Delta\lambda_{\text{FBG-1}}, \Delta\lambda_{\text{FBG-2}}$), $\lambda_{\text{FBG-1}}$ and

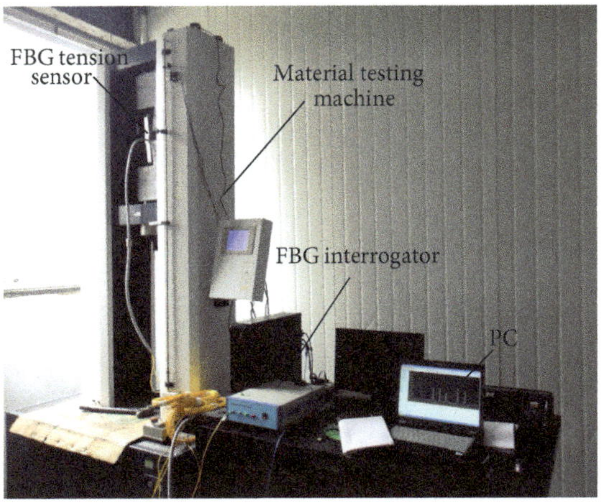

FIGURE 4: Setup of the calibration experiment.

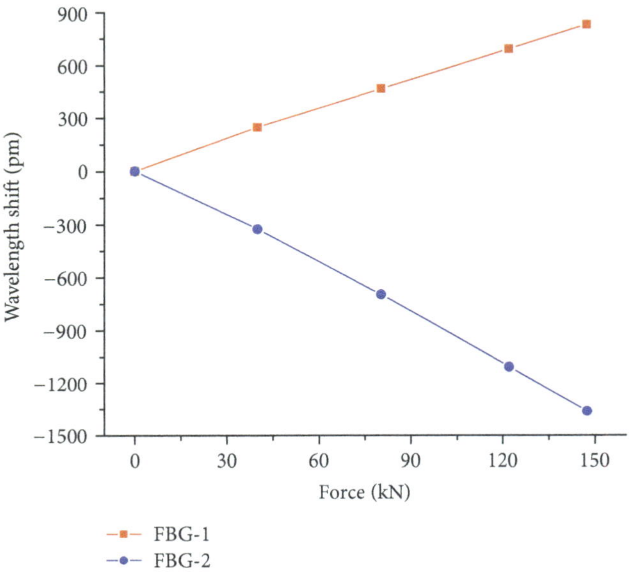

FIGURE 5: Wavelength shifts of the two FBGs versus force.

$\lambda_{FBG\text{-}2}$ could be replaced by an equivalence value (λ). Assuming that the two FBGs undergo the same temperature variation, the difference between the two wavelength shifts could be estimated by subtracting (2) from (3):

$$\Delta\lambda_{FBG\text{-}1} - \Delta\lambda_{FBG\text{-}2} = (1 - p_e)\,\lambda\,(\Delta\varepsilon_1 - \Delta\varepsilon_2). \tag{4}$$

For an FBG of central wavelength of 1550 nm, typical strain sensitivity is approximately 1.2 pm/microstrain. Equation (4) can be adapted as

$$\Delta\lambda_{FBG\text{-}1} - \Delta\lambda_{FBG\text{-}2} = 1.2\,(\Delta\varepsilon_1 - \Delta\varepsilon_2). \tag{5}$$

Figure 3 shows the strain values of ε_1 and ε_2 calculated by the finite element analysis (FEA) with ANSYS, when the spring element is applied by different values of force F. The material of the spring element is 1045 steel and the structure parameters of the spring element used in the calculation and simulation are as follows: Young's modulus $E = 1.93\times10^{11}$ Pa,

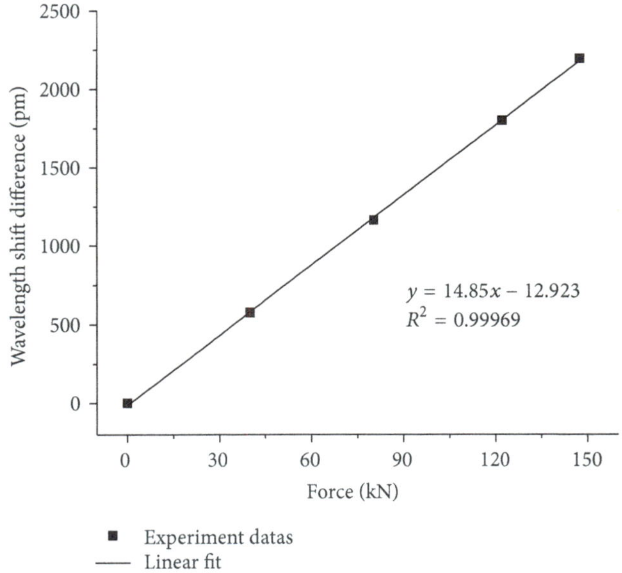

FIGURE 6: Variation of the wavelength shift difference of the two FBGs versus force.

FIGURE 7: General view of the test site.

Poisson's ratio $\mu = 0.31$, radius $R = 33$ mm, diameter $D = 55$ mm, thickness $H = 30$ mm, $h = 10$ mm, length $A = 270$ mm, width $B = 106$ mm, and distance $c = 39$ mm.

It could be found from Figure 3 that there is positive strain ε_1 at point 1 and negative strain ε_2 at point 2. The linear relation between the strain $(\varepsilon_1, \varepsilon_2)$ and the applied force F can be, respectively, expressed as

$$\varepsilon_1 = 5.56F, \qquad \varepsilon_2 = -9F. \tag{6}$$

Assuming that the strain transmission ratio from the spring element to FBGs is 1, regardless of the effect of adhesive, according to (5) and (6), the relation of wavelength

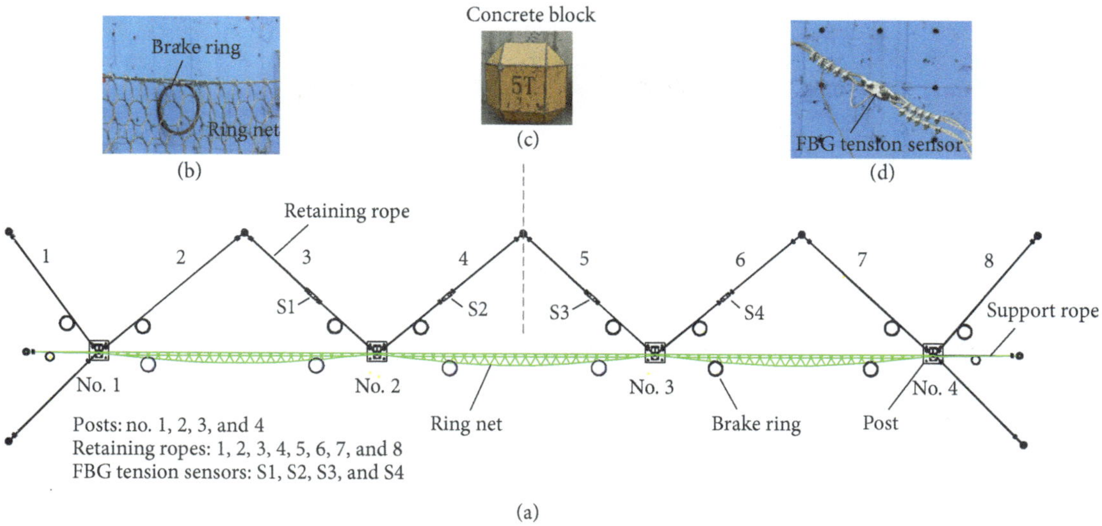

FIGURE 8: Details of the impact test: (a) sketch of the tested barrier, (b) brake ring and ring net, (c) concrete block, and (d) FBG tension sensor installed on the rope.

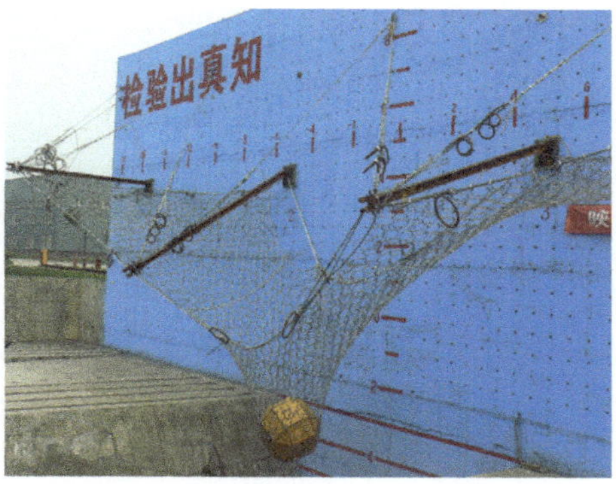

FIGURE 9: The tested barrier after the impact.

shift difference of the two FBGs and the applied force can be obtained:

$$\Delta\lambda_{\text{FBG-1}} - \Delta\lambda_{\text{FBG-2}} = 17.472\Delta F. \tag{7}$$

Equation (7) shows that the difference in wavelength shifts of the two FBGs is a linear function of force and independent of temperature variation. In addition, the force sensitivity calculated is 17.472 pm/kN, and it is enhanced compared to the wavelength shift of one single FBG as sensing signal.

2.3. Calibration Experiment. The calibration experiment has been carried out on a material testing machine as shown in Figure 4. The wavelength interrogation system is used to monitor the FBGs' wavelength shifts and mainly composed of an FBG interrogator and related ancillary components such as multiplexers, switches, and software. The FBG interrogator shown in Figure 4 is based on diffractive grating and charge-coupled device (CCD) array technology with minimum wavelength resolution of 0.4 pm and maximum data acquisition rate of 8 kHz, which could suit most measurement applications of dynamic force. During the experiment, the force applied to the FBG tension sensor is controlled by the material testing machine and changed from 0 to 150 kN.

Figure 5 shows the wavelength shifts of the two FBGs versus the applied force. The wavelength of FBG-1 presents a red shift, indicating that FBG-1 senses the positive strain, while the wavelength of FBG-2 has a blue shift which verifies that FBG-2 measures the negative strain. Both variation patterns are linear, and the slope is 5.57 and −9.28, respectively.

Figure 6 lists the difference values in the wavelength shifts of the two FBGs at each force value. The fitting results are also presented and the force sensitivity obtained is 14.85 pm/kN with a fitting linear correlation coefficient of 99.969%. The measured sensitivity is in agreement with the theoretical calculation value, and the difference is mainly caused by the strain transmission and simulation calculation error.

3. Application for Dynamic Force Measurement

Rockfall protection barriers have been extensively used in civil engineering as passive countermeasures to prevent a rockfall event to reach the surrounding roads, railways, and built-up areas as discussed elsewhere [16, 17]. Full-scale impact tests are very important to evaluate the behaviour of rockfall protection barriers and design procedures that still mostly rely on the results of impact rest as discussed elsewhere [18, 19]. The FBG tension sensors presented in this paper have been utilized to measure the dynamic forces acting on the anchor ropes of the barriers in impact test.

The test site depicted in Figure 7 located in Sichuan, China, was built by Sichuan OST Slope Protection Engineering Co., Ltd, in order to verify and assess whether a rockfall protection barrier is effectively able to withstand the designed impact energy. The test block, a concrete polyhedron, is lifted

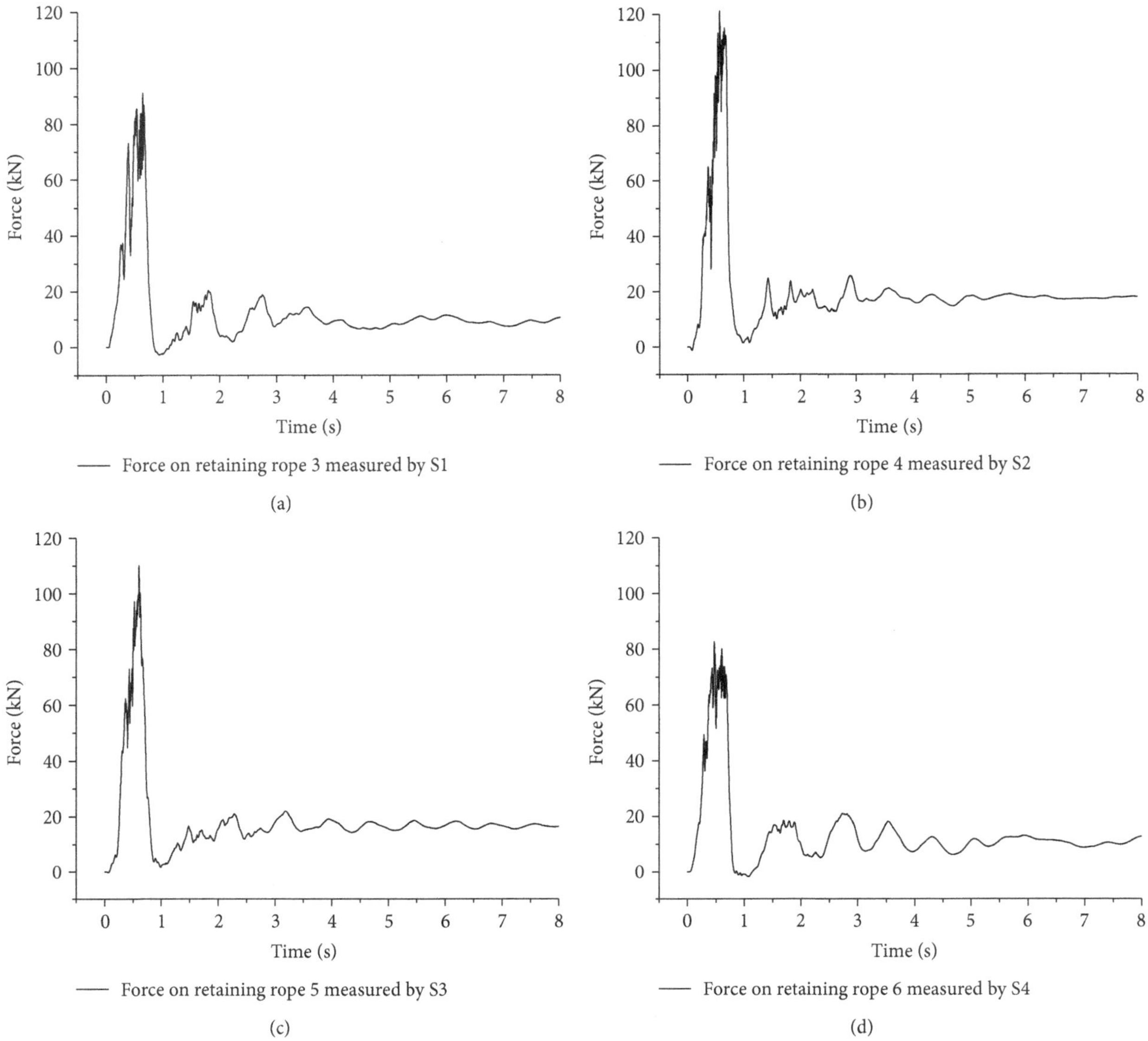

FIGURE 10: The impact forces measured by the FBG tension sensors.

by a travelling crane and then released to vertically impact into the barrier in freefall motion with known kinetic energy.

The tested barrier, a full-scale flexible rockfall protection barrier manufactured by the company, has a capacity to protect against rockfall impact energies of up to 1000 kJ. Figure 8(a) shows the sketch of the tested barrier, which is mainly composed of ring nets, posts, anchor ropes (support rope, retaining rope, and other ropes), clamps, studs, and so forth. The ring nets and brake rings illustrated in Figure 8(b) are the dissipating devices in a barrier.

Four FBG tension sensors (S1, S2, S3, and S4) were installed on the main force-bearing retaining ropes (3, 4, 5, and 6) in the barrier to measure the dynamic tensile forces during the impact test as shown in Figures 8(a) and 8(d). After free falling from a height of 20 m, the 5 ton concrete block smashed into the ring net of the barriers with an impact kinetic energy of 1000 kJ, which is the maximum designed absorbing energy value of the tested barrier.

4. Results and Discussion

In the impact test, the tested barrier successfully caught the falling concrete block and effectively withstood the impact energy of 1000 kJ. The barrier after the impact with the block in the net is shown in Figure 9.

As depicted in Figure 10, the FBG tension sensors clearly recorded the impact forces acting on the retaining ropes during the test with a sampling rate of 1 kHz. The peak values of impact forces on retaining cope 3, 4, 5, and 6 were, respectively, 91.2 kN, 119.5 kN, 110.2 kN, and 82.7 kN, and the overall braking time was approximately 0.87 s. As the block was still in the net, the forces on the ropes did not return to zero after the impact as shown in Figure 10.

The impact forces recorded in the initial first second as shown in Figure 11 indicate that the variation of the impact forces on the retaining ropes could be mainly divided into three stages: smooth increasing to a value, seesaw rising until

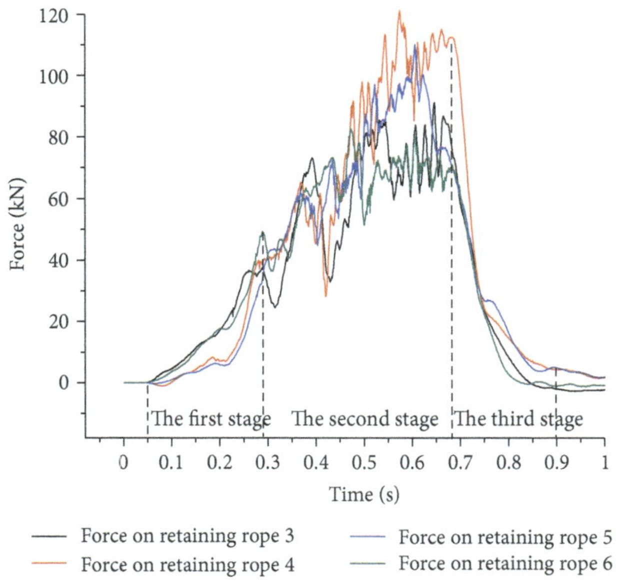

FIGURE 11: The impact forces recorded in the initial first second.

reaching the peak value, and rapidly decreasing. This impact process measured by the FBG tension sensors has a good agreement with the results obtained by Gottardi and Govoni using traditional load cells [20].

Considering the barrier's structure characteristics and the measurement results, it could be found that

(1) in the first stage, the tensile forces on the retaining ropes are lower than the start thresholds of the brake rings on the ropes, and the impact kinetic energy is at first absorbed through plastic deformation of the ring bundles of the ring nets;

(2) in the second stage, the tensile forces on the ropes reach and exceed the start thresholds, and the brake rings are added to dissipating the kinetic energy;

(3) finally in the third stage, the impact energy has been absorbed by the energy dissipating devices.

The presented FBG tension sensors are extremely useful in the impact tests to provide accurate and reliable experimental data for investigating the behaviour of the rockfall protection barriers. Furthermore, due to the characteristics of remote sensing, environmental ruggedness, and massive multiplexing capabilities, they could be applied to monitor the security conditions of the barriers at a remote distance.

5. Conclusions

An FBG tension sensor for anchor rope has been presented and applied to measure the dynamic forces on anchor ropes in impact test. The temperature cross-sensitivity is effectively avoided by using the wavelength shift difference between the two FBGs adhered on the spring element as the force sensitivity parameter. The experiment results indicate that the force sensitivity is 14.85 pm/kN in a range from 0 to 150 kN. The dynamic forces on anchor ropes in the full-scale impact test of a rockfall protection barrier have been clearly recorded by the FBG tension sensors. The measurement range and sensitivity of the tension sensor can be easily adjusted by optimizing the size and material of the spring element to adapt different measure demands, and multiplexing could be realized by adding the FBGs in a single optical fiber. This kind of force sensor is supposed to have many potential applications for quasi-distributed force measurement and remote online monitoring in industrial and mechanical fields.

Acknowledgments

This work was supported by the National Science and Technology Major Project of China under Grant no. 2012ZX04001-012-05 and the Key Project of National Natural Science Foundation of China under Grant no. 50935005. The authors are very grateful to Sichuan OST Slope Protection Engineering Co., Ltd. for the use of test site.

References

[1] I. Muller, R. de Brito, C. E. Pereira, and V. Brusamarello, "Load cells in force sensing analysis—theory and a novel application," *IEEE Instrumentation and Measurement Magazine*, vol. 13, no. 1, pp. 15–19, 2010.

[2] H. Kumar, C. Sharma, and A. Kumar, "Design and development of precision force transducers," *Journal of Scientific and Industrial Research*, vol. 70, no. 7, pp. 519–524, 2011.

[3] D. J. Park, M. K. Kang, J. H. Lee, and S. S. Lee, "Load cell design using fiber Bragg grating sensors," *International Journal of Modern Physics: Conference Series*, vol. 6, pp. 203–208, 2012.

[4] B. Lee, "Review of the present status of optical fiber sensors," *Optical Fiber Technology*, vol. 9, no. 2, pp. 57–79, 2003.

[5] S. J. Mihailov, "Fiber bragg grating sensors for harsh environments," *Sensors*, vol. 12, no. 2, pp. 1898–1918, 2012.

[6] G. R. C. Possetti, R. C. Kamikawachi, M. Muller, and J. L. Fabris, "Metrological evaluation of optical fiber grating-based sensors: an approach towards the standardization," *Journal of Lightwave Technology*, vol. 30, no. 8, pp. 1042–1052, 2012.

[7] J. M. López-Higuera, L. R. Cobo, A. Q. Incera, and A. Cobo, "Fiber optic sensors in structural health monitoring," *Journal of Lightwave Technology*, vol. 29, no. 4, pp. 587–608, 2011.

[8] S. Z. Tian, X. F. Zhao, Z. Zhou, and J. P. Ou, "Application of fiber Bragg grating sensors in civil engineering," in *Fundamental Problems of Optoelectronics and Microelectronics II*, vol. 5851 of *Proceedings of SPIE*, pp. 182–189, Sevilla, Spain, September 2004.

[9] T. H. Loutas, A. Panopoulou, D. Roulias, and V. Kostopoulos, "Intelligent health monitoring of aerospace composite structures based on dynamic strain measurements," *Expert Systems with Applications*, vol. 39, no. 9, pp. 8412–8422, 2012.

[10] G. C. Kahandawa, J. Epaarachchi, H. Wang, and K. T. Lau, "Use of FBG sensors for SHM in aerospace structures," *Photonic Sensors*, vol. 2, no. 3, pp. 203–214, 2012.

[11] M. S. Milczewski, J. C. da Silva, C. Martelli, L. Grabarski, I. Abe, and H. J. Kalinowski, "Force monitoring in a maxilla modal and dentition using optical fiber Bragg gratings," *Sensors*, vol. 12, no. 9, pp. 11957–11965, 2012.

[12] G. M. Ma, C. R. Li, J. Jiang, Y. T. Luo, and Y. C. Cheng, "A novel optical load cell used in icing monitoring on overhead

transmission lines," *Cold Regions Science and Technology*, vol. 71, pp. 67–72, 2012.

[13] J. M. Kim, H. W. Kim, Y. H. Park, I. H. Yang, and Y. S. Kim, "FBG sensors encapsulated into 7-wire steel strand for tension monitoring of a prestressing tendon," *Advances in Structural Engineering*, vol. 15, no. 6, pp. 907–917, 2012.

[14] J. G. Liu, L. F. Xue, G. Y. Kai, Q. D. Zhao, and T. Guo, "A pull sensor with temperature self-compensated based on loop thin-wall section beam," *Journal of Optoelectronics Laser*, vol. 17, no. 9, pp. 1031–1034, 2006.

[15] A. D. Kersey, M. A. Davis, H. J. Patrick et al., "Fiber grating sensors," *Journal of Lightwave Technology*, vol. 15, no. 8, pp. 1442–1463, 1997.

[16] A. Trad, A. Limam, and P. Robit, "Real scale experiments on rockfall protection barriers," *Applied Mechanics and Materials*, vol. 82, pp. 734–739, 2011.

[17] D. F. Castro, L. Q. Lopez, L. A. F. Sanudo, and J. P. G. Mery, "Research and development of tests for the certification of dynamic rockfall barriers," *Revista de la Construction*, vol. 8, no. 2, pp. 51–59, 2009.

[18] S. Margreth and A. Roth, "Interaction of flexible rockfall barriers with avalanches and snow pressure," *Cold Regions Science and Technology*, vol. 51, no. 2-3, pp. 168–177, 2008.

[19] O. Buzzi, M. Spadari, A. Giacomini, S. Fityus, and S. W. Shoan, "Experimental testing of rockfall barriers designed for the low range of impact energy," *Rock Mechanics and Engineering*, vol. 46, no. 4, pp. 701–712, 2013.

[20] G. Gottardi and L. Govoni, "Full-scale modelling of falling rock protection barriers," *Rock Mechanics and Rock Engineering*, vol. 43, no. 3, pp. 261–274, 2010.

Nomographs and Feasibility Graphs for Enumeration of Ravigneaux-Type Automatic Transmissions

Essam L. Esmail

College of Engineering, University of Qadisiyah, Al Qadisiyah, Iraq

Correspondence should be addressed to Essam L. Esmail; dr.essamesmail@yahoo.com

Academic Editor: Marco Ceccarelli

Although there are many epicyclic-type automatic transmissions in production, the related configuration design methods are still tedious and borne to human error. A simple methodology for the systematic design of the Ravigneaux-type epicyclic gear transmissions needs to be developed. First, fundamentals and gear-shifting of four-speed and six-speed epicyclic-type automatic transmissions are illustrated to establish the design requirements. Second, based on the kinematic nomographs of the corresponding basic gear ratios, a simple clutching-sequence method is proposed and illustrated. Next, a planar-graph representation is presented to arrange the desired clutches for each possible clutching sequence into the epicyclic gear mechanism. Then, with the above methods, the systematic designs of the epicyclic gear mechanisms are given for demonstrating the feasibility of the proposed methodology. The result of this work shows that the seven-, eight-, and nine-link two-DOF Ravigneaux-type epicyclic gear mechanisms could reach four-, six-, and eight-forward speeds at most, respectively. New five-, six-, seven-, and eight-velocity automatic transmissions are enumerated from the two-ring eight- and nine-link Ravigneaux gear mechanisms. It is a major breakthrough to design completely satisfactory eight-speed automatic transmissions from the nine-link Ravigneaux gear mechanism.

1. Introduction

For a long-time automatic transmissions with planetary gear trains are used in the automotive industry. Ravigneaux in 1940 proposed seven- and eight-link two-degree-of-freedom (DOF) epicyclic gear mechanisms [1, 2]. These epicyclic gear mechanisms are called the Ravigneaux gear mechanisms. Figure 1 shows an automatic transmission which provides three forward speeds and one reverse speed [3]. It consists of a seven-link two-DOF Ravigneaux gear mechanism, two rotating clutches C_1 and C_2, and two band clutches B_1 and B_3. In the associated clutching sequence Table, X_i indicates that the corresponding clutch is activated on the ith link for that gear. The ranges of output velocities are classified into two kinds: under drive (UD) and reverse drive (RD) according to whether the velocity is between zero and the input velocity, or less than zero. A "direct drive" (DD) is equal to the input velocity.

The seven-link two-DOF-Ravigneaux gear train has been developed by nearly all automotive manufacturers as three- or four-velocity automatic transmission [3]. It can be found in Ford C_3, Ford C_5, Mercedes Benz, Toyota A40, and Nissan, to name, a few three-velocity automatic transmissions. It can also be found in KM 175 and 176, Ford AOD, ZF 4 HP 14, and Borg-Warner [4], to name, a few four-velocity automatic transmissions [5]. Figure 2 shows the ZF 4 HP 14 automatic transmission [6], which can provide four forward speeds and one reverse speed.

Sometimes, the seven-link two-DOF Ravigneaux gear train is integrated with a simple epicyclic gear train to form ten-link three-DOF epicyclic gear mechanisms to enhance the number of speeds [6, 7], providing six forward speeds. Figure 3 shows Lepelletier automatic transmission and its clutching sequence table [8]. A widespread gear set concept is that of Lepelletier. This design is based on a single planetary gear set with rear-mounted Ravigneaux gear set. In 2003, ZF [9] used this gear set design to bring the first 6-speed passenger car transmission 6 HP 26 on the market (Figure 3).

Two-DOF eight-link Ravigneaux gear mechanism, shown in Figure 4(b), consists of a long pinion and two

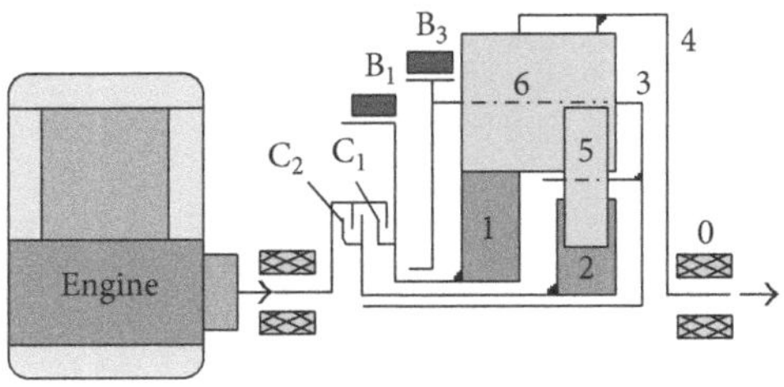

Range	Activated clutches			
	C_1	C_2	B_1	B_3
UD_1		X		X
UD_2		X	X	
DD	X	X		
RD	X			X

FIGURE 1: A typical automatic transmission mechanism and its clutching sequence table.

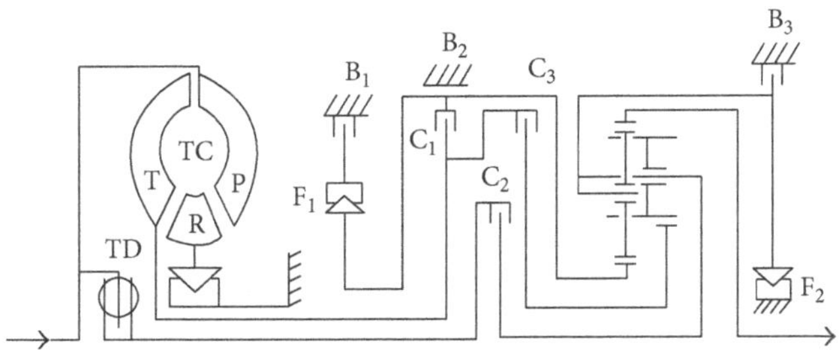

FIGURE 2: Gearbox diagram for 2 ZF 4 HP 14 4-speed automatic transmission with Ravigneaux gear set (TC: Trilok converter, P: pump, T: turbine, R: reactor with freewheel, TD: torsion damper, F: freewheels, B: brakes, C: clutches) [6].

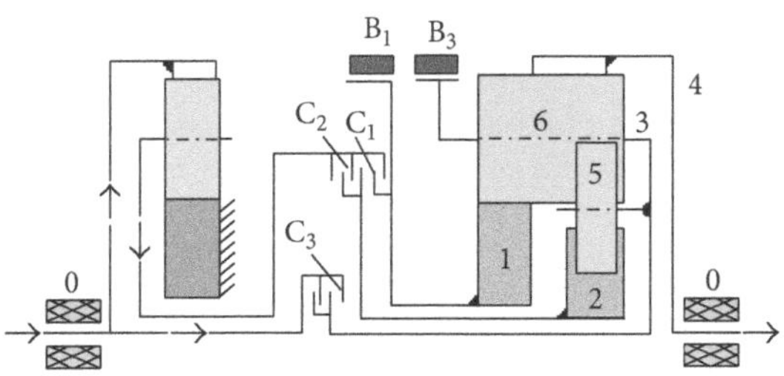

	C_1	C_2	C_3	B_1	B_3	Velocity ratio
UD_1		X			X	4.135
UD_2		X		X		2.363
UD_3	X	X				1.508
UD_4		X	X			1.14
OD_1	X		X			0.86
OD_2			X	X		0.675
RD	X				X	−3.127

FIGURE 3: Six-velocity Lepelletier automatic transmission and its clutching sequence table [8].

or more short pinions connecting two sun gears and two ring gears. Unfortunately, two-DOF eight-link Ravigneaux gear trains have been developed as four-velocity automatic transmission [10] while they can be used to produce automatic transmissions having more than four speeds. This work in part will attempt to attain maximum sequential velocity ratios for any given epicyclic gear train.

A seven-link 2-DOF Ravigneaux gear mechanism, an eight-link 2-DOF Ravigneaux gear mechanism, and a nine-link 2-DOF stepped Ravigneaux gear mechanism are shown in Figure 4, respectively.

The literature on the design of planetary gear trains includes conceptual designs, kinematic analysis, power flow and efficiency analysis, and configuration designs. However, relatively little work has been done on the enumeration of clutching sequences and configuration design of EGMs. This paper presents a systematic procedure to enumerate clutching sequences and to find feasible clutch layouts for Ravigneaux planetary gear trains.

2. Literature Review

The selection of an optimal clutching sequence cannot be solved analytically. Nadel et al. [11–13] formulated the task as a constraint satisfaction problem. Hsieh and Tsai [14, 15], Hwang and Huang [16], and Hsu and Huang [17] used algorithmic techniques. Ross and Route [18] and Esmail [19] introduced graphic techniques. Hattori et al. [20] used phase geometry method.

Nadel et al. applied an artificial intelligence technique to enumerate clutching sequences for EGMs made up of two basic epicyclic gear trains. The artificial intelligence technique is a powerful tool for solving the transmission design problem. However, the technique assumes that the design variables have discrete values in prescribed domains. Furthermore, it requires a search over the entire feasible solution space. This methodology is suitable for PGTs in which two simple PGTs are combined. These shortcomings inevitably reduce the efficiency of the algorithm.

Hattori et al. [20] proposed twenty-three phase geometric patterns for five-speed automatic transmissions; each could provide four clutching sequences. Each feasible clutching sequence obtained could be used to construct a clutch layout. However, this approach is suitable only for PGTs consisting of two sun gears, two ring gears, and one to three meshed planet gears mounted on a common arm. Ross and Route [18] introduced a design tool based on a lever analogy. It includes calculation of gear ratios, gear trains selection,

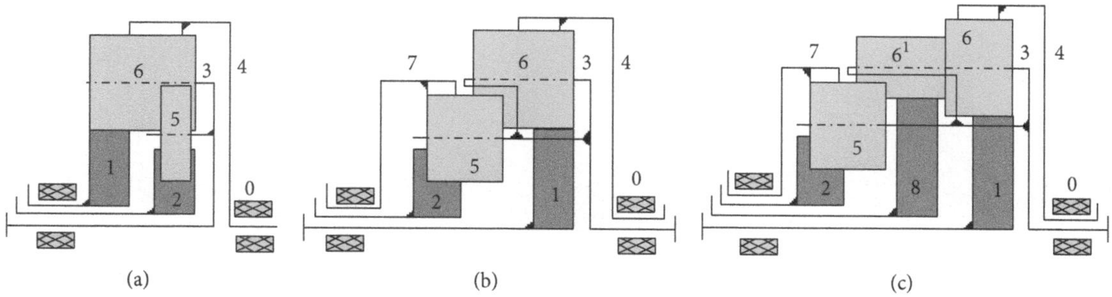

FIGURE 4: Seven-, eight-, and nine-link Ravigneaux gear mechanisms.

and the construction of clutch layouts for parallel-connected PGTs. Hwang and Huang [16] and Hsu and Huang [17] used similar methods to that used by Hsieh and Tsai [14] to enumerate the clutching sequences of two-DOF eight-link Ravigneaux gear mechanism. Moreover, the five studies cited above are restricted to constructing clutch layouts of automatic transmissions for specific types of PGTs.

Hsieh and Tsai [14] used the concept of fundamental gear entities (FGEs) proposed by Chatterjee and Tsai [21] in conjunction with their earlier kinematic study [22] to determine the most efficient clutching sequence associated with automatic transmission [15]. They applied combinatorial enumeration procedure to arrange the velocity ratios in a descending sequence. Then, they used an algorithm to enumerate clutching sequences for EGMs composed of two or more FGEs. A computer algorithm for the enumeration of clutching sequences is given by Hsieh [23]. The algorithm needs information containing the approximate gear sizes arranged in a descending order.

Most combinatorial enumeration procedures are done through the process of generating and testing. The procedure is thus divided into two parts: a generator of all possible solutions and a tester that selects only those solutions that meet the constraints. An important issue in using a generating and testing technique is the distribution of knowledge between the generator and tester. The generator produces solutions satisfying some of the constraints. The tester then selects those solutions that satisfy the rest of the constraints. While this technique is valid for solving transmission design problems, it limits the solutions to the knowledge (information) contained in the generator and tester. This inevitably reduces the efficiency of this solution technique and needs complicated computer algorithm. Usually, putting more knowledge in the generator results in a more efficient procedure. Nevertheless, Hsieh and Tsai [14] did not develop any effective method of arranging clutch layouts for the synthesized clutching sequences. However, the elimination of invalid clutching sequences was conducted by inspection. Identifying invalid clutching sequences by inspection is not always reliable.

Esmail [19] proposed a methodology, based on nomographs, for the enumeration of the associated clutching sequence table for an epicyclic gear mechanism. This method simplifies the synthesis of the clutching sequence of an epicyclic gear mechanism efficiently.

Without careful examination, a clutching sequence may be mistaken for a usable clutch layout for an automatic transmission. Hsu and Huang [17] proposed a planar-graph representation to arrange the desired clutches for each possible clutching sequence into the Ravigneaux gear mechanism. Hwang and Huang [24] introduced coded sketches for connecting clutch elements to planetary gear trains for automotive automatic transmissions.

The literature survey reveals that although some studies are made on the clutching sequence synthesis, they are all tedious. A simple methodology for the design of planetary automatic transmission needs to be developed. In this paper nomographs are used to synthesize the clutching sequences of Ravigneaux gear trains; the nomograph method is described in a series of previous papers [19, 25, 26]. Only the related topics needed in this paper will be reviewed wherever appeared.

In this paper, more efficient solution techniques are developed to overcome those shortcomings. This paper applies kinematic nomographs [19, 25] and feasibility graphs [17] of Ravigneaux gear trains to achieve the goal. By virtue of these solution techniques, completely satisfactory six- and eight-speed automatic transmissions are designed from the eight- and nine-link Ravigneaux gear mechanism, respectively.

3. Enumeration of All Feasible Clutching Sequences with the Aid of Nomograph

The first stage of designing epicyclic-type automatic transmissions is to synthesize all possible clutching sequences from the epicyclic-gear mechanism. Nomographs are used for this purpose.

3.1. Nomographs and Geometry Relations for Seven-Link Ravigneaux Gear Train. A nomograph is defined as three or more axes, or scales, arranged such that problems of three or more variables can be solved using a straightedge. In the particular case of EGTs, a nomograph can be constructed using three or more vertical parallel axes [19, 27].

Traditionally, the velocity ratio is used to study the velocity relations among the different links of an EGT [3].

Let the symbol $R^z_{x,y}$ denote the velocity ratio between links x and y, with reference to link z where x, y, and z are any three links in the gear train, then

$$R^z_{x,y} = \frac{\omega_x - \omega_z}{\omega_y - \omega_z}. \tag{1}$$

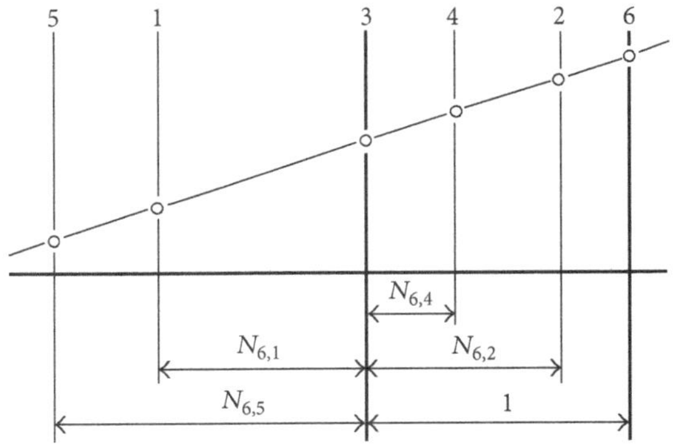

FIGURE 5: Nomograph for the double-planet FGT shown in Figure 4(a), in terms of planet gear 6, as given in [19].

Since the gear mechanism, shown in Figure 4(a), is a double-planet FGE, then its nomograph can be drawn in terms of planet gear 6 as shown in Figure 5.

The term "gear ratio" is used in this paper to denote the ratio of a meshing gear pair. It is defined by the ratio of a planet gear p with respect to a sun or ring gear x:

$$N_{p,x} = \mp \frac{Z_p}{Z_x}, \quad (2)$$

where Z_p and Z_x denote the numbers of teeth on the planet and the sun or ring gear, respectively, and the positive or negative sign depends on whether x is a ring or sun gear.

Considering the kinematics of a fundamental circuit, the fundamental circuit equation can be written as [19]

$$\frac{(\omega_x - \omega_c)}{(\omega_p - \omega_c)} = N_{p,x}. \quad (3)$$

Equation (3) can be rewritten for ring, sun, planet gears, and carrier to obtain $N_{p,r}$, $N_{p,s}$, $N_{p,p}$, and $N_{p,c}$, respectively. The gear ratio for small sun gear s_s that is not meshing directly with the first planet gear p on which the nomograph is drawn and is meshing with the second planet gear p_1 can be found in terms of the gear ratio of the two planets $N_{p,p1}$ as

$$N_{p,s_s} = N_{p,p_1} \cdot N_{p_1,s_s}. \quad (4)$$

The values of the gear ratios are used to place the axes of the nomograph shown in Figure 5. The ω_c axis passes at the origin, and the ω_p axis is one unit apart from it.

The gear ratios for the Ravigneaux gear train are

$$N_{6,5} = -\frac{Z_6}{Z_5}, \quad (5)$$

$$N_{5,2} = -\frac{Z_5}{Z_2}, \quad (6)$$

$$N_{6,1} = -\frac{Z_6}{Z_1}, \quad (7)$$

$$N_{6,4} = \frac{Z_6}{Z_4}. \quad (8)$$

Considering the geometry relations of the Ravigneaux gear mechanism given in (5) through (8), the values of the gear ratios can be deduced as follows:

$$\begin{gathered} -\infty \le N_{6,5} \le 0, \\ -\infty \le N_{5,2} \le 0, \\ -\infty \le N_{6,1} \le 0, \\ 0 \le N_{6,4} \le 1. \end{gathered} \quad (9)$$

From (4), we can write

$$N_{6,2} = N_{6,5} \cdot N_{5,2}, \quad (10)$$

$$N_{6,2} = \frac{Z_6}{Z_2}. \quad (11)$$

Since Z_4 is greater than Z_2, then from (8) and (11) it can be shown that

$$N_{6,2} > N_{6,4}. \quad (12)$$

3.2. Enumeration of All Feasible Clutching Sequences. Kinematic relationships among the links of this FGE can easily be visualized by observation from the nomograph shown in Figure 5. Any straight line through the input operating velocities of the EGT will intersect other axes at the operating velocities of the links representing to those axes. From the nomograph $R^z_{x,y}$ can be written as [19]

$$R^z_{x,y} = \frac{N_{p,x} - N_{p,z}}{N_{p,y} - N_{p,z}}. \quad (13)$$

Because there are four coaxial links and link four is preassigned as the output link, this gear train can provide six overall velocity ratios [27, 28]. Figure 6 shows the clutching sequence nomograph for this mechanism.

In arranging a clutching sequence, it is highly desirable to achieve a single-shift transition [19]. In order to achieve single-shift transitions, the UDs can be further classified into two sets. The ODs and RDs are used with both of the UD sets.

A direct drive is obtained by simultaneously clutching two coaxial links of an EGT to the input power source. A reverse drive can be obtained by applying one or two of the clutches designed for the forward drives to the reverse drive.

As a result, we obtain two descending sequences of velocities as shown in Figures 7(a) and 7(b), which result in two three-velocity and two four-velocity clutching sequences.

3.3. Feasibility Graphs for an Epicyclic Gear Train to Form an Automatic Transmission. A graphical representation is proposed by [17] to easily and quickly detect the possibility of the arrangement of all the rotating clutches and band clutches into the gear mechanism to form an automatic transmission.

A graph that can be drawn on a plane such that none of two of its edges intersect is called planar; meaning it can form an automatic transmission. A graph that cannot be drawn on a plane without crossover between its edges is called nonplanar; meaning there is no possibility to form an automatic transmission. Here we shall call such graphs as

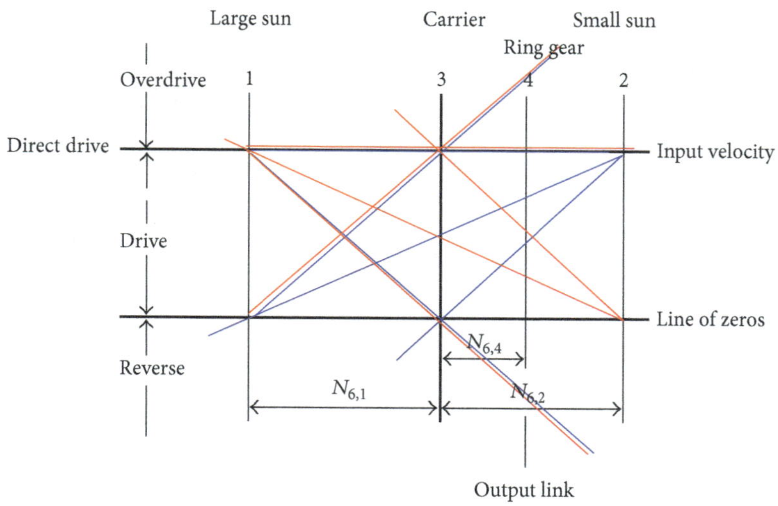

FIGURE 6: Clutching sequence nomograph for the Ravigneaux gear train.

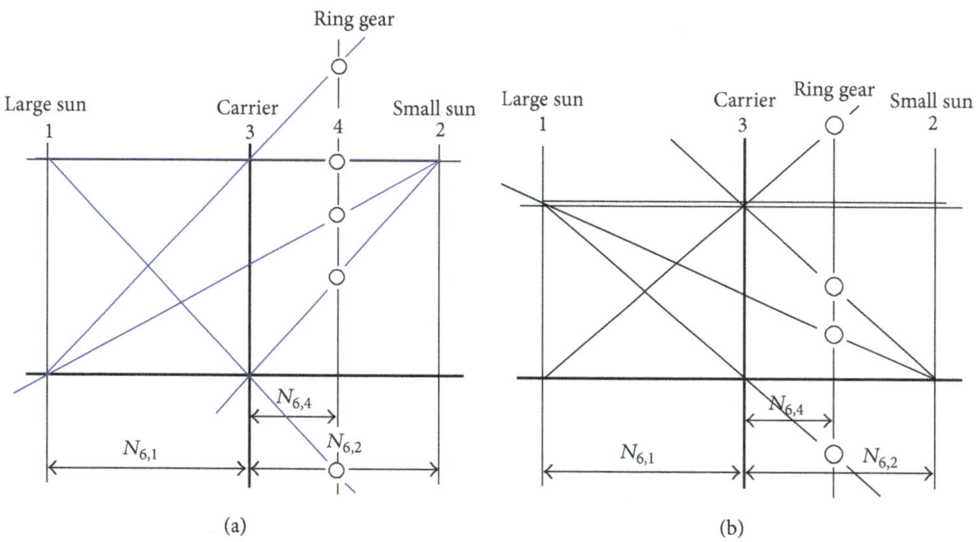

FIGURE 7: Clutching sequence nomograph for (a) the first reduction set and (b) the second reduction set.

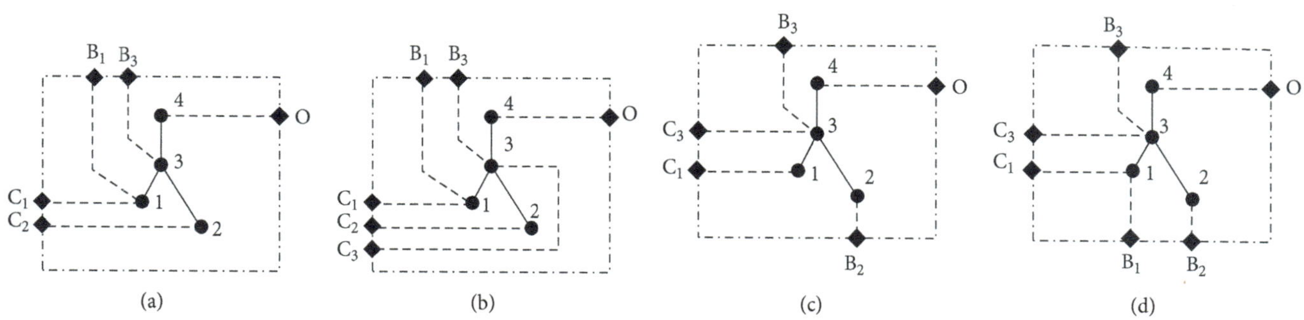

FIGURE 8: Feasibility graphs for the Ravigneaux gear train to form three- or four-velocity automatic transmissions.

the feasibility graphs of a gear train to form an automatic transmission. They are shown in Figure 8(a) through 8(d).

As a result, all of the four sets of clutching sequences are feasible to form automatic transmissions, and the corresponding four-velocity automatic transmissions are drawn, as shown in Figures 9(a) through 9(d), respectively.

Figure 9(a) shows one feasible clutching sequence with rotating clutches attached to links 1 and 2 and band clutches attached to links 1 and 3. This clutching sequence has been applied in most three-velocity Ravigneaux automatic transmissions [3, 6]. The other clutching sequences obtained are in agreement with those reported in the literature [5, 14, 23],

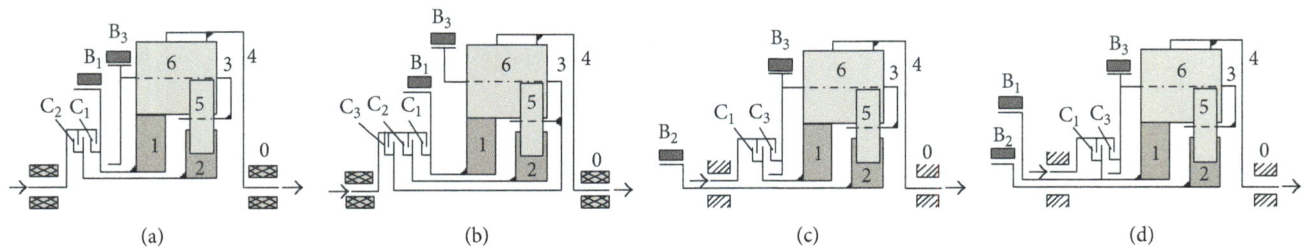

FIGURE 9: Functional representation of the three- and four-velocity Ravigneaux automatic transmissions.

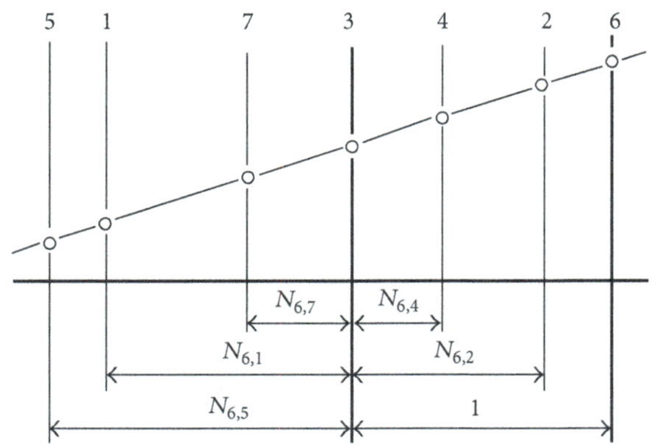

FIGURE 10: Nomograph for the double-planet FGT shown in Figure 4(b), in terms of planet gear 6 [19].

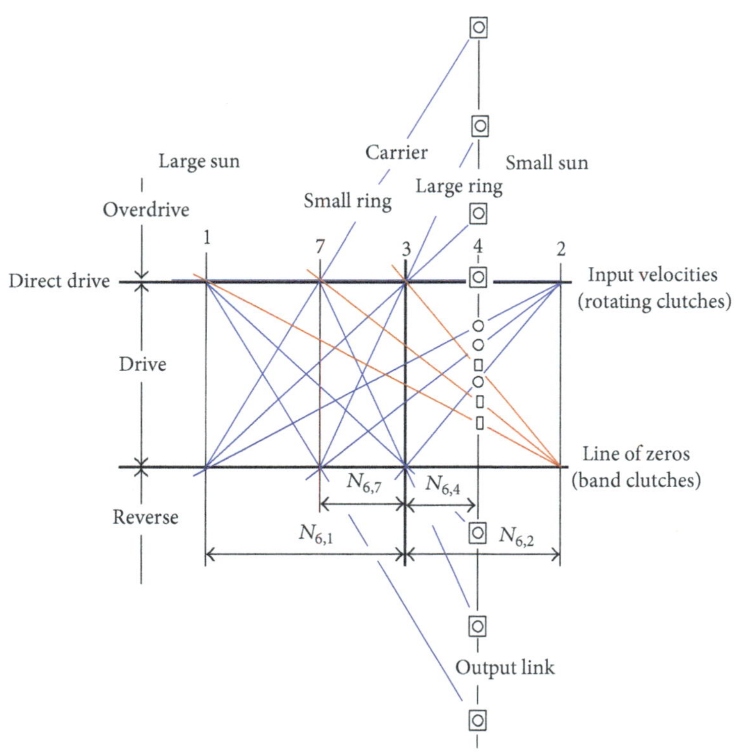

FIGURE 11: Clutching sequence nomograph for the Ravigneaux gear train shown in Figure 4(b).

except that there is no need to any information containing the approximate gear sizes arranged in a descending order.

3.4. Nomographs and Geometry Relations for Eight-Link Ravigneaux Gear Train. Figure 10 shows the nomograph of the mechanisms shown in Figure 4(b).

The gear ratios for the eight-link Ravigneaux gear train are the same as that for the seven-link given before. In addition the gear ratio for the added small ring gear can be written as follows:

$$N_{5,7} = \frac{Z_5}{Z_7}. \tag{14}$$

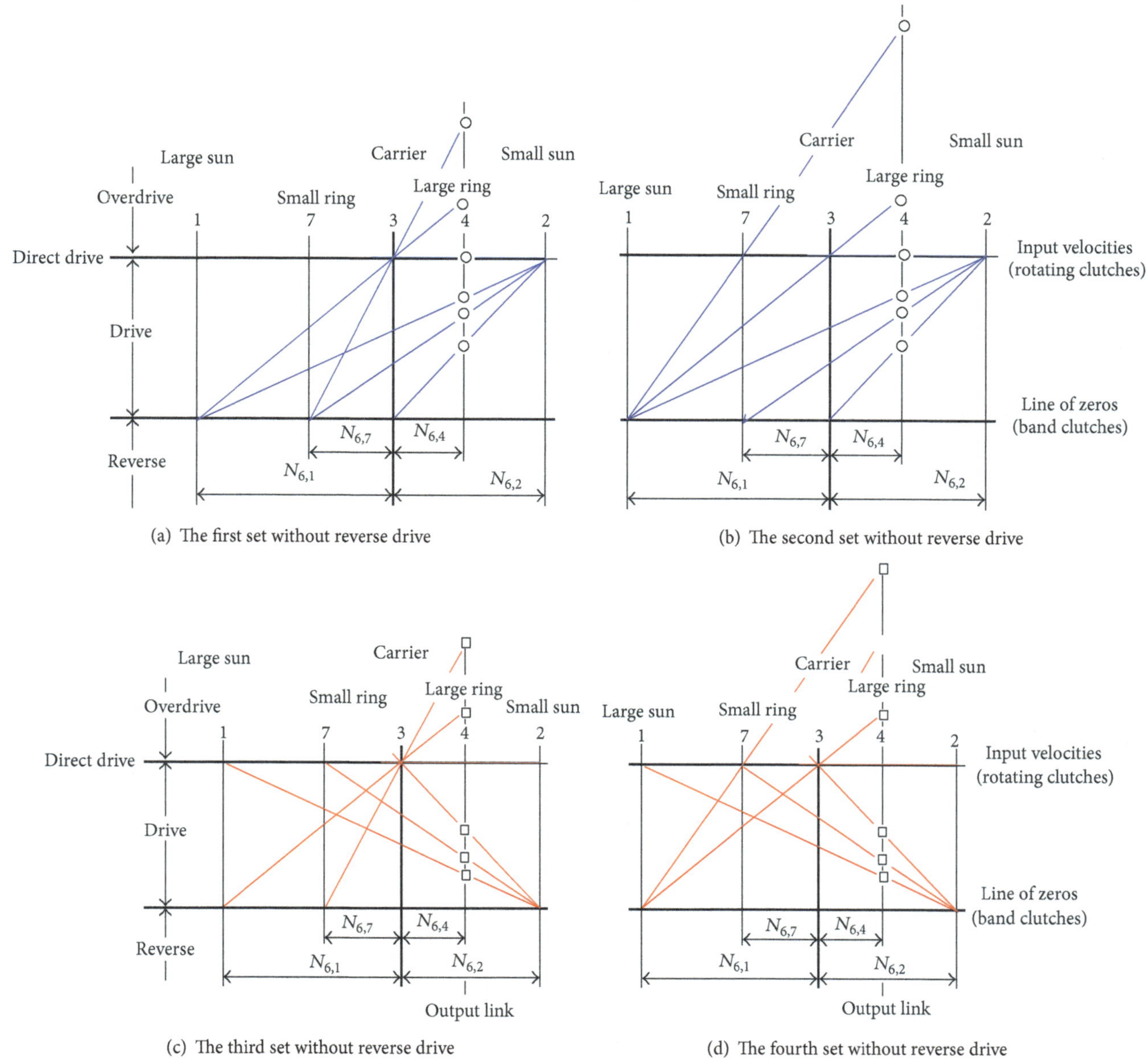

FIGURE 12: Clutching sequence nomographs for the first, second, third, and fourth sets.

It can be concluded that $0 \leq N_{5,7} \leq 1$. From (4), we can write

$$N_{6,7} = N_{6,5} \cdot N_{5,7}. \quad (15)$$

Therefore,

$$N_{6,7} = -\frac{Z_6}{Z_7}. \quad (16)$$

Since Z_7 is greater than Z_1, then from (7) and (16)

$$N_{6,7} > N_{6,1}. \quad (17)$$

3.5. Enumeration of All Feasible Clutching Sequences. Because there are five coaxial links and link four or seven can be assigned as the output link, this gear train can provide twelve overall velocity ratios for each assignment. Figure 11 shows the clutching sequence nomograph for this mechanism.

In order to achieve single-shift transitions, the UDs can be further classified into two sets. By adding a direct drive and two ODs, a total of four six-velocity sets are obtained. A direct drive is obtained by simultaneously clutching two coaxial links of an EGT to the input power source. As a result, we obtain four descending sequences of velocities as shown in Figure 12.

A reverse drive can be obtained by applying one or two of the clutches designed for the forward drives to the reverse drive.

If a reverse drive is added, three six-velocity clutching sequences can be enumerated for each set which result in twelve six-velocity clutching sequences. For the small ring

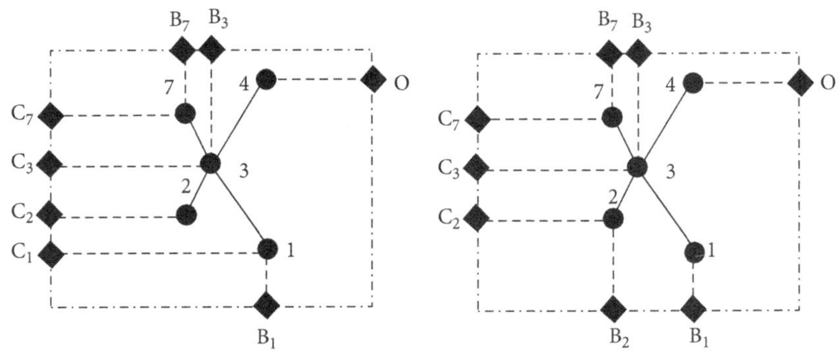

(a) Feasibility graph for the first and second sets to form six-velocity automatic transmission

(b) Feasibility graph for the third and fourth sets to form five-velocity automatic transmission

FIGURE 13: Feasibility graphs for the Ravigneaux gear train to form five- or six-velocity automatic transmissions.

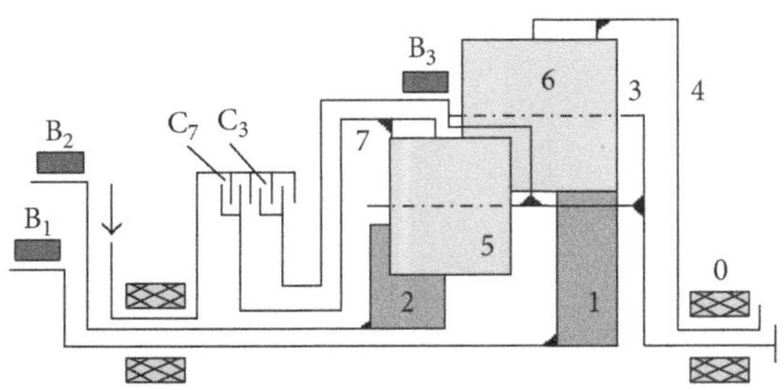

FIGURE 14: Functional representation of five-velocity Ravigneaux gear train based on the fourth reduction set shown in Figure 12(d).

to be an output link another twelve six-velocity clutching sequences can be enumerated.

The first set can realize six-speed automatic transmission, while having three clutches and three brakes only.

3.6. Feasibility Graphs for the Two-Ring Epicyclic Gear Train to Form an Automatic Transmission. The feasibility graphs of the eight-link Ravigneaux gear train to form automatic transmissions are shown in Figures 13(a) and 13(b).

Figure 13(a) reveals that only a clutch can be attached to link 2. For the first and second sets of clutching sequences shown in Figures 12(a) and 12(b), they are controlled by $C_1, C_2, C_3, C_7, B_1, B_3$, and B_7. As a result, all of the six clutching sequences that can be obtained from the first and second sets are feasible to form automatic transmissions.

Figure 13(b) reveals that only a band clutch can be attached to link 1. The third and fourth sets of clutching sequences shown in Figures 12(c) and 12(d) are controlled by $C_1, C_2, C_3, C_7, B_1, B_2, B_3$, and B_7. As a result by excluding C_1 from the clutching sequences nomograph, two clutching sequences are feasible to form five-velocity automatic transmissions.

Figure 12(d) shows a clutching sequence nomograph without reverse drive based on the fourth reduction set. By excluding C_1 from the clutching sequence nomograph, a five-velocity automatic transmission is possible. Figure 14 shows the corresponding functional representation, with a band clutch B_3 attached to link 3 for reverse drive. The input clutches C_7 and band clutch B_3 are applied for the reverse drive.

3.7. Nomographs and Geometry Relations for Nine-Link Ravigneaux Gear Train. Figure 15 shows the nomograph of the mechanisms shown in Figure 4(c).

The gear ratios for the stepped nine-link Ravigneaux gear train are the same as those for the seven- and eight-link gear trains given before, except that the gear ratios are written in terms of the plant to which it is meshing, either gear 6 or 6^1 of the stepped plant:

$$N_{6^1,7} = -\frac{Z_{6^1}}{Z_7}. \quad (18)$$

In addition, the gear ratio for the added third sun gear can be written as, $N_{6^1,8} = -Z_{6^1}/Z_8$.

Since Z_7 is greater than Z_1 and Z_6 is greater than Z_{6^1}, then from (7) and (18) it can be shown that

$$N_{6^1,7} > N_{6,1}. \quad (19)$$

Similarly $N_{6^1,8} > N_{6,1}$ and $N_{6^1,7} > N_{6^1,8} > N_{6,1}$. Moreover

$$N_{6^1,2} = \frac{Z_{6^1}}{Z_2}. \quad (20)$$

Since Z_4 is greater than Z_2 and Z_6 is greater than Z_{6^1}, then from (8) and (20) it can be shown that these gear ratios work oppositely; while one increases the gear ratio, the other decreases it, but, without loss of generality, we can select the most probable case where $N_{6^1,2} > N_{6,4}$.

Because there are six coaxial links and link four or eight can be assigned as the output link, this gear train can provide twenty overall velocity ratios for each assignment. The velocity ratios are classified into three groups; underdrive,

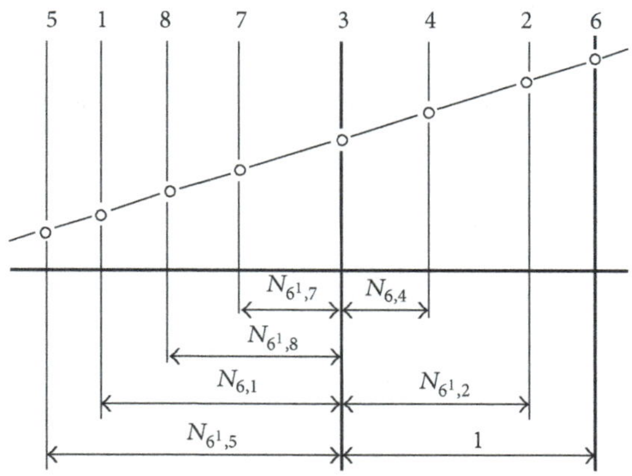

FIGURE 15: Nomograph for the double-planet FGT shown in Figure 4(c), in terms of planet gear 6.

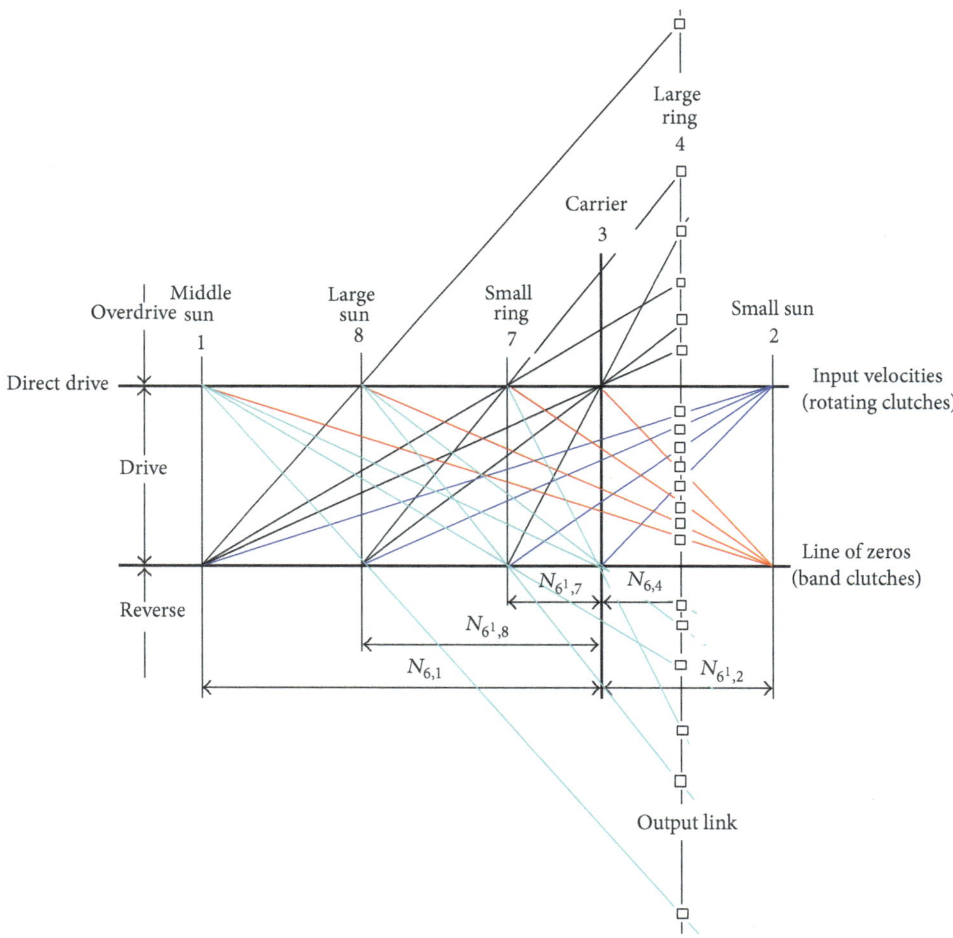

FIGURE 16: Clutching sequence nomograph for the nine-link Ravigneaux gear train shown in Figure 4(c).

overdrive, and reverse drive. If there are more than two sets of velocity ratios in any group, they are further classified into all possible sets or subgroups based on the constraint that only one clutch can be shifted in each set. The velocity ratios in any set are arranged in a descending order, and then the corresponding clutching sequence is generated. Possible clutching sequences are generated by combining the UD, DD, OD, and RD subgroups together. Figure 16 shows the clutching sequence nomograph for the nine-link Ravigneaux gear mechanism shown in Figure 4(c).

In order to achieve single-shift transitions, the UD velocities are classified into two sets (shown in Figure 17).

For the first set, the UD clutching sequences shown in Figure 17(a) are controlled by C_2, B_1, B_3, B_7 and B_8, while

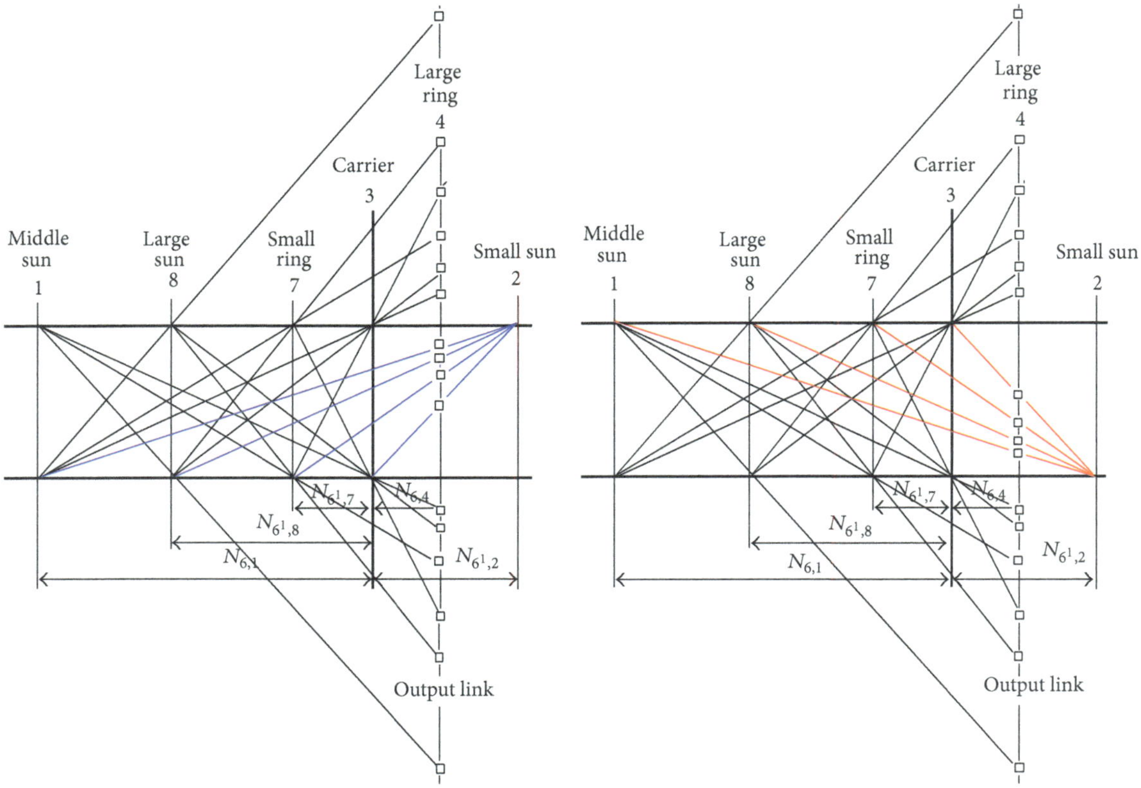

(a) First UD set (b) Second UD set

FIGURE 17: Clutching sequence nomograph for the two UDs of the nine-link Ravigneaux gear train.

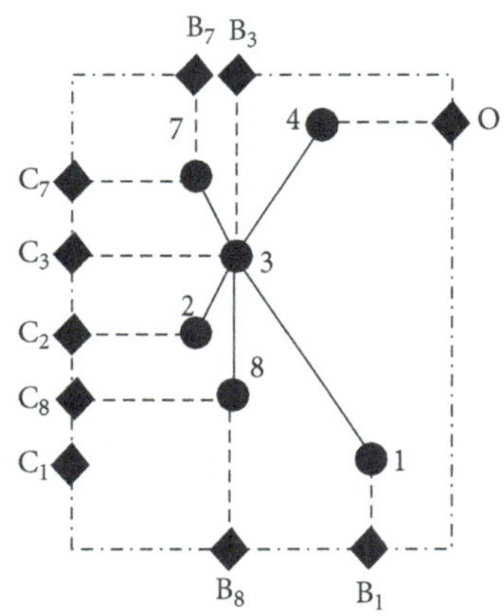

FIGURE 18: Feasibility graph for the first UD set to form eight-velocity automatic transmission.

those for the second set shown in Figure 17(b) are controlled by B_2, C_1, C_3, C_7, and C_8.

The feasibility graphs of the nine-link Ravigneaux gear train to form automatic transmissions are shown in Figures 18 and 23.

For the first UD set, the clutching sequences, shown in Figure 18, are controlled by C_2, C_3, C_7, C_8, B_1, B_3, B_7, and B_8. As a result, all of the UD and OD clutching sequences that can be obtained from the first set are feasible to form automatic transmissions. The RDs are only obtained from one of the rotating clutches C_7 or C_8 and one of the band clutches B_3 or B_7.

Figure 18 reveals that a clutch cannot be attached to link 1; thus, the last three reverse clutching sequences shown in Figure 17(a) are inadmissible. As a result, only thirteen clutching sequences shown in Figure 19 are feasible at this design stage.

Based on the first UD set of feasibility graph, the general clutching sequence layout for the nine-link Ravigneaux gear train is shown in Figure 20. All of the eight-velocity feasible

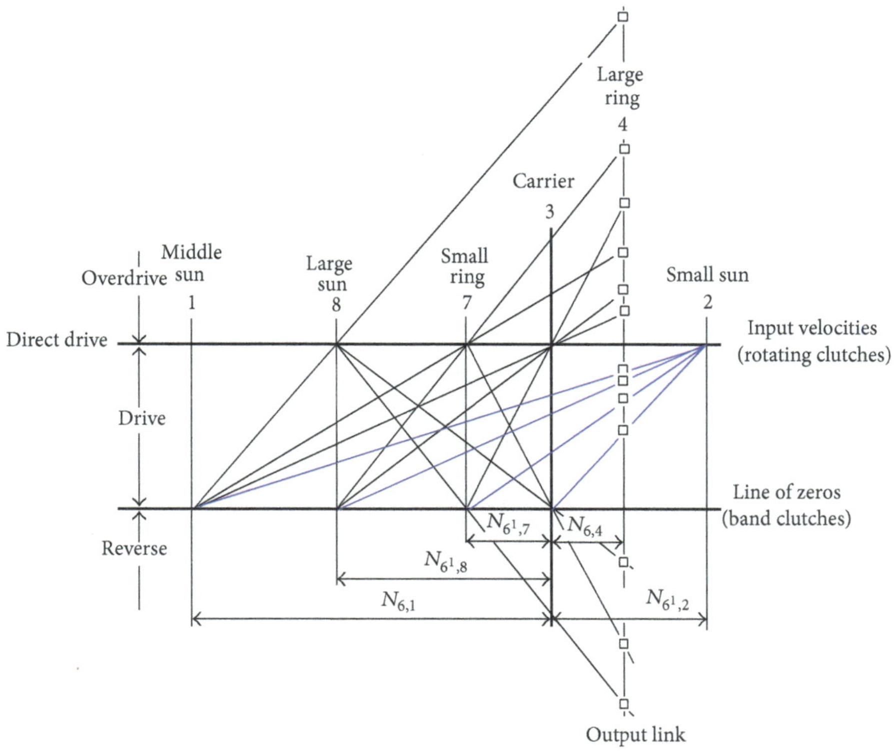

FIGURE 19: Clutching sequence nomograph for the nine-link Ravigneaux gear train shown in Figure 4(c) based on the feasibility graph.

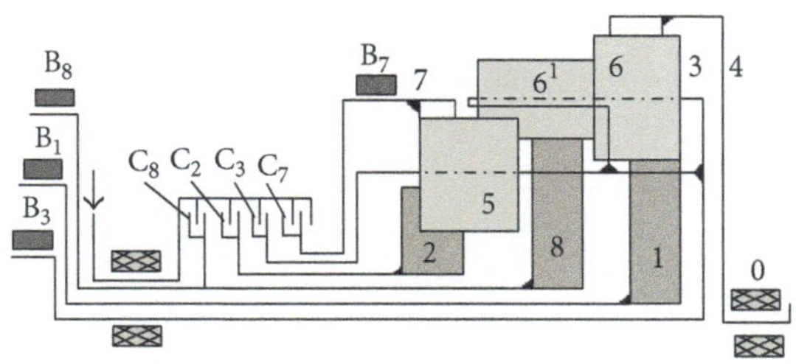

FIGURE 20: Clutching sequence layout for the nine-link Ravigneaux gear train shown in Figure 4(c) based on the feasibility graph.

clutching sequence layouts can be extracted from this general layout.

In order to achieve single-shift transitions, the ODs can be further classified into five sets; three sets consist of three-overdrives and two sets consist of two overdrives.

By combining the first set of UD clutching sequences with three sets of three-velocity OD clutching sequences and three sets of RDs, nine possible clutching sequences for eight-velocity automatic transmissions are feasible at this design stage as shown in Figure 21.

Moreover, the first set of UD clutching sequences can also be combined with two sets of two-velocity OD clutching sequences and three sets of RDs, and six possible clutching sequences for seven-velocity automatic transmissions are feasible at this design stage, as shown in Figure 22.

For the second set, since B_2 is essential for the UD clutching sequences, then only band clutches can be attached to links 1 and 8 as shown in Figure 23. Therefore, the clutching sequences are controlled by C_2, C_3, C_7, B_1, B_2, B_3, B_7, and B_8. The clutch C_2 has no role in the clutching sequence. As a result by excluding C_2, three clutching sequences are feasible to form six-velocity automatic transmissions as shown in Figure 24.

In order to achieve single-shift transitions, the ODs can be further classified into three sets consisting of three-overdrives.

By combining the second set of UD clutching sequences with the three sets of three-velocity OD clutching sequences and the one RD, three possible clutching sequences for six-velocity automatic transmissions are feasible at this design stage as shown in Figure 25.

Based on the second UD set of feasibility graph, the general clutching sequence layout for the nine-link Ravigneaux gear train to form six-velocity Ravigneaux gear mechanisms is shown in Figure 26.

4. Conclusions

A novel feature of the suggested approach for constructing multiaxis nomographs is the ability to enumerate the feasible

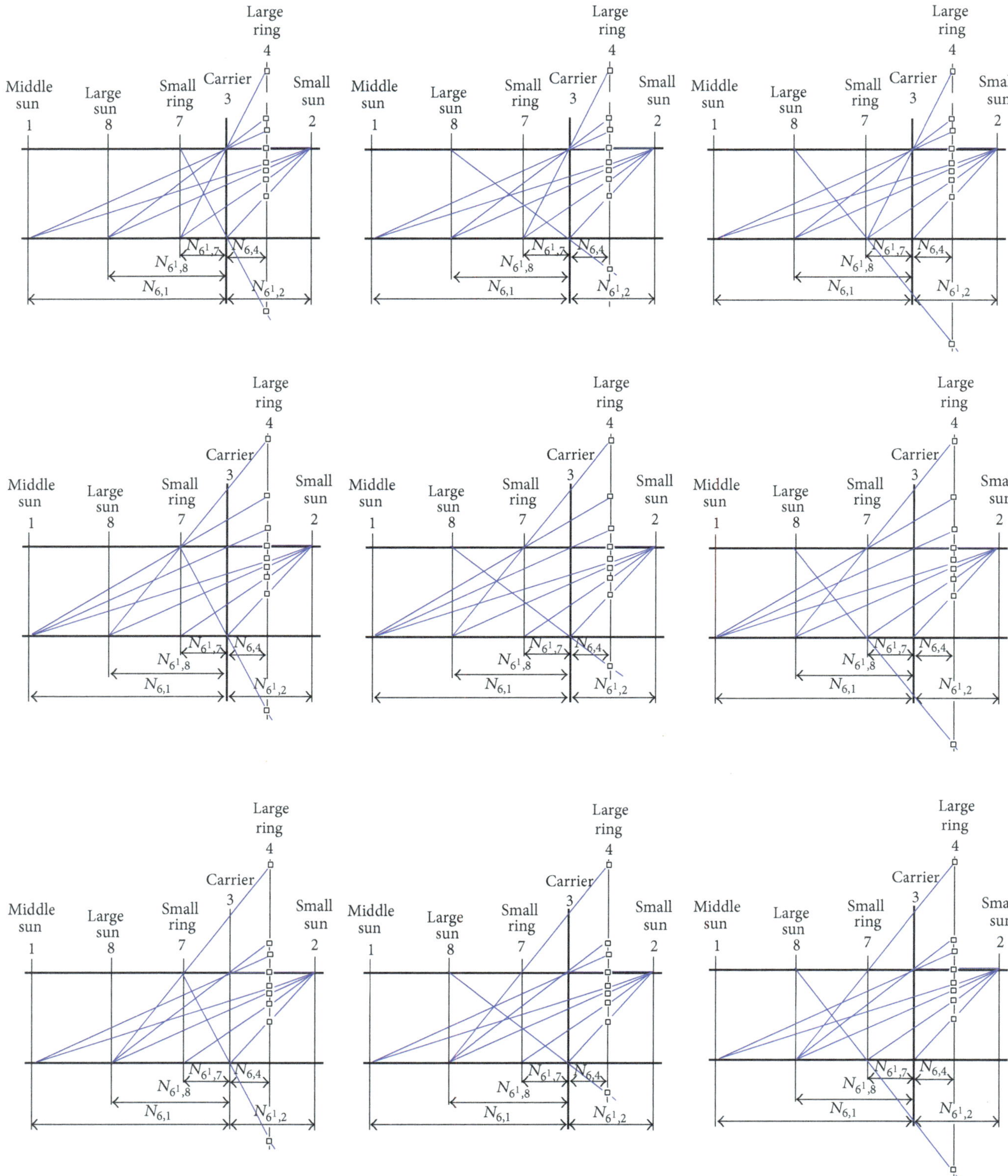

FIGURE 21: Clutching sequence nomographs for eight-velocity transmissions.

clutching sequences associated with two-ring eight-, and nine-link Ravigneaux-type automatic transmissions. It is a major breakthrough to design a completely satisfactory six-speed automatic transmission from the two-ring Ravigneaux gear train since it has only eight links. Moreover, new eight-speed automatic transmissions are enumerated from nine-link gear mechanisms. The new designs make use of the benefits of the Ravigneaux gear train and overcome the previous art difficulties. The completeness of the results cannot be confirmed since no publications exist for eight-velocity

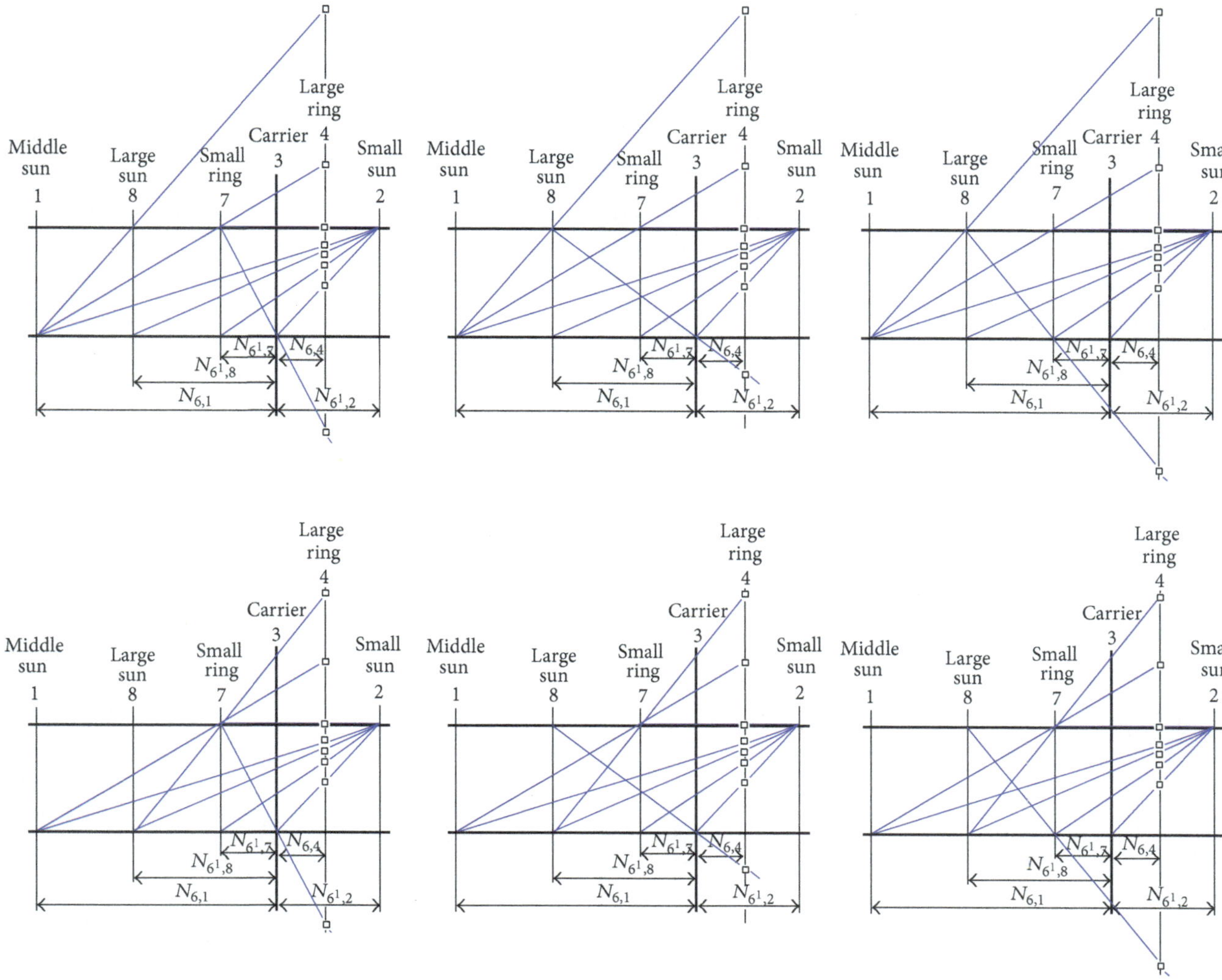

FIGURE 22: Clutching sequence nomographs for seven-velocity transmissions.

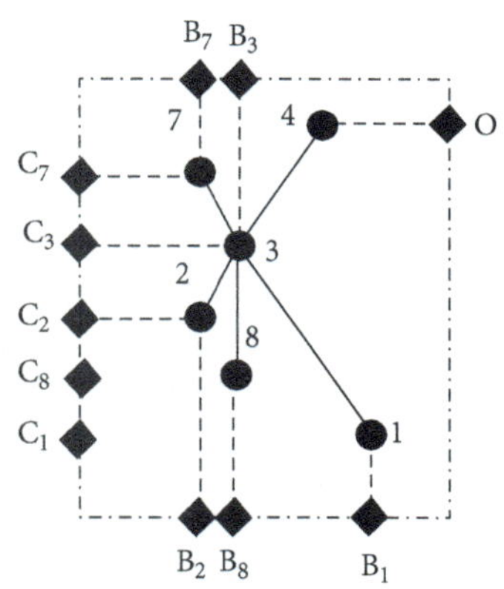

FIGURE 23: Feasibility graphs for the Ravigneaux gear train.

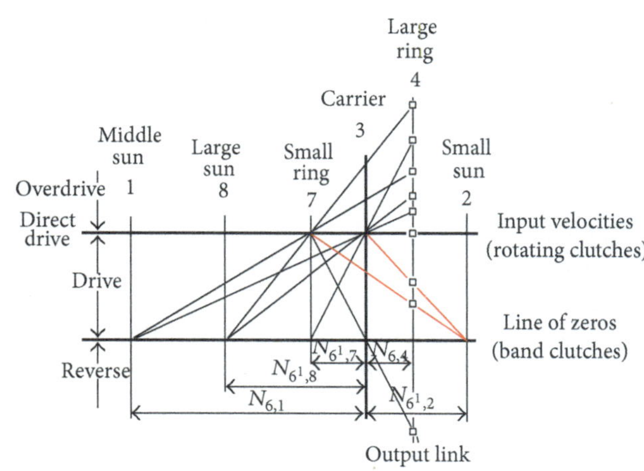

FIGURE 24: Clutching sequence nomograph for the nine-link Ravigneaux gear train based on the second UD set feasibility graph.

Ravigneaux gear mechanisms. The proposed methodology can be used for the systematic design of any epicyclic-type automatic transmission. Feasibility graphs have been also used as a design tool, allowing the designer to quickly select the most viable clutching sequence.

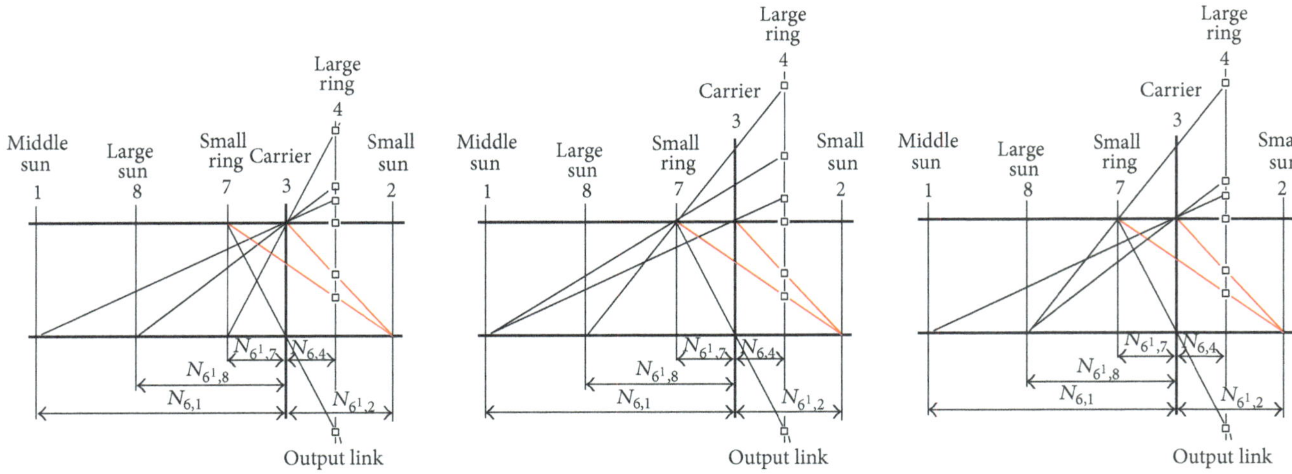

FIGURE 25: The three feasible clutching sequences for six-velocity automatic transmissions.

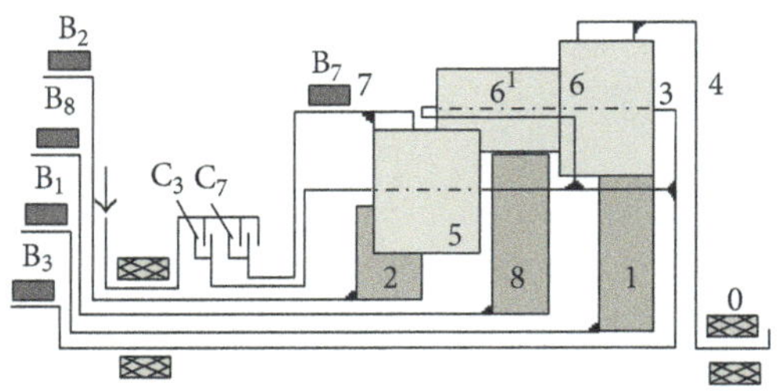

FIGURE 26: Clutching sequence layout for the nine-link Ravigneaux gear train based on the second UDs, ODs and RDs.

References

[1] P. Ravigneaux, "Gear box with freewheeling gear," US patent no. 2195783, 1940.

[2] P. Ravigneaux, "Speed changing device," US patent no. 2220174, 1940.

[3] L. W. Tsai, *Mechanism Design: Enumeration of Kinematic Structures According to Function*, CRC Press, Boca Raton, Fla, USA, 2001.

[4] E. Wilfinger and J. Thompson, *Borg-Warner Australia Model 85 Automatic Transmission*, SAE Paper no. 880480, Society of Automotive Engineers, 1988.

[5] C. H. Hsu and R. H. Huang, "Systematic design of four-speed ravigneaux-type automatic transmissions," *Journal of Chinese Society of Mechanical Engineers*, vol. 30, no. 3, pp. 201–209, 2009.

[6] H. Naunheimer, B. Bertsche, J. Ryborz, and W. Novak, "Vehicle transmission systems: basic design principles," in *Automotive Transmissions: Fundamentals, Selection, Design and Application*, Springer, Berlin, Germany, 2nd edition, 2011.

[7] N. Katou, T. Taniguchi, and K. Tsukamoto, *Aisin AW New Six-Speed Automatic Transmission for FWD Vehicles*, SAE Paper no. 2004-01-0651, 2004.

[8] P. A. G. Lepelletier, "Multispeed automatic transmission for automobile vehicles," US patent no. 5106352, 1992.

[9] H. Scherer, *ZF 6-Speed Automatic Transmission for Passenger Cars*, SAE Paper no. 2003-01-0596, Society of Automotive Engineers, 2003.

[10] P. Ravigneaux, "Epicyclic change-speed gear," US patent no. 2761333, 1956.

[11] B. A. Nadel and J. Lin, "Automobile transmission design as a constraint satisfaction problem: first results," in *Proceedings of the 7th IEEE Conference on Artificial Intelligence Applications*, vol. 1, pp. 248–256, February 1991.

[12] B. A. Nadel and J. Lin, "Automobile transmission design as a constraint satisfaction problem: modeling the kinematic level," *Artificial Intelligence for Engineering Design, Analysis and Manufacturing*, vol. 5, no. 3, pp. 137–171, 1991.

[13] B. A. Nadel, X. Wu, and D. Kagan, "Multiple abstraction levels in automobile transmission design: constraint satisfaction formulations and implementations," *International Journal of Expert Systems*, vol. 6, no. 4, pp. 489–559, 1993.

[14] H. I. Hsieh and L. W. Tsai, "A methodology for enumeration of clutching sequences associated with epicyclic-type automatic transmission mechanisms," *SAE Transactions*, vol. 105, no. 6, pp. 928–936, 1996.

[15] H. I. Hsieh and L. W. Tsai, "The selection of a most efficient clutching sequence associated with automatic transmission mechanisms," *Journal of Mechanical Design*, vol. 120, no. 4, pp. 514–519, 1998.

[16] W. M. Hwang and Y. L. Huang, "Configuration design of six-speed automatic transmissions with two-degree-of-freedom planetary gear trains," *Transactions of the Canadian Society for Mechanical Engineering*, vol. 29, no. 1, pp. 41–55, 2005.

[17] C. H. Hsu and R. H. Huang, "Systematic design of six-speed automatic transmissions with an eight-link two-DOF Ravigneaux gear mechanism," *Journal of Mechanical Design*, vol. 131, no. 1, Article ID 011004, 8 pages, 2009.

[18] C. S. Ross and W. D. Route, "A method for selecting parallel-connected, planetary gear train arrangements for automotive automatic transmissions," *SAE Transactions*, vol. 100, no. 6, pp. 1765–1774, 1991.

[19] E. L. Esmail, "Nomographs for enumeration of clutching sequences associated with epicyclic type automatic transmission mechanisms," in *Proceedings of the ASME International Mechanical Engineering Congress and Exposition (IMECE '08)*, Boston, Mass, USA, October-November 2008, paper no. 66409, also at *Emirates Journal for Engineering Research*, vol. 14, no.1, pp. 29–38, 2009.

[20] N. Hattori, T. Oshidari, and Y. Morimono, "Application of a new complex planetary gear set to five-speed automatic transmission gear train," *Transactions of the Society of Automotive Engineers of Japan*, vol. 26, no. 1, pp. 79–82, 1995.

[21] G. Chatterjee and L. W. Tsai, "Enumeration of epicyclic-type transmission gear trains," *Transactions of SAE*, vol. 103, no. 6, pp. 1415–1426, 1995, paper no. 941012.

[22] H. I. Hsieh and L. W. Tsai, "Kinematic analysis of epicyclic-type transmission mechanisms using the concept of fundamental geared entities," *Journal of Mechanical Design*, vol. 118, no. 2, pp. 294–299, 1996.

[23] H. I. Hsieh, *Enumeration and selection of clutching sequences associated with epicyclic-type transmission mechanisms [Ph.D. thesis]*, Department of Mechanical Engineering, University of Maryland, College Park, Md, USA, 1996.

[24] W. M. Hwang and Y. L. Huang, "Connecting clutch elements to planetary gear trains for automotive automatic transmissions via coded sketches," *Mechanism and Machine Theory*, vol. 46, no. 1, pp. 44–52, 2011.

[25] E. L. Esmail, "Kinematic nomographs of epicyclic-type transmission mechanisms," *Emirates Journal for Engineering Research*, vol. 12, no. 3, pp. 47–55, 2007.

[26] E. L. Esmail, "Nomographs for analysis of power circulation through closed epicyclic gear trains," *Emirates Journal For Engineering Research*, vol. 16, no. 2, pp. 1–9, 2011.

[27] E. L. Esmail and H. A. Hussen, "Nomographs for kinematics, statics and power flow analysis of epicyclic gear trains," in *Proceedings of the ASME International Mechanical Engineering Congress and Exposition (IMECE '09)*, vol. 13, pp. 631–640, Lake Buena Vista, Fla, USA, November 2009, paper no. 10789.

[28] E. L. Esmail, "Teaching planetary gear trains with the aid of nomographs," *Advances in Mechanical Engineering*, vol. 2013, Article ID 978418, 9 pages, 2013.

4

Investigation into the Impact of n-Decane, Decalin, and Isoparaffinic Solvent on Elastomeric Sealing Materials

Yue Liu and Christopher W. Wilson

Department of Mechanical Engineering, University of Sheffield, Mappin Street, Sheffield S1 3JD, UK

Correspondence should be addressed to Yue Liu, mep09yl@sheffield.ac.uk

Academic Editor: Ali Fatemi

One of the crucial aspects in the adoption of alternative fuels (e.g., GtL fuel) in aviation industry is to investigate their compatibility with elastomeric materials used in current gas turbine engines. This study employed stress relaxation technique to investigate the effects of three solvents, namely, decalin (cycloparaffin), n-decane (normal paraffin), and ShellSol T solvent (isoparaffins) on O-rings made from different materials. Results indicated that both fluorosilicone and fluorocarbon O-rings showed excellent compatibility with all 25 blends tested. The stress relaxation characteristic of nitrile O-ring was highly dependent on the composition of the solvents; the more decalin (n-decane) is in the blend, the better (worse) its sealing performance becomes. Effects of the three solvents presented in the tests indicated aromatics are not the only compounds that can swell nitrile O-ring. It is also important to notice that although decalin presents good O-ring swelling ability, it does not mean all cycloparaffins have the same property. n-decane also showed certain O-ring swelling ability but its main effect during the polymer-fuel interaction process is to extract materials out of nitrile O-ring. Isoparaffins do not participate in the O-ring swelling process. They only extract polymer materials; however, its extraction ability is relatively weaker than n-decane.

1. Introduction

Global petroleum crisis and environmental stress have boosted the research and development of alternative fuels in aviation industry. Due to more restrictions in alternative fuel usage than in any other sector [1], development of "drop-in" fuels has been the main interest in aviation. One of the crucial "fit-for-purpose" properties of "drop-in" fuels is their compatibility with elastomeric sealing components, most commonly O-rings, in an engine fuel system.

An O-ring is an elastomeric material that is squashed between two mating faces in a fuel system. The material deforms into the gap between the faces and thus provides a seal between them [2]. To guarantee a good sealing service, it is important to understand the interaction between O-ring material and fuel which is being sealed. Physically speaking, two processes occur when a new O-ring material is exposed to fuel for the first time. First, the material may absorb components from the fuel (such as alkanes, aromatics, etc.), which would generally cause the material to swell and soften. Second, the fuel may extract components from the material (such as plasticizers, processing aids, etc.), which would generally cause the material to shrink and harden. The overall effect the fuel has on the material will be the balance of these two processes. Once the material has been in service for some time, the fuel-extractable components will have been removed and all subsequent changes in physical properties will result from a shifting equilibrium between the material and the overlying fuel, which in turn will depend on the composition of the fuel [3].

Previous experiences indicated O-rings that are being used, particularly nitrile O-rings create a good seal because they swell in the presence of aromatics in the kerosene. When alternative fuels are used, normally trace or no aromatics in their compositions, O-rings either shrink or no longer swell to the volume that could provide a good seal [4–6]. Gas-to-Liquid (GtL) kerosene produced via the Fisher-Tropsch process is one of the alternative fuels which could potentially be adopted. This type of kerosene mainly consists of normal and isoparaffins, with no aromatic content. As aromatics

FIGURE 1: Elastocon Relaxation Tester EB 02 (source: Elastocon AB).

TABLE 1: Solvent matrix for testing.

Blend number	Composition (%)		
	Decalin	ShellSol T	n-decane
B1	100	0	0
B2	75	25	0
B3	50	50	0
B4	41.5	41.5	17
B5	33.3	33.3	33.3
B6	66	17	17
B7	61.1	27.8	11.1
B8	25	75	0
B9	0	100	0
B10	17	66	17
B11	27.8	61.1	11.1
B12	0	75	25
B13	11.1	61.1	27.8
B14	0	50	50
B15	17	41.5	41.5
B16	0	25	75
B17	11.1	27.8	61.1
B18	0	0	100
B19	17	17	66
B20	25	0	75
B21	27.8	11.1	61.1
B22	41.5	17	41.5
B23	50	0	50
B24	61.1	11.1	27.8
B25	75	0	25

are responsible for the soot formation from the emission, GtL kerosene has the benefits of reducing its footprint on environment. However, as paraffins are relatively large and nonpolar molecules, their ability to penetrate into the polymer structure in order to swell the O-rings are much worse than aromatic compounds, which are more compact and exhibit polar and hydrogen bonding character [3]. To overcome this drawback and also to keep its benefit on emissions, attention has been drawn to look for alternative ways to promote its O-ring swelling property, without adding aromatic species. Cycloparaffins are a class of compound which has the potential because research has been done on Coal-to-Liquid (CtL) kerosene (mainly cycloparaffins) suggested it has the same swelling characteristics for nitrile O-ring as JP-8 does (contains 17% aromatics and diaromatics and significant levels of polar compounds) [7]. It would be beneficial if it can be proved that cycloparaffins can be blended into the composition of GtL fuel to promote its O-ring swelling property without sacrificing other promising fuel properties. Meanwhile, further understanding on the roles of normal and isoparaffins play in the O-ring swelling process is also needed.

The aim of this paper is to investigate the impact of normal, iso-, and cycloparaffins on O-ring materials commonly used in the fuel system of an aircraft engine. n-decane and decalin were chosen as the candidate solvents to represent normal and cycloparaffins, as they are commonly found in the composition of jet fuel. ShellSol T, a synthetic isoparaffinic hydrocarbon solvent provided by Shell, was used as candidate isoparaffins. Three types of O-rings were tested, which were nitrile, fluorocarbon, and fluorosilicone. Rather than only immerge the O-rings in these solvents, stress relaxation technique was employed to look at how O-rings would perform under simulated service conditions. A "normal-iso-cycloparaffins" triangle sample matrix was prepared to look at what proportions of the three solvents could be possibly adopted to provide effective O-ring swelling property.

2. Experimental Details

2.1. Procedure. The equipment employed for the stress relaxation test was the Elastocon Relaxation Tester EB 02 (Figure 1). It has four rigs in a cell oven which all have the same specification for compression test. According to the test standard ISO 3384, test method "A" was used; therefore, the samples were compressed and all counterforce measurements were made at the test temperature [8]. The sample O-ring was inserted between two compression plates and compressed by 25% of its original size. Then, 150 mL of fuel was injected into the container to immerse the sample O-ring. Each rig contained a different type of solvent. The test temperature was set at 30°C in order to keep the rigs running safely. Counterforce and temperature of each rig were monitored and recorded continually throughout the experiment on a computer connected to the Elastocon. Normally the test period was one week, but may vary depending on how long it took to illustrate different relaxation characteristics between the various solvents.

2.2. Solvent Matrix. n-decane, decalin (both were purchased from VWR), and ShellSol T solvent (provided from Shell) were blended to prepare 25 solvents within the triangle shown in Figure 2. Each vertex of the triangle represented one neat solvent. The big triangle was divided into six small triangles by the medians on each side. Then, the vertexes, intersection point of the medians, and the midpoints on all sides of each small triangle were set as the sample points

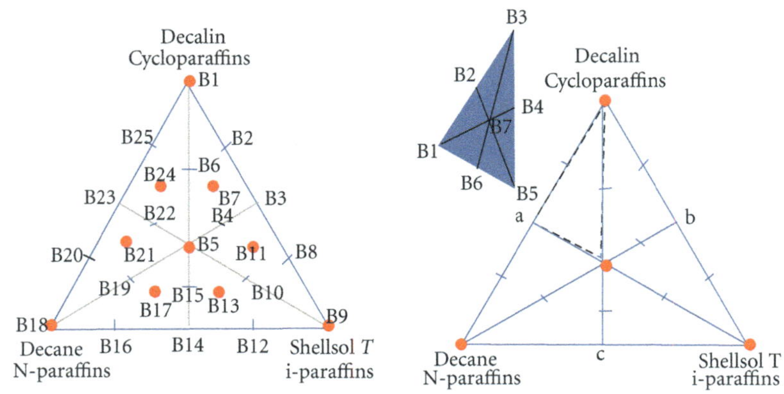

Figure 2: "Normal-iso-cycloparaffins" solvent triangle.

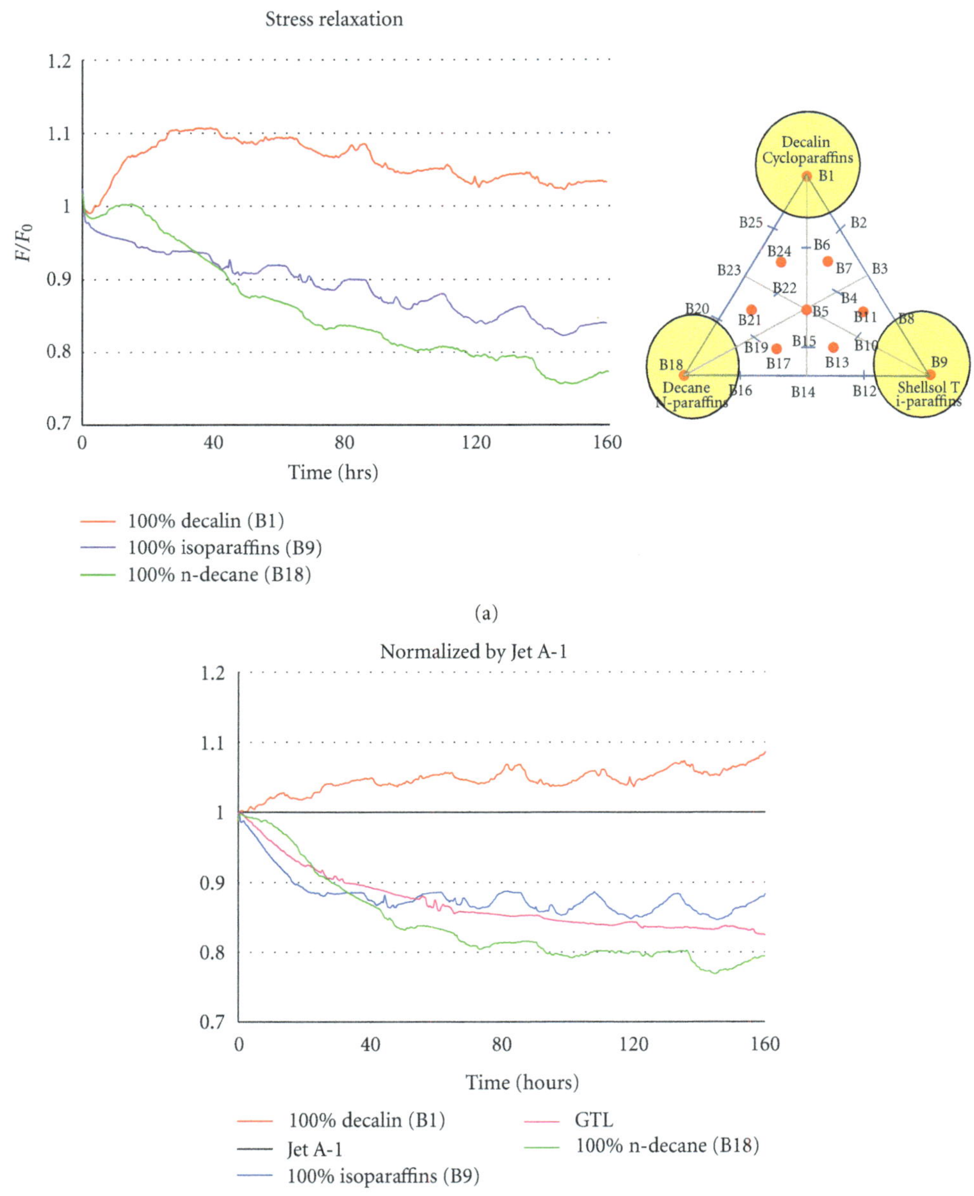

Figure 3: Relaxation process of nitrile O-ring: (a) in three single solvents, respectively; (b) normalized by Jet A-1, with comparison to GtL kerosene.

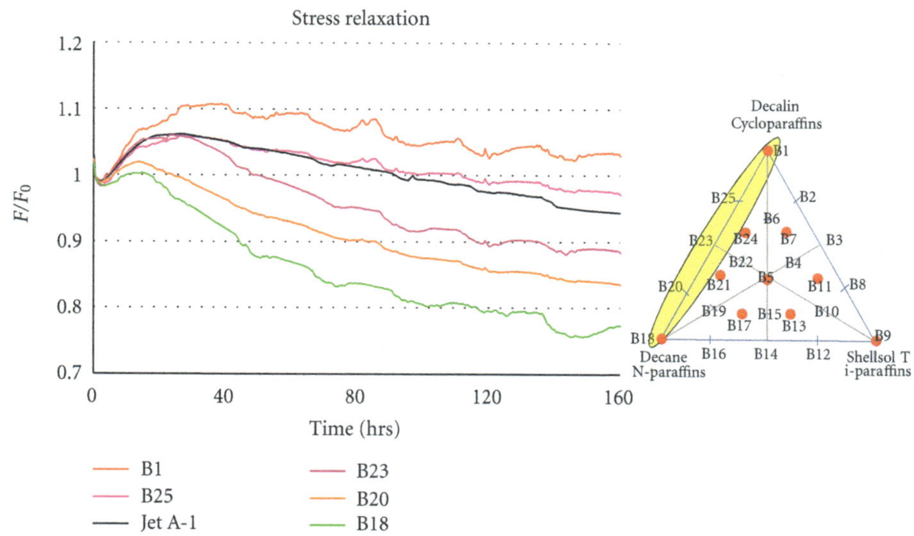

FIGURE 4: Relaxation process of nitrile O-rings in the blends of decalin and n-decane solvents.

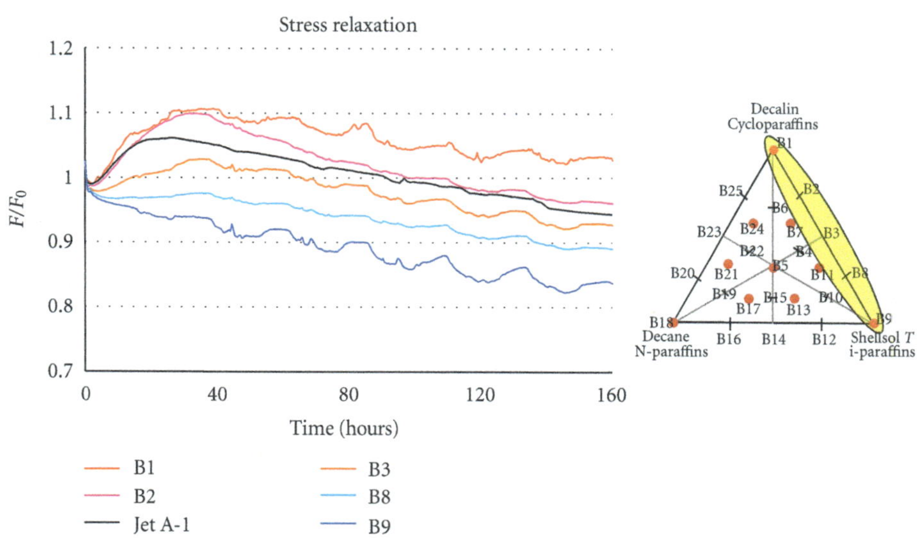

FIGURE 5: Relaxation process of nitrile O-rings in the blends of decalin and ShellSol T solvents.

TABLE 2: Specification of sample O-rings.

Specification	Fluorosilicone	Nitrile	Fluorocarbon
Part no.	L1218 2-113	100-113-1109	100-113-2326
Color	Blue	Black	Black
ID (mm)		13.94 ± 0.18	
CS (mm)		2.62 ± 0.08	

for blending. Detailed proportions of each solvent in these blends are in Table 1.

2.3. Sample O-Rings. Three types of O-ring materials were chosen for testing: nitrile, fluorocarbon, and fluorosilicone. They were chosen not only because they represent the types of O-rings used in the current market, but also the trend of changing standard of O-ring materials employed in the aviation industry. Fluorosilicone O-rings were supplied by Parker Hannifin Corporation and both nitrile and fluorocarbon O-rings were provided by Trelleborg. All O-rings have the standard AS568 113 size [9], and each type of O-rings was from the same batch to ensure the similar specification and quality. Specifications of each material are provided in Table 2.

3. Results and Discussion

When raw data was obtained from the stress relaxation tests, F_t/F_0 was generated to normalize each relaxation force by its original compression in order to compare the relative relaxation behavior. The undulating shape of the relaxation curves was caused by ambient temperature fluctuation.

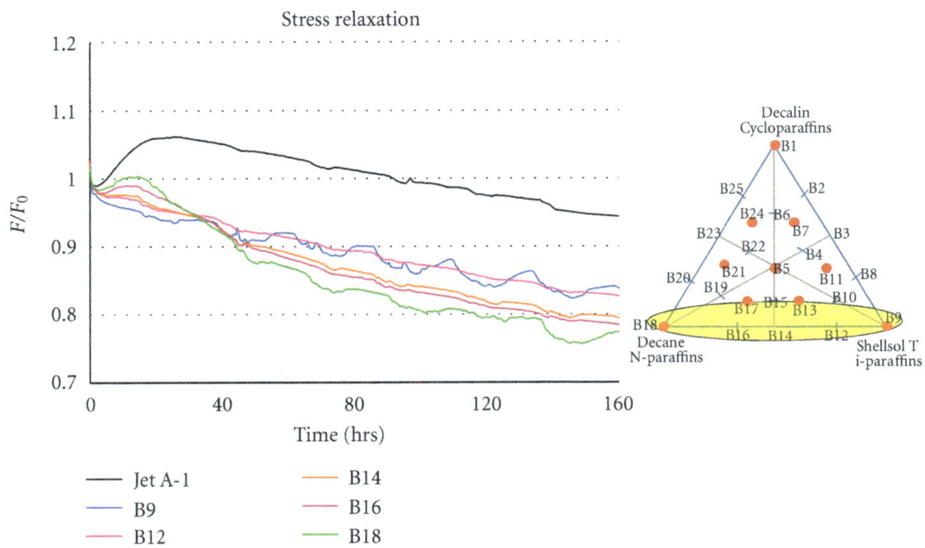

FIGURE 6: Relaxation process of nitrile O-rings in the blends of ShellSol T and n-decane solvents.

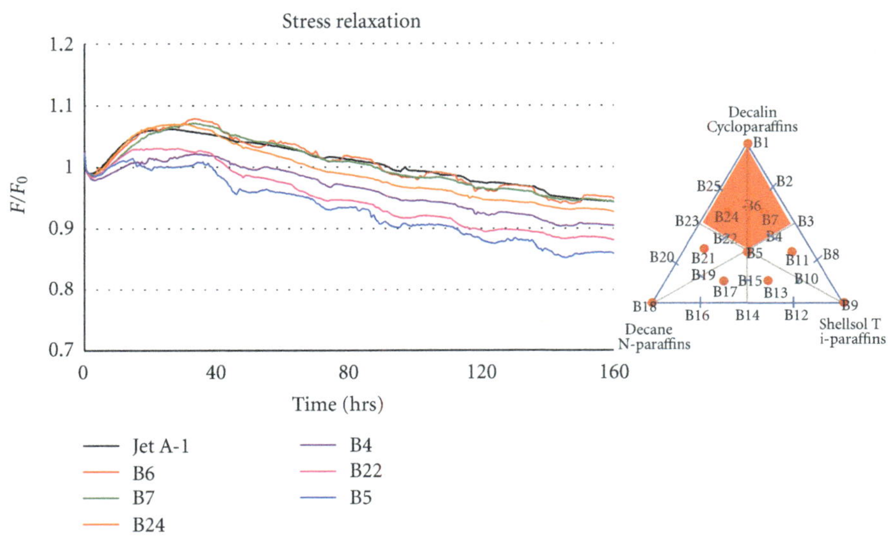

FIGURE 7: Relaxation process of nitrile O-rings in the blends of "decalin region" (red).

3.1. Nitrile

3.1.1. Vertexes of the Triangle. Nitrile O-rings performed quite differently in these three solvents. In decalin and n-decane, its stress relaxation process could be divided into three phases (see Figure 3(a)).

Phase 1 (Reduction). Counterforce decreased sharply with time during this phase as elastomeric material is viscoelastic, which means it has both the properties of viscous and elastic materials and, as such, exhibits time-dependant strain [10]. Time periods of this phase for decalin and n-decane lasted for 3 hours since compressions were applied.

Phase 2 (Increase). Instead of continuous relaxation after the initial counterforce reduction, counterforce increased in this phase which was due to O-ring swelling. This observation proved that both decalin and n-decane have the ability to diffuse into the polymer structure of nitrile O-ring due to their relatively small molecular size. Decalin may also benefit from its ring-shape molecular geometry which is similar to naphthalene (diaromatics). Previous research suggested that alkanes are relatively "inactive" in swelling O-rings and that is a main reason why a certain percentage of aromatics are needed in the composition of jet fuel. Result here indicated alkanes are also capable to participate in the O-ring swelling process if their sizes are small enough. The load cell of the relaxation rig sensed the pressure imposed by the volume swell of the O-ring which resulted in the increase of compression force. The relaxation curves reached their peak points at around 40 and 15 hours for decalin and n-decane, respectively, which also indicated the maximum volume swell points.

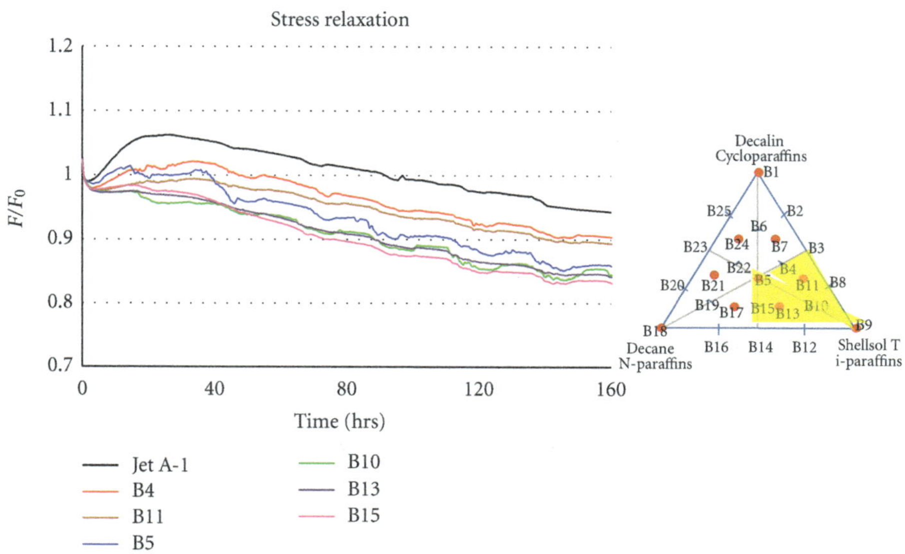

FIGURE 8: Relaxation process of nitrile O-rings in the blends of "isoparaffin region" (yellow).

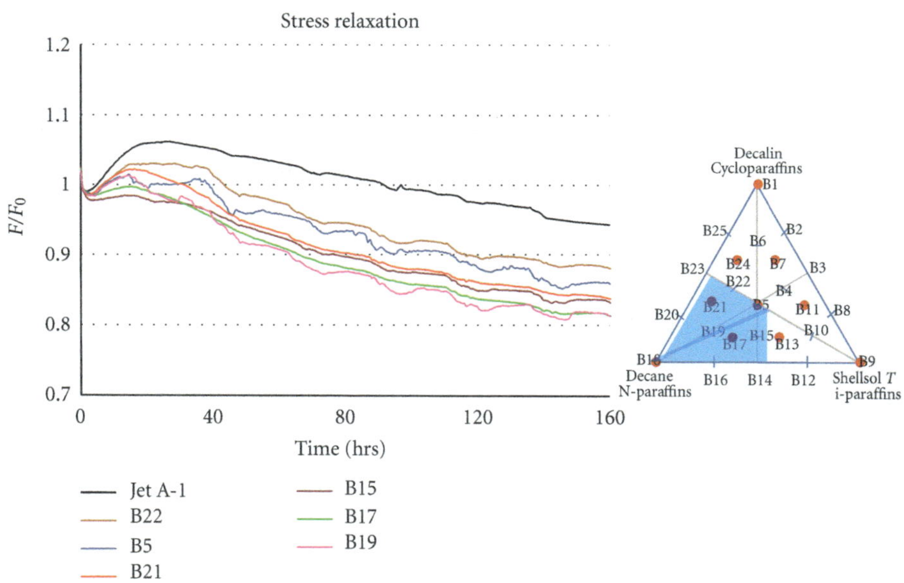

FIGURE 9: Relaxation process of nitrile O-rings in the blends of "n-decane region" (blue).

Phase 3 (Reduction). When the volume swell and the compression force reached equilibrium, counterforce began to decrease again gradually due to the solvents extracting chemical components from the O-rings [6]. The speed of relaxation in n-decane was much faster than that in decalin, indicating n-decane's ability to extract component out of nitrile O-ring is greater than decalin.

By contrast, nitrile O-ring tested in isoparaffins did not present any swelling behavior (see Figure 3(a)). After the initial counterforce quick reduction, O-ring relaxed gradually, with similar relaxation speed as that in decalin in its phase 3. This suggests isoparaffins might not participate in the O-ring swelling process, which means this class of hydrocarbon compound is not a contributor with regard to fuel's O-ring swelling property.

When normalized by the stress relaxation characteristics in Jet A-1 fuel with comparison to GtL [11] (Figure 3(b)), it can be seen that nitrile O-ring in decalin showed even better relaxation performance than that in Jet A-1. O-ring swelled to a larger amount of volume and relaxed slower afterwards, which suggests that pure decalin has good nitrile O-ring swelling property. Isoparaffins, GtL kerosene, and n-decane solvents presented weaker O-ring swelling ability than Jet A-1 fuel. GtL kerosene, whose main components are normal and isoparaffins, presented a combined characteristic of both n-decane and isoparaffins as its relaxation curve lied between those of the two solvents. Its ability to swell nitrile O-ring

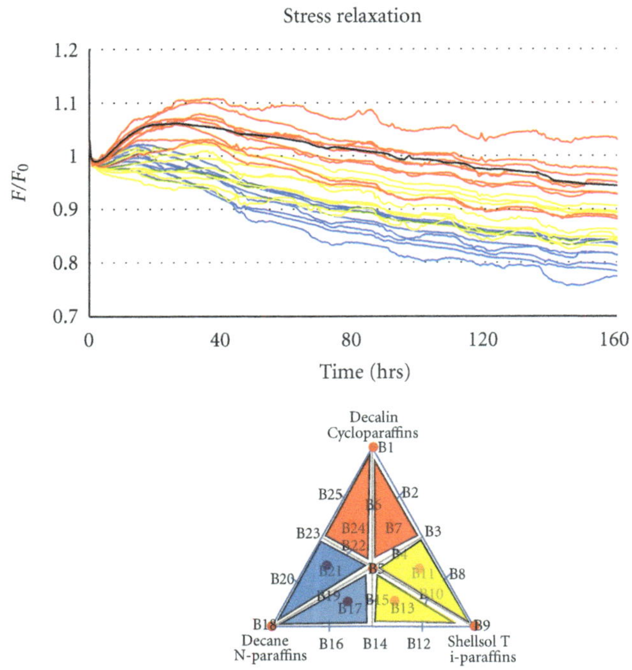

FIGURE 10: Summary of the relaxation behavior of nitrile O-rings in the triangle.

TABLE 3: O-ring swelling property (OSP)* of five solvents.

Solvent	Jet A-1	GtL	Decalin	n-decane	ShellSol T
OSP	+	−	+	−	−

*+: contribute to the O-ring swelling process.
−: weaken the O-ring swelling process.

was weaker but relaxed the O-ring slower than pure n-decane solvent, due to the effect of isoparaffins. Regarding the ability to swell nitrile O-rings, these five types of fuels or solvents could be graded in Table 3.

3.1.2. Edges of the Triangle

Decalin ↔ n-Decane. Figure 4 showed the stress relaxation characteristics of nitrile O-rings in decalin blended with n-decane by 0%, 25%, 50%, 75%, and 100% (v/v). It can be seen clearly that as the proportion of decalin (n-decane) increased (decreased) in the blends, nitrile O-ring relaxed slower, or its sealing performance was improved dramatically. Without considering the influence by the temperature fluctuation, its degree of relaxation was reduced roughly proportional to the percentage of decalin in the blends. It was proved again that the OSP of decalin is much better than that of n-decane. The relaxation behavior of the blend could be comparable to that of Jet A-1 when the proportion of decalin reached 75%.

Decalin ↔ ShellSol T (Isoparaffins). Figure 5 showed the stress relaxation behaviors of nitrile O-rings in decalin blended with ShellSol T solvent by 0%, 25%, 50%, 75%, and 100% (v/v). As the proportion of decalin (ShellSol T) increased (decreased) in the blends, nitrile O-ring's sealing performance was improved dramatically. The degree of relaxation was also reduced roughly proportional to the percentage of decalin in the blends, when not taking the influence by the temperature fluctuation in consideration. It could be said that the OSP of decalin is also better than that of isoparaffins. The relaxation behavior of the blend could also be comparable to that of Jet A-1 when the proportion of decalin reached 75%.

ShellSol T (Isoparaffins) ↔ n-Decane. Figure 6 showed the stress relaxation behaviors of nitrile O-rings in ShellSol T blended with n-decane solvent by 0%, 25%, 50%, 75%, and 100% (v/v). As the proportion of ShellSol T (n-decane) increased (decreased), the relaxation curves raised slightly. Although isoparaffins do not participate in the O-ring swelling process, their ability of extracting material out of the O-ring is weaker than n-decane, which makes them an "improver" when blended with the latter. So overall, their OSP is slightly better than that of n-decane. However, only mixing isoparaffins and n-decane cannot provide enough O-ring swelling property to the blended solvents, no matter by what proportions; as O-ring relaxed much faster than that in Jet A-1.

From the results obtained from the "edge of the triangle," it can be concluded that when blending two of the three solvents together, only decalin can promote the OSP of the final blends. However, in order to be comparable to Jet A-1, the proportion of decalin in the blend need to be around 75%. The OSP of the three solvents in two component blends can be compared as

Decalin > Isoparaffins > n-decane.

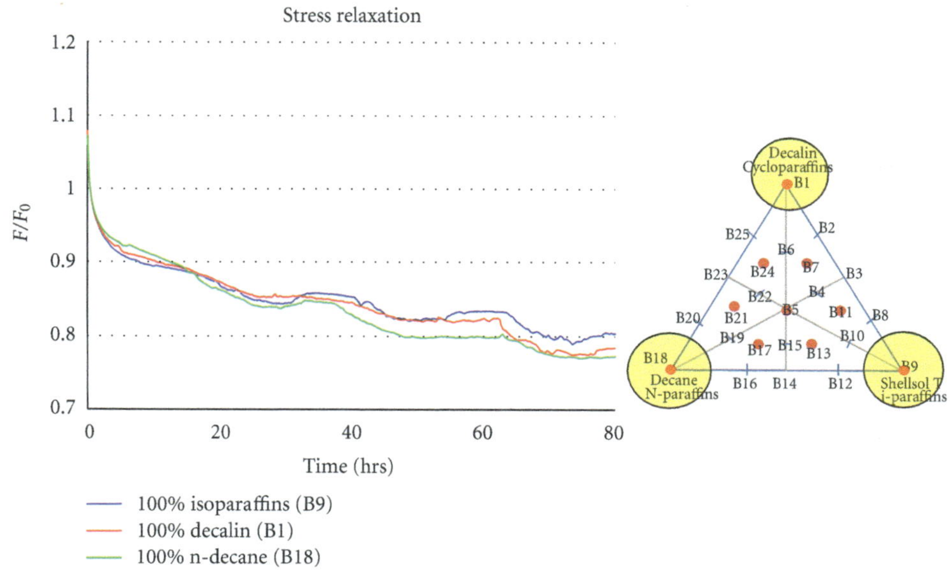

(a)

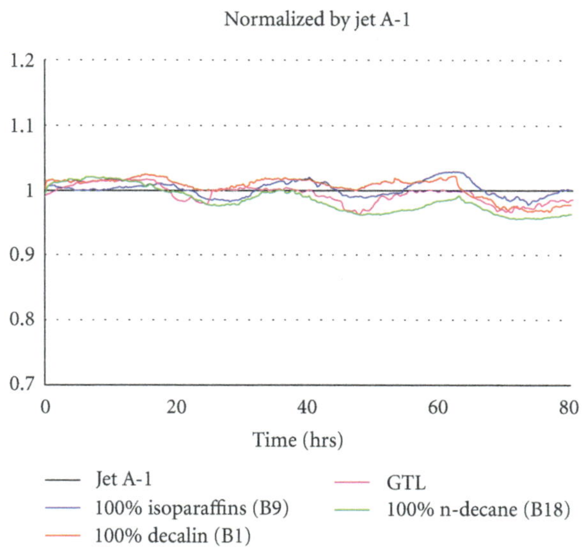

(b)

Figure 11: Relaxation behavior of fluorosilicone O-rings: (a) in three single solvents, respectively; (b) normalized by Jet A-1, with comparison to GtL kerosene.

3.1.3. Insides of the Triangle. In order to compare the respective effect of the three solvents in three-component blends, the insides of the triangle were divided into three regions, based on their relative positions towards each vertex.

Decalin Region. The "decalin region" consisted of six blends, which were B4, B5, B6, B7, B22, and B24. The percentages of decalin in the blends ranged from 33.3% to 66%, while the contents of both isoparaffins and n-decane were between 11.1% and 41.5%. O-rings in these blends all presented a certain degree of swelling; and as the proportion of decalin increased, the relaxation characteristics were improved (Figure 7). When the amount of decalin was certain, the more n-decane was in the blends, the worse the relaxation behavior became. It seemed that more than 60% of decalin was needed in the composition of the blends (B6 and B7) in order to be comparable to Jet A-1.

Isoparaffin Region. The "isoparaffin region" consisted of six blends, which were B4, B5, B10, B11, B13, and B15. The percentages of isoparaffins in the blends ranged from 33.3% to 66%, while the contents of both decalin and n-decane were between 11.1% and 41.5%. Compared with the "decalin region," the whole spectrum of relaxation curves shifted downwards dramatically (Figure 8). Slight difference in the composition of the blends seemed not change the relaxation behavior much (e.g., B10 and B13), though generally it was still improved with the increase of decalin content.

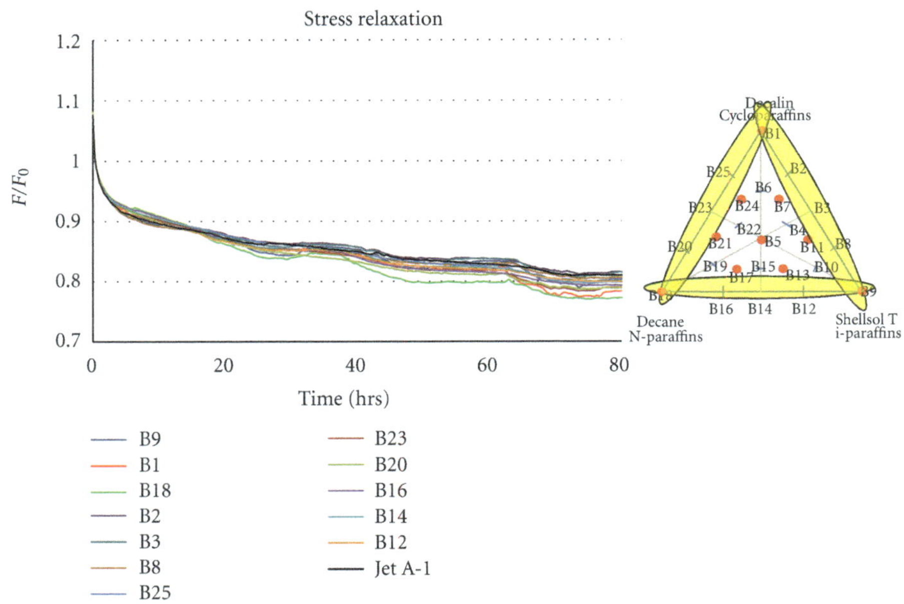

FIGURE 12: Relaxation behavior of fluorosilicone O-rings in the blends on the edges of the triangle.

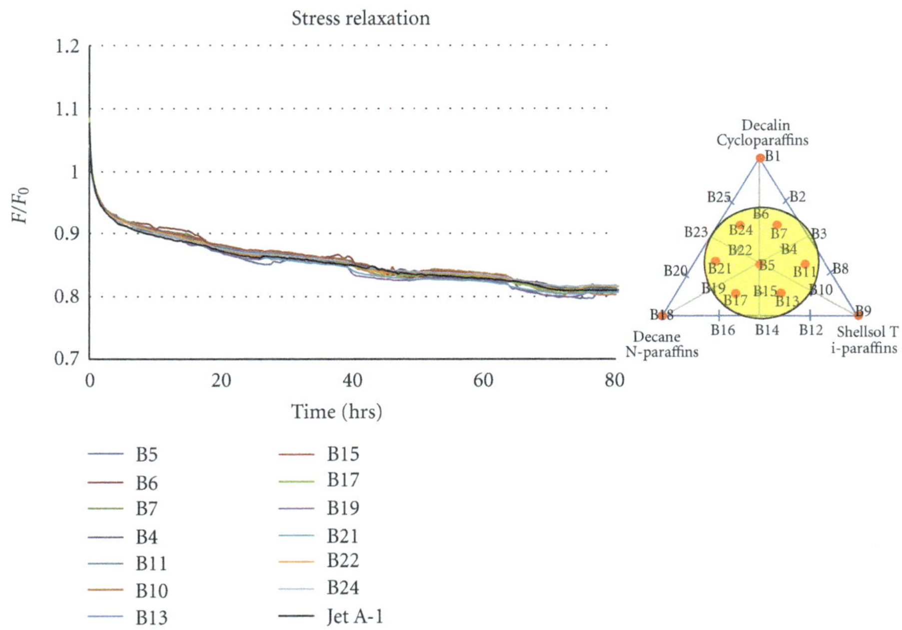

FIGURE 13: Relaxation behavior of fluorosilicone O-rings in the blends inside the triangle.

n-Decane Region. The "n-decane region" also consisted of six blends, which were B5, B15, B17, B19, B21, and B22. The percentages of n-decane in the blends ranged from 33.3% to 66%, while both decalin and isoparaffins were between 11.1% and 41.5%. Compared with the "isoparaffin region," the whole spectrum of relaxation curves shifted further downwards (Figure 9). Generally, the closer the blend was towards the vertex of n-decane, the lower the relaxation curve would be; or the quicker O-ring relaxed in this blend. Although n-decane is able to swell the O-ring during the initial stage of relaxation, its main effect is "negative" during the whole process.

Results obtained from the "insides of the triangle" indicated the same conclusion as that of the "edges of the triangle," which further proved decalin's ability to promote the OSP of the final blend. The proportion of decalin in the blend needs to be more than 60% to be considered comparable to Jet A-1 fuel. The OSP of the three solvents in three component blends were still as

$$\text{Decalin} > \text{Isoparaffins} > \text{n-decane}.$$

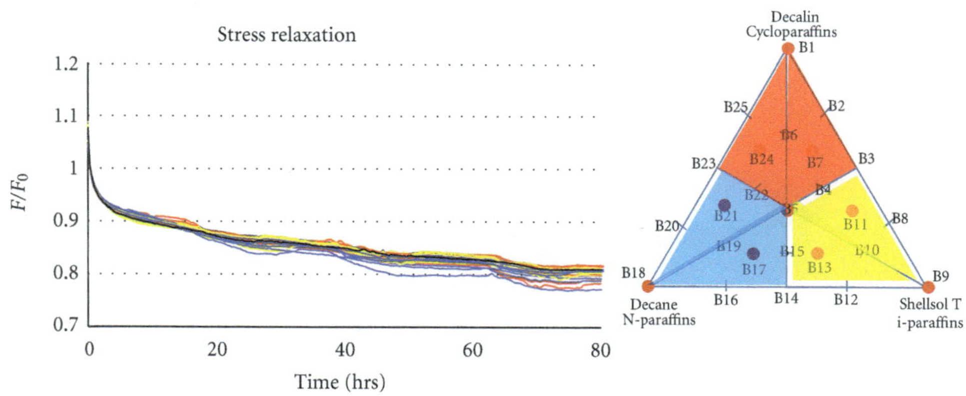

Figure 14: Summary of the relaxation behavior of fluorosilicone O-rings in the triangle.

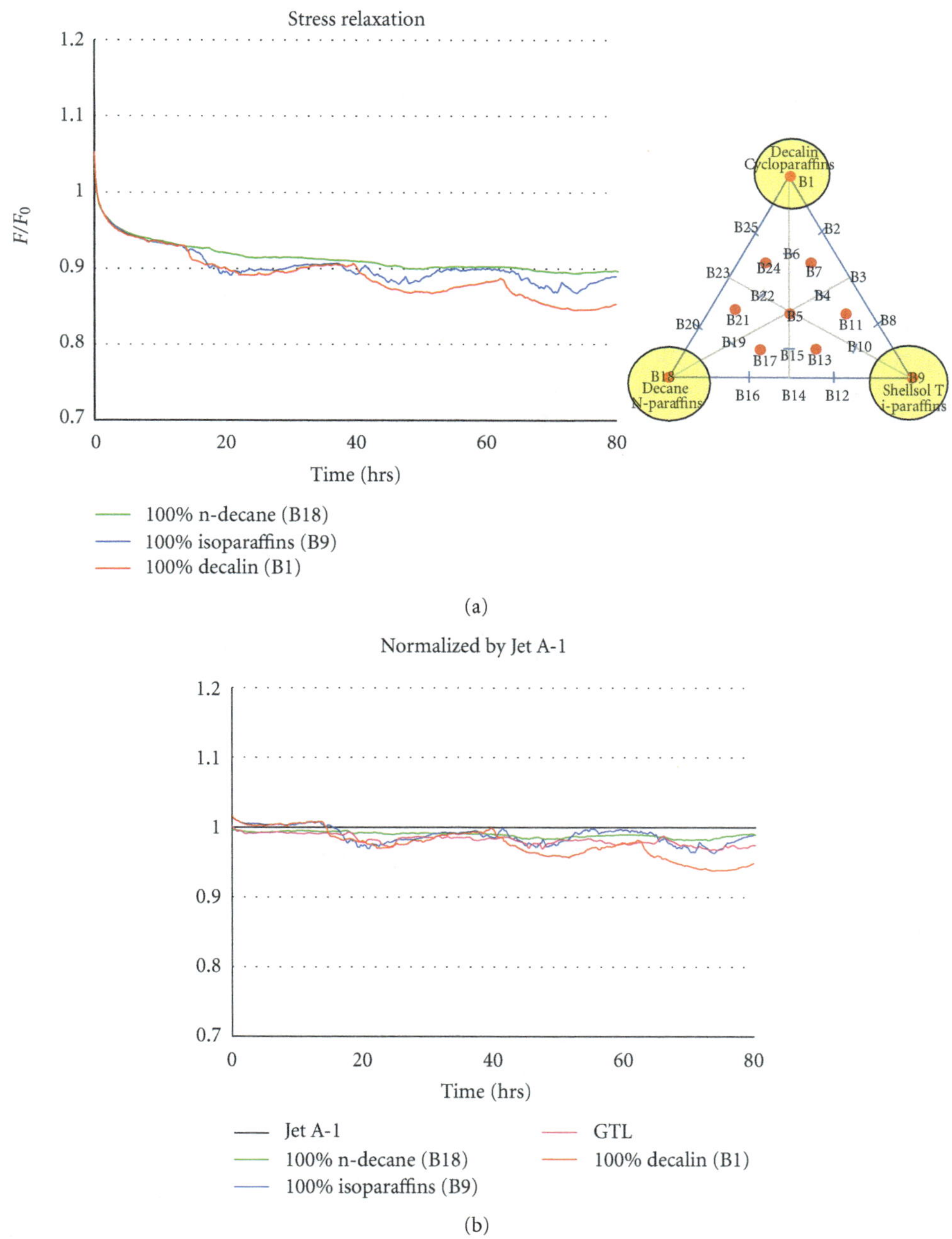

Figure 15: Relaxation behavior of fluorocarbon O-rings: (a) in three single solvents, respectively; (b) normalized by Jet A-1, with comparison to GtL kerosene.

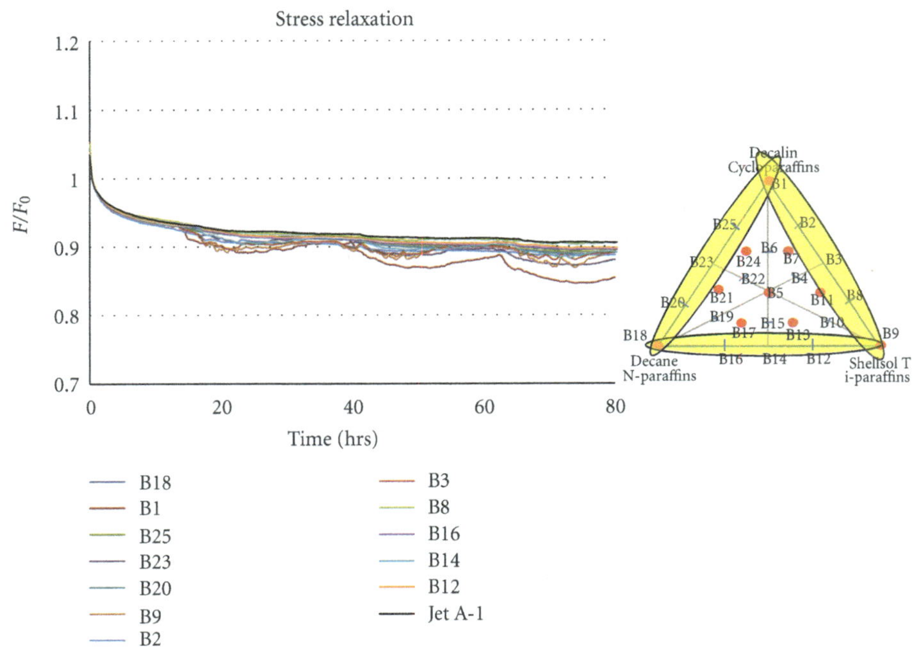

Figure 16: Relaxation behavior of fluorocarbon O-rings in the blends on the edges of the triangle.

If all the stress relaxation curves were drawn in one graph, the effects of the three solvents during the relaxation process of nitrile O-ring could be seen more clearly as shown in Figure 10. Generally, the relaxation curves shifted upwards with the increase of decalin in the blends, while moved downwards if the amount of n-decane became significant. Isoparaffins played a similar role as n-decane in this process but its effect was slightly weaker. To make the blend comparable to Jet A-1 fuel, over 60% (v/v) of decalin is needed in the composition, while the proportions of other two solvents are not as significant.

3.2. Fluorosilicone

3.2.1. Vertexes of the Triangle. Fluorosilicone O-rings presented very similar relaxation characteristics in all three solvents (Figure 11(a)). No swelling process or obvious acceleration in the relaxation was observed. When compared with Jet A-1 and GtL kerosene (Figure 11(b)), the relaxation curves almost overlapped each other, indicating excellent compatibility with all solvents.

3.2.2. Edges of the Triangle. Relaxation behaviors of the sample blends on all three edges of the triangle were plotted in Figure 12, with Jet A-1 as a reference. All relaxation curves lied closely to that of Jet A-1 which meant fluorosilicone O-ring exhibited quite similar relaxation characteristics in all these two-component solvents. It is relatively inert to the changes of fuel composition.

3.2.3. Insides of the Triangle. When three solvents were blended together, fluorosilicone O-ring presented even closer relaxation characteristics to that in Jet A-1 fuel (Figure 13). No solvent preference was observed as the changes of blend composition did not affect O-ring's relaxation process.

Results from the triangle solvents indicated excellent compatibility of fluorosilicone O-rings with three solvents (Figure 14).

3.3. Fluorocarbon

3.3.1. Vertexes of the Triangle. Fluorocarbon O-ring also presented good compatibility with the solvents (Figure 15(a)). A slight difference was observed in decalin solvent as O-ring relaxed more than those in the other two solvents. By the end of the 80-hour test, fluorocarbon O-ring relaxed less than fluorosilicone O-ring, with the F/F_0 value ranged from 0.9–0.85 for the former and around 0.8 for the latter. When compared with Jet A-1, with comparison to GtL kerosene, all solvents showed excellent compatibility with fluorocarbon O-ring (Figure 15(b)). Although in decalin, the relaxation curve was below all the others, but the F/F_0 value was still higher than 9.3.

3.3.2. Edges of the Triangle. Relaxation behaviors of the sample blends on all three edges of the triangle were plotted in Figure 16, with Jet A-1 as a reference. All relaxation curves lied closely to that of Jet A-1 which meant fluorocarbon O-ring exhibited quite similar relaxation characteristics in all these two-component solvents.

3.3.3. Insides of the Triangle. When three solvents were blended together, fluorocarbon O-ring presented quite similar relaxation characteristics to that in two-component blends (Figure 17). No solvent preference was observed as

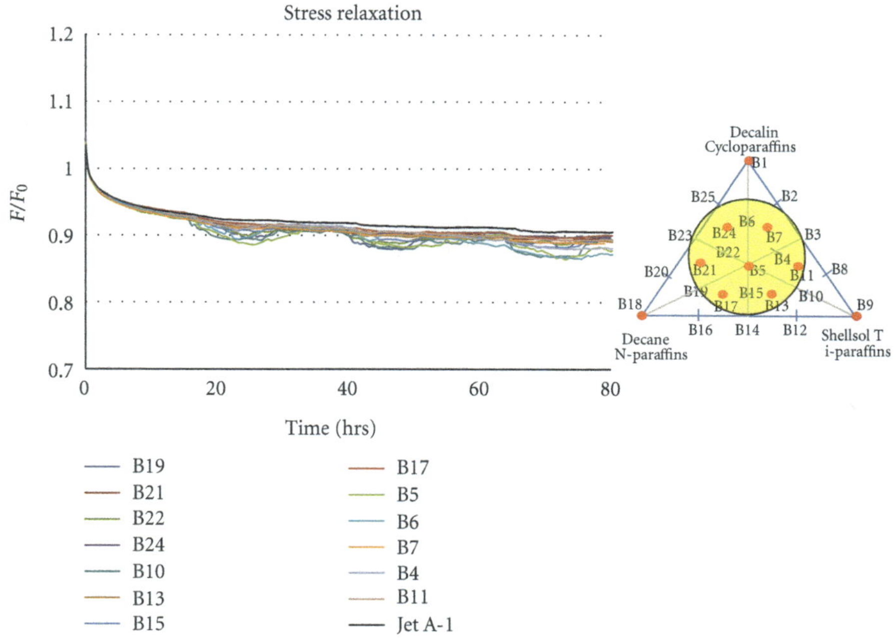

FIGURE 17: Relaxation behavior of fluorocarbon O-rings in the blends inside the triangle.

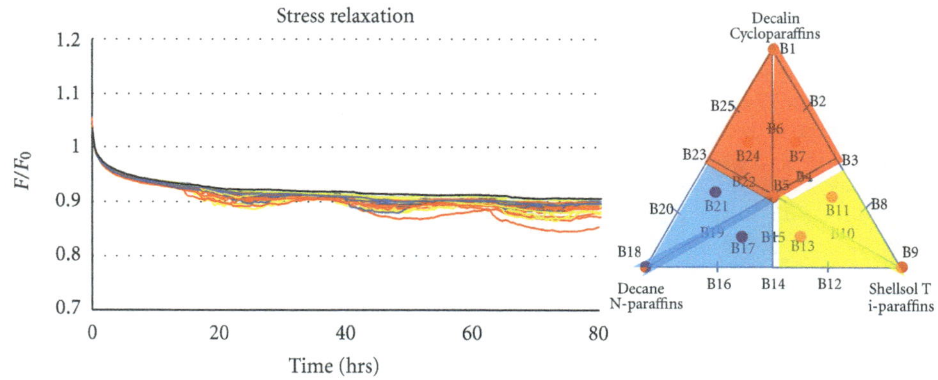

FIGURE 18: Summary of the relaxation behavior of fluorocarbon O-rings in the triangle.

the changes of blend composition did not affect O-ring's relaxation process.

Generally, fluorocarbon O-ring also has excellent compatibility with all three solvents tested (Figure 18).

4. Conclusions

Results obtained in this paper further proved excellent compatibility of both fluorocarbon and fluorosilicone O-rings with alternative fuels. Solvents consist of n-decane, isoparaffins, and decalin have very limited impact on the relaxation characteristics of these two polymer materials, no matter what composition they are in the blends. The relaxation behavior or sealing performance of nitrile O-ring is highly dependent on the composition of the fuel being sealed. As the composition shifts within the "triangle", its relaxation characteristics change dramatically. Generally, the more decalin is in the blends, the better its sealing performance becomes.

Effects of three solvents presented in the tests indicate aromatics are not the only compounds that can swell the O-ring in order to enhance its sealing performance. Decalin, as a representative of cycloparaffins, is also capable to swell nitrile O-rings. However, the proportion of decalin in a fuel blend has to be no less than 60% to be comparable to Jet A-1 fuel, which is too high to be an ideal additive to GtL kerosene. It is also important to notice that although decalin presents good O-ring swelling ability, it does not mean all cycloparaffins have the same property. Further tests are needed on other cycloparaffinic compounds to see if they have similar or different properties. n-decane also showed certain O-ring swelling ability but its main effect during the polymer-fuel interaction process is to extract materials out of nitrile O-rings. Isoparaffins do not participate in the O-ring swelling process but only extract polymer materials; however, its extraction ability is relatively weaker than n-decane.

Nomenclature

F_t: Counterforce measured after the specified time (t) of the test (N)

F_0: Initial counterforce measured 30 min after completing the compression (N).

Acknowledgments

The research described in this paper was a part of an international research program funded by Qatar Science and Technology Park (QSTP). The authors would like to acknowledge the support of QSTP in carrying out this work.

References

[1] S. Blakey, L. Rye, and C. W. Wilson, "Aviation gas turbine alternative fuels: a review," in *Proceedings of the Combustion Institute*, 2010.

[2] J. Whitlock and EPM, *The Seal Man's O-Ring Handbook*, EPM, Atlanta, Ga, USA, 2004.

[3] J. Graham, T. F. Rahmes, M. C. Kay et al., "Alternative fuel, final report," Tech. Rep. FAA OTA DTFAWA-10-C-0030.

[4] J. L. Graham, R. C. Striebich, K. J. Myers, D. K. Minus, and W. E. Harrison, "Swelling of nitrile rubber by selected aromatics blended in a synthetic jet fuel," *Energy and Fuels*, vol. 20, no. 2, pp. 759–765, 2006.

[5] D. D. Link, R. J. Gormley, J. P. Baltrus, R. R. Anderson, and P. H. Zandhuis, "Potential additives to promote seal swell in synthetic fuels and their effect on thermal stability," *Energy and Fuels*, vol. 22, no. 2, pp. 1115–1120, 2008.

[6] R. J. Gormley, D. D. Link, J. P. Baltrus, and P. H. Zandhuis, "Interactions of jet fuels with nitrile o-rings: petroleum-derived versus synthetic fuels," *Energy and Fuels*, vol. 23, no. 2, pp. 857–861, 2009.

[7] X. Zabarnick, D. K. Phelps, Z. J. West et al., "Evaluation and testing of the suitability of a coal-based jet fuel," Tech. Rep. AFRL-RZ-WP-TR-2009-2063, 2009.

[8] S. Ronan, T. Alshuth, S. Jerrams, and N. Murphy, "Long-term stress relaxation prediction for elastomers using the time-temperature superposition method," *Materials and Design*, vol. 28, no. 5, pp. 1513–1523, 2007.

[9] Society of Automotive Engineers (SAE)-AS568-08 Aerospace Size Standard for O-ring.

[10] M. A. Meyers and K. K. Chawla, *Mechanical Behavior of Materials*, Cambridge University Press, New York, NY, USA, 2009.

[11] Y. Liu, C. W. Wilson, and S. Blakey, "Elastomer compatibility test of alternative fuels using stress relaxation test and FTIR spectroscopy," in *Proceedings of the ASME Turbo Expo*, Vancouver, Canada, 2011, GT2011-46100.

5

Analysis of Dual Heat Sources in a Partially Open Enclosure

M. McGarry and James G. Kohl

University of San Diego, Department of Engineering, 5998 Alcala Park, Loma Hall, San Diego, CA 92110, USA

Correspondence should be addressed to M. McGarry; mmcgarry@sandiego.edu

Academic Editor: C. T. Nguyen

In this study a partially open enclosure with two heat sources is examined to determine the Nusselt number along the hot wall. The first part of the study examines the effect of the Rayleigh number ratio on the Nusselt number. The largest Rayleigh number ratio results in the highest magnitude of the Nusselt number. The second part of the study examines what happens to the Nusselt number along the hot wall as the Rayleigh number ratio for the two blocks is different. This situation can arise when the two blocks have unequal heat source values. The largest Nusselt numbers correspond to the situation when the block along the top wall has a Rayleigh number ratio that is ten times higher than the Rayleigh number ratio for the block along the hot wall. The largest Nusselt number magnitudes can be attributed to the strong vortex in the corner near the hot wall.

1. Introduction

Natural convection plays an important role in many applications where heat must be removed from objects that generate heat. Often, these objects are inside of an enclosure that is partially open. The practical applications for this situation have been well documented by previous researchers [1] and include nuclear reactors, solar energy collectors, electronic equipment, and the design of indoor environments.

Several studies have been conducted regarding natural convection in partially open enclosures. Angirasa et al. [2] examined transient natural convection in an enclosure with three heated walls and one open wall. Other studies involving open enclosures were performed with large domains outside of the enclosure [3–5]. All of these studies did not involve heated objects in the enclosure.

Xia and Zhou [6] explored a partially open enclosure interacting with an external heat generating source. Mariani and Da Silva [1] examined the local and average Nusselt number along a hot and cold wall for a single internal heat source inside a partially open enclosure. They varied the position of the heat source along the insulated bottom and hot side wall. In addition, they varied the ratio of the Rayleigh number for the internal heat source (internal Rayleigh number) and the Rayleigh number based on the temperature difference between the hot and cold wall (external Rayleigh number).

Recently, a new study examined natural convection in vented cavities using a restricted domain approach [7]. In addition, Muftuoglu and Bilgen [8] and Muftuoglu and Bilgen [9] studied a single heat source in an open enclosure to determine the optimal heat source location. Liu and Phan-Thien determined the optimal spacing for three chips mounted on a vertical substrate in an enclosure [10]. Also, Da Silva et al. found the optimal position on a wall and a plate undergoing natural convection [11, 12]. Other studies have examined heat sources in vertical channels and L-shaped channels [13, 14].

All of the above studies examined enclosures that contained a single heat source or multiple heat sources along a plate or an open channel. This study will examine multiple heat sources in a partially open enclosure. This study will characterize the Nusselt number along the hot wall as a function of the external Rayleigh number for various ratios between the internal Rayleigh number, Ra_i, and the external Rayleigh number, Ra_e. In addition, this study will examine the Nusselt number along the hot wall as a function of the external Rayleigh number for various ratios of the internal

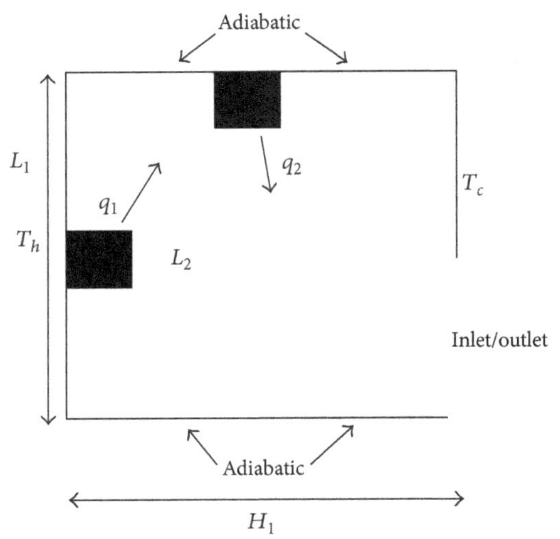

FIGURE 1: The geometry for the problem.

TABLE 1: The results from the grid refinement study.

Number of elements	Nu_h
25,000	−11.89
40,000	−11.22
60,000	−11.17

Rayleigh numbers for the two heat sources. Streamlines and isotherms will be presented for the two cases listed above.

2. Mathematical Model

The geometry for the model consists of a partially open enclosure with two heated blocks inside of the enclosure, Figure 1. The enclosure is a square with a length of L_1 and an opening of one half of L_1. The left side of the enclosure is at a constant high temperature, T_h, and the right wall of the enclosure is held at a lower temperature, T_c. The upper and the lower walls of the enclosure are adiabatic. The two heat sources are located in the middle of the left and upper walls. Both heat sources are square and have a length L_2 that is one tenth of L_1. The heat sources produce an amount of heat equal to q_1 and q_2, respectively.

A computational mesh is constructed for the geometry with a high density of elements placed near the walls where the gradients are the largest. A grid refinement study is performed at a Rayleigh number of 5,000, and the results can be seen in Table 1. The optimal number of elements is found to be 40,000 across all Rayleigh numbers examined. In addition to the grid refinement study, our model is compared to Mariani and Da Silva for two different external Rayleigh numbers, Figure 2. Figure 2 shows excellent agreement between the two models.

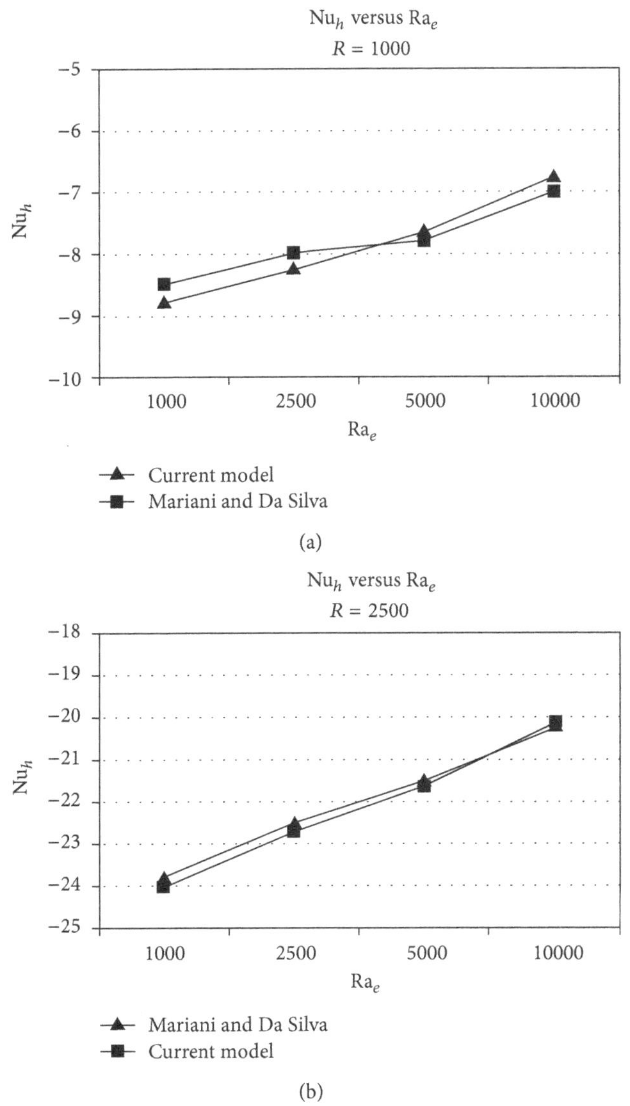

FIGURE 2: The comparison between Mariani and Da Silva and our model for the external Rayleigh number of (a) 1000 and (b) 2500.

The dimensionless form of the steady mass, momentum, and energy equations in the fluid region for the problem is defined as follows:

$$\begin{gathered}\frac{\partial U}{\partial X}+\frac{\partial V}{\partial Y}=0,\\ U\frac{\partial U}{\partial X}+V\frac{\partial U}{\partial Y}=-\frac{\partial P}{\partial X}+\Pr\left(\frac{\partial^2 U}{\partial X^2}+V\frac{\partial^2 U}{\partial Y^2}\right),\\ U\frac{\partial V}{\partial X}+V\frac{\partial V}{\partial Y}=-\frac{\partial P}{\partial Y}+\Pr\left(\frac{\partial^2 V}{\partial X^2}+V\frac{\partial^2 V}{\partial Y^2}\right)+\mathrm{Ra}\Pr\theta,\\ U\frac{\partial \theta}{\partial X}+V\frac{\partial \theta}{\partial Y}=\left(\frac{\partial^2 \theta}{\partial X^2}+V\frac{\partial^2 \theta}{\partial Y^2}\right).\end{gathered} \tag{1}$$

In the solid conducting block, the dimensionless energy equation is

$$k^* \left(\frac{\partial^2 \theta_s}{\partial X^2} + \frac{\partial^2 \theta_s}{\partial Y^2} \right) = 0, \quad (2)$$

where the nondimensional variables used in (1) and (2) are defined as

$$U = \frac{uL}{\alpha}, \quad V = \frac{vH}{\alpha}, \quad X = \frac{x}{H},$$
$$Y = \frac{y}{L}, \quad P = \frac{(p - p_\infty) L^2}{\rho \alpha^2}, \quad \theta = \frac{(T - T_c)}{(T_H - T_c)},$$
$$\theta_s = \frac{(T_s - T_c)}{(T_H - T_c)},$$
$$\text{Ra} = \frac{gB(T_h - T_c) L^3}{\nu \alpha}, \quad (3)$$
$$\text{Pr} = \frac{\nu}{\alpha},$$
$$k^* = \frac{k_{\text{solid}}}{k_{\text{fluid}}}.$$

u and v are the fluid velocity in the x and y directions, α the thermal diffusivity, ρ the fluid density, k the thermal conductivity, T the temperature, ν the kinematic viscosity, p_∞ the reference pressure, p the static pressure, and B the thermal expansion coefficient.

The boundary conditions in the fluid domain are as follows:

$$X = 0, \quad U = V = 0, \quad \theta = 1,$$
$$X = 1, \quad U = V = 0, \quad \theta = 0,$$
$$Y = 0, \quad U = V = 0, \quad \frac{\partial \theta}{\partial Y} = 0, \quad (4)$$
$$Y = 1, \quad U = V = 0, \quad \frac{\partial \theta}{\partial Y} = 0.$$

At the surface of the conducting block, the boundary conditions are

$$U = V = 0,$$
$$\theta_s = \theta, \quad \frac{\partial \theta}{\partial n} = k^* \frac{\partial \theta_s}{\partial n}, \quad (5)$$

where n is the normal direction to the surface. The solid block and fluid were initialized to have the median temperature between the hot and cold walls for all the simulations. The equations above were solved using the finite volume software FLUENT. The pressure-velocity coupling was accomplished by the SIMPLE algorithm, and a power law representation was used for the convective flux terms. The Boussinesq approximation was used to model the buoyancy effects.

The key dimensionless groups that characterize the problem are developed by Mariani and Da Silva and are the average Nusselt number, Nu, the external Rayleigh number, Ra_e, the internal Rayleigh number, Ra_i, and the ratio between the internal and external Rayleigh number, R. In addition to the above parameters, this study will examine the effect of different R values for each of the heat sources, R^*. The dimensionless groups are defined below

$$\text{Nu} = \frac{hL}{k},$$
$$\text{Ra}_e = \frac{gB(T_h - T_c) L^3}{\nu \alpha},$$
$$\text{Ra}_i = \frac{gBqL^5}{\nu \alpha k},$$
$$R = \frac{\text{Ra}_i}{\text{Ra}_e}, \quad (6)$$
$$R_1 = \frac{\text{Ra}_{i,1}}{\text{Ra}_{e,1}},$$
$$R_2 = \frac{\text{Ra}_{i,2}}{\text{Ra}_{e,2}},$$
$$R^* = \frac{R_1}{R_2}.$$

To explore how the average and local Nusselt numbers are affected by these dimensionless groups, simulations are performed by varying the external Rayleigh number, which is consistent with the method by Mariani and da Silva. Table 2

TABLE 2: The range of dimensionless parameters in the study.

Dimensionless group	Range
Ra_e	1,000–10,000
R	250, 1000, 2500
R^*	0.1, 1, 10
Ra_i	0.05–10

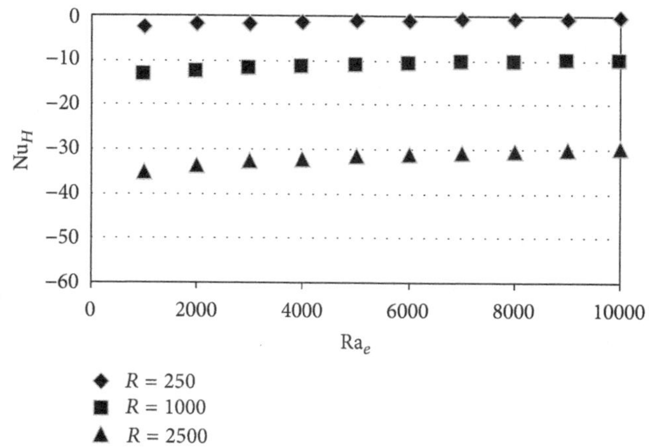

FIGURE 3: The average Nusselt number along the hot wall versus the external Rayleigh number for a Rayleigh number ratio of 250, 1000, and 2500.

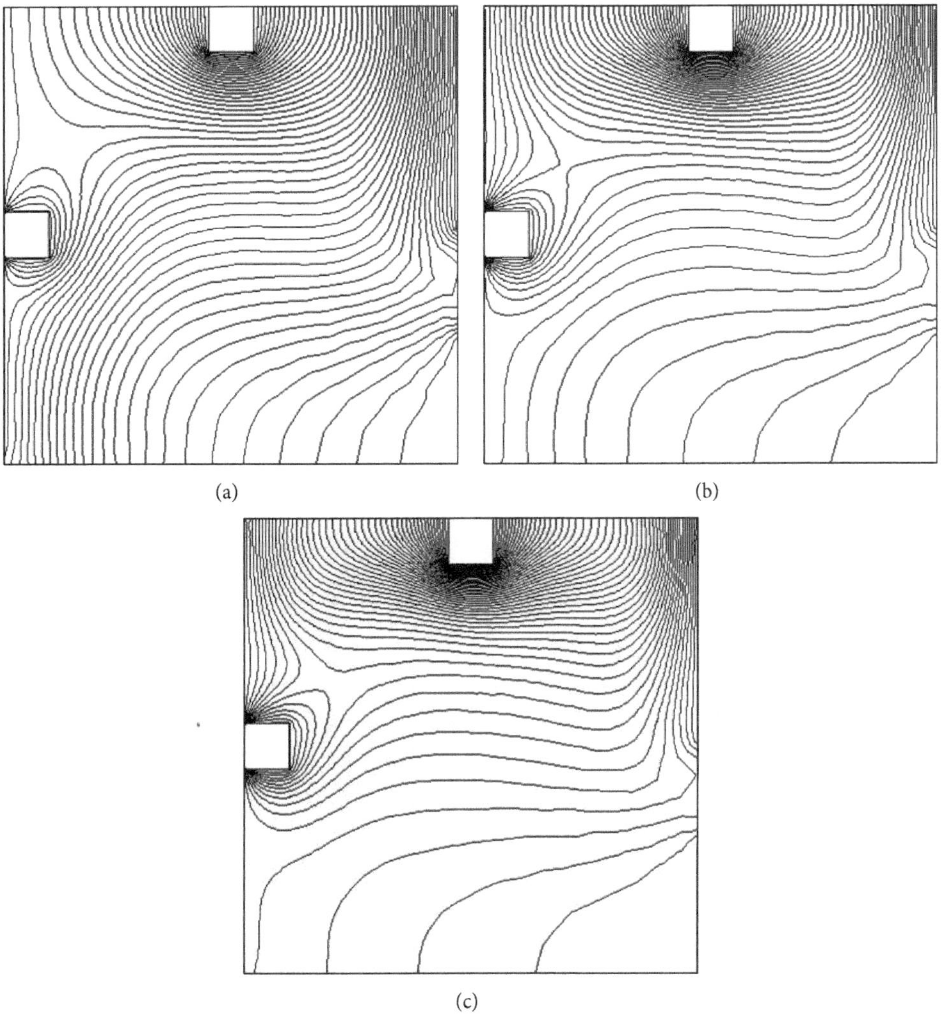

FIGURE 4: Isotherms for an external Rayleigh number ratio of 6000 for a Rayleigh number ratio of (a) 250, (b) 1000, and (c) 2500.

TABLE 3: The comparison between the average Nusselt number along the hot wall for three Rayleigh number ratios.

Rayleigh number ratio, R	Magnitude of the average Nusselt number (Mariana and Da Silva)	Magnitude of the average Nusselt number (current study)
250	2	2.5
1000	9	13
2500	22	35

displays the range of the dimensionless groups used. The subsequent sections detail the results of how the Nusselt number is affected by each group.

3. Effect of the Rayleigh Number Ratio

To determine the effect of the Rayleigh number ratio, R, on the local and average Nusselt numbers, Figure 3 is created. In Figure 3, the average Nusselt number along the hot wall is plotted versus the external Rayleigh number for three different Rayleigh number ratios, 250, 1000, and 2500. From the figure, one can see that the most negative Nusselt numbers correspond to the largest Rayleigh number ratio of 2500. Also, as the external Rayleigh number is increased, the average Nusselt number increases for all Rayleigh number ratios, and the rate of change decreases. The negative Nusselt numbers are due to the direction of heat flow from the air inside the enclosure to the wall. These trends are consistent with the works by Mariana and Da Silva. However, the average Nusselt number for our Rayleigh number ratios is much larger in magnitude compared to the average Nusselt number from Mariana and Da Silva at the same Rayleigh number ratios. Table 3 shows the maximum magnitude of the average Nusselt number along the hot wall for each Rayleigh number ratio for the current study compared with the one by Mariana and Da Silva. From Table 3, one can see that the

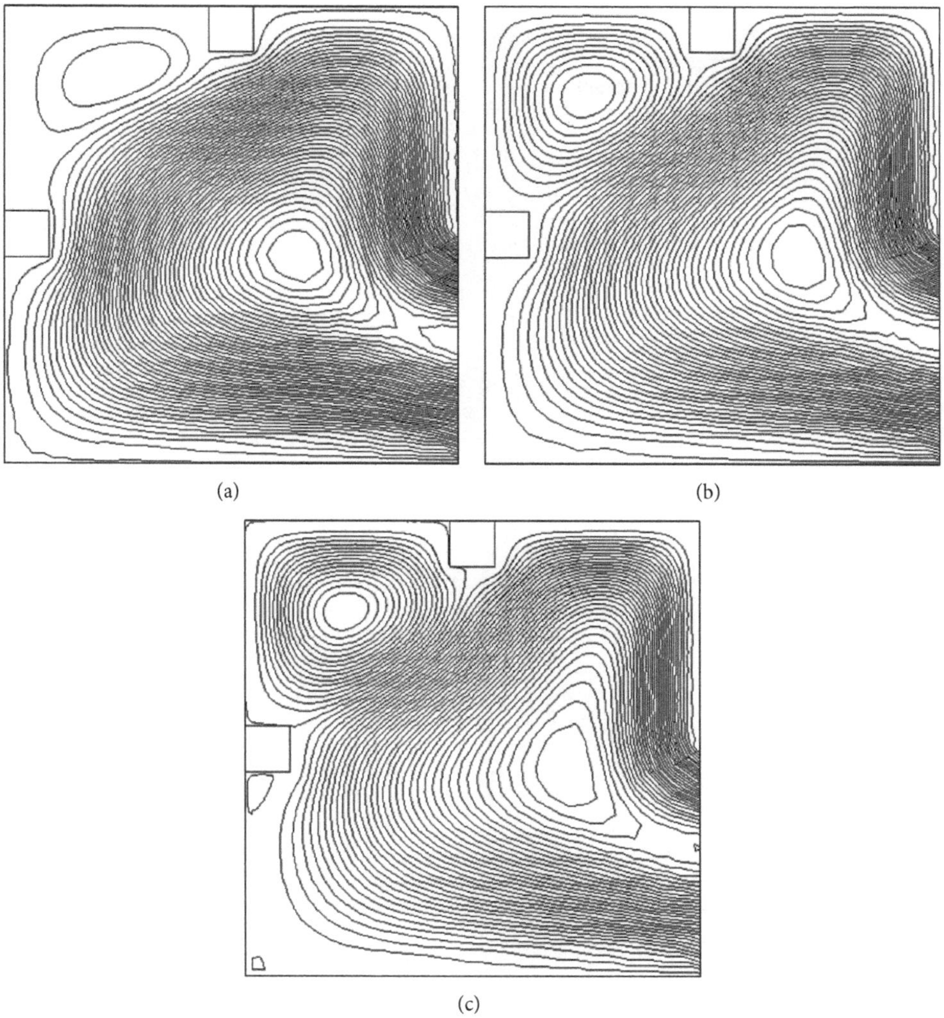

FIGURE 5: Streamlines for an external Rayleigh number ratio of 6000 for a Rayleigh number ratio of (a) 250, (b) 1000, and (c) 2500.

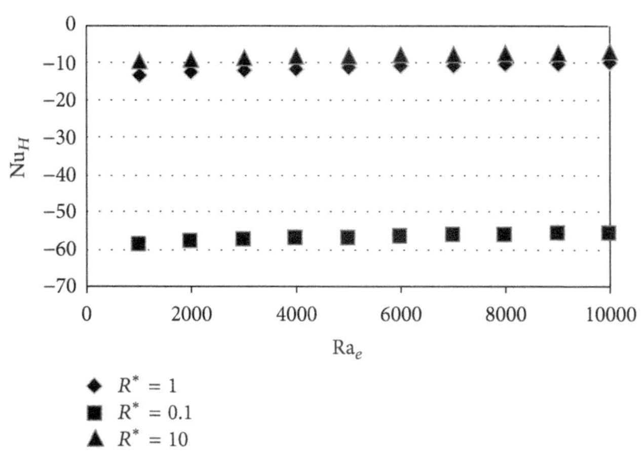

FIGURE 6: The average Nusselt number along the hot wall versus the external Rayleigh number for an R^* ratio of 0.1, 1.0, and 10.0.

addition of the second heat source results in higher Nusselt numbers.

The reason for this increase in the magnitude of the Nusselt number along the hot wall can be seen with the aid of Figure 4. Figure 4 shows the isotherms for an external Rayleigh number of 6000 for the three Rayleigh number ratios. Notice the large gradient in the isotherms along the upper part of the hot wall and the left side of the heated block along the insulated wall for the Rayleigh number ratio of 2500.

In addition to the large gradients, the temperature of the air in that corner is higher for this Rayleigh number ratio than the other two Rayleigh number ratios. These large gradients affect the flow patterns.

Figure 5 shows the streamlines for an external Rayleigh number of 6000 for the three Rayleigh number ratios. The streamlines indicate two distinct vortices in the flow. The main vortex is centered near the middle of the enclosure, and the second vortex is located in the corner of the enclosure between the two heat sources. The second vortex increases in strength as the Rayleigh number ratio is increased. This can be seen by the increased gradient in the streamlines as the Rayleigh number ratio is increased. This large temperature

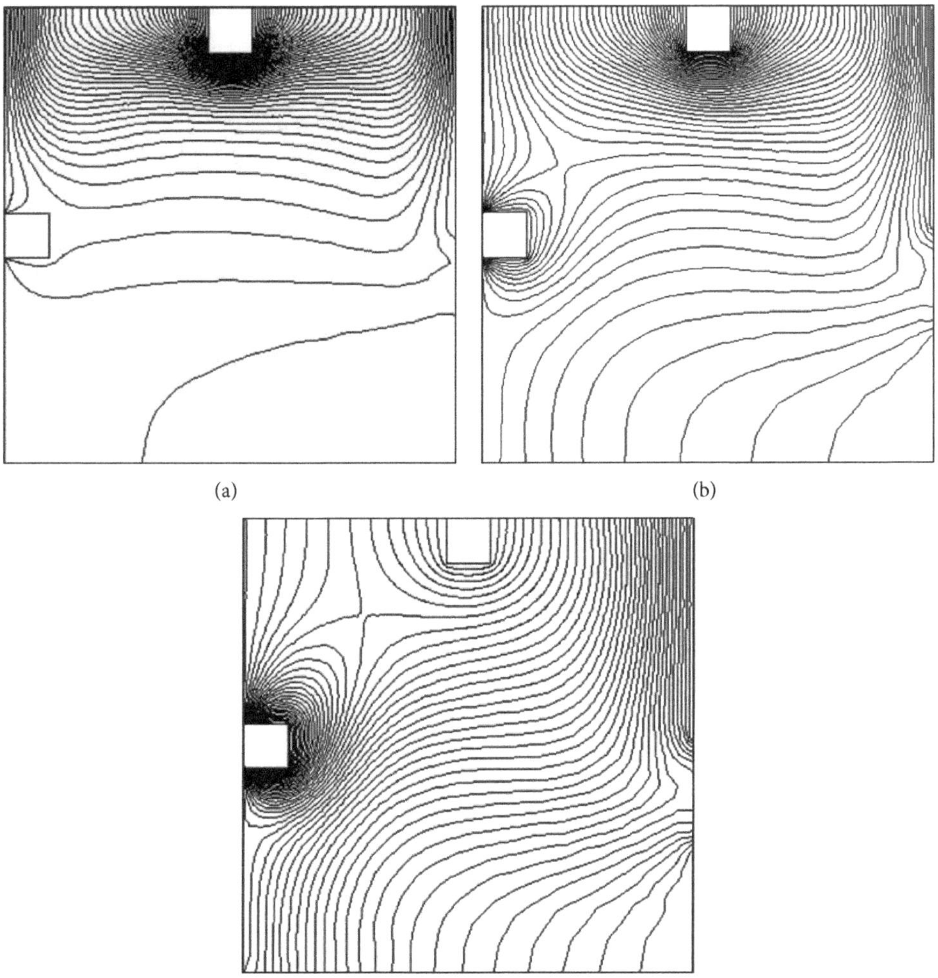

(a) (b) (c)

FIGURE 7: Isotherms for an external Rayleigh number ratio of 6000 for a Rayleigh number ratio of (a) 250, (b) 1000, and (c) 2500.

gradient causes a strong vortex to form in this corner associated with this Rayleigh number ratio. Mariani and Da Silva also reported a second vortex near the heat source; however, the size of the vortex is larger in the current study. This increase in the isotherm gradients and the resulting vortex size contribute to the larger Nusselt numbers in the current study.

4. Effect of Unequal Rayleigh Number Ratios for the Heated Blocks

In order to determine the effect of different heat generation rates for the two blocks, the Rayleigh number ratio for each block is set to a different value. The ratio between the Rayleigh number ratios for each block is defined as the parameter R^*. The block on the left-hand hot wall is designated block one with a Rayleigh number ratio of R_1, and the block along the top wall is called block two with a Rayleigh number ratio of R_2. Figure 6 shows how the Nusselt number along the hot wall is affected by the external Rayleigh number for three different R^* values, 0.1, 1, and 10.0. The R^* value of 1.0 corresponds to the two blocks having equal Rayleigh ratios and is the value in the previous section. From Figure 6, one can see that the magnitude of the Nusselt number is the largest for the value of R^* equal to 0.1, and the magnitude of the Nusselt numbers for the other two R values, 1.0 and 10.0, are close to each other. The Nusselt numbers for the R^* value of 10.0 are slightly less than the Nusselt numbers for R^* value of 1.0.

The larger magnitude of the Nusselt numbers for the R^* value of 0.1 can be explained with the aid of Figure 7. Figure 7 shows the resulting isotherms for the three R^* values for an external Rayleigh number of 6000. Notice the large gradient in the isotherms associated with the R^* value of 0.1. In contrast, the isotherms for the R^* value of 10.0 are spaced relatively wide apart. The resulting streamlines for the three R^* values can be seen in Figure 8. One can see that a vortex exists in the upper corner near the hot wall for the case where R^* equals 0.1. In comparison, no vortex appears in that same corner for the R^* value of 10.0. The strong vortex in the corner of the hot wall for the R^* value of 0.1 is associated with large

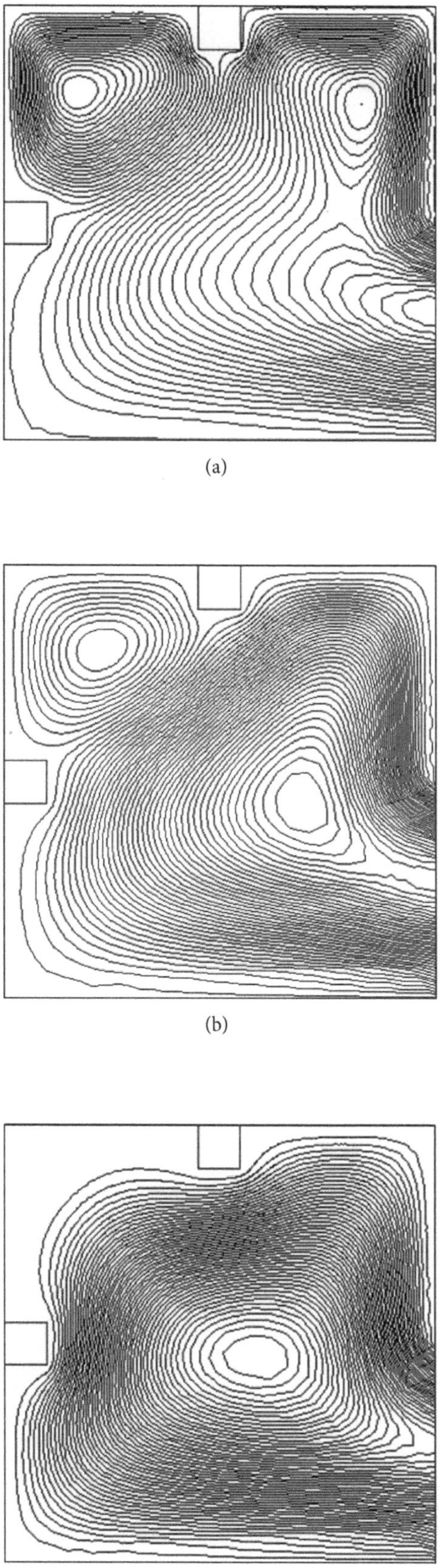

(a)

(b)

(c)

FIGURE 8: Isotherms for an external Rayleigh number ratio of 6000 for a Rayleigh number ratio of (a) 250, (b) 1000, and (c) 2500.

gradients in the isotherms. A vortex does exist in the upper corner for the case where R^* equals 1.0. However, the strength of this vortex is much smaller than the vortex associated with the R^* value of 0.1. This can be seen by the gradient between the streamlines for these two cases. In addition to the vortex in the upper corner near the hot wall, a second vortex appears in the upper corner near the cold wall for the case where R^* equals 0.1.

5. Conclusions

In this study, a partially open enclosure with two heat sources is examined to determine the Nusselt number along the hot wall. The study is divided into two parts. The first part examines the effect of the Rayleigh number ratio on the Nusselt number. The largest Rayleigh number ratio results in the highest magnitude of the Nusselt number. When compared to the study by Mariani and Da Silva, the current study shows much larger magnitudes for the Nusselt number for the same Rayleigh number ratios. This increase is attributed to the large vortex that forms of the corner near the hot wall. The strength and size of this vortex are due to the large gradient in the isotherms with higher Rayleigh number ratios.

The second part of the study examines what happens to the Nusselt number along the hot wall as the Rayleigh number ratio for the two blocks is different. This situation can arise when the two blocks have unequal heat source values. The largest Nusselt number magnitudes are associated with the R^* value of 0.1. This corresponds to the situation when the block along the top wall has a Rayleigh number ratio that is ten times higher than the Rayleigh number ratio for the block along the hot wall. The larger Nusselt number magnitudes can be attributed to the strong vortex in the corner near the hot wall. In contrast, there is no vortex in the same corner for an R^* value of 10.0. A second vortex also exists in the corner near the cold wall for the R^* value of 0.1.

Nomenclature

c_p: Specific heat
g: Acceleration due to gravity
k: Thermal conductivity
L: Length of the square
Nu: Average Nusselt number
Pr: Prandtl number
p: Static pressure
q: Heat generation per unit volume of the heat source
Ra_e: External Rayleigh number
Ra_i: Internal Rayleigh number
R: Rayleigh number ratio
R_1: Rayleigh number ratio for block 1
R_2: Rayleigh number ratio for block 2
R^*: Rayleigh number ratio of the two heat sources
T: The temperature
v: Fluid velocity
α: Thermal diffusivity
β: Volumetric coefficient of thermal expansion
μ: Dynamic viscosity
ρ: Fluid density.

Subscripts

1: Block 1
2: Block 2
c: Cold
e: External
i: Internal.

References

[1] V. C. Mariani and A. Da Silva, "Natural convection: analysis of partially open enclosures with an internal heated source," *Numerical Heat Transfer A*, vol. 52, no. 7, pp. 595–619, 2007.

[2] D. Angirasa, J. G. M. Eggels, and F. T. M. Nieuwstadt, "Numerical simulation of transient natural convection from an isothermal cavity open on a side," *Numerical Heat Transfer A*, vol. 28, no. 6, pp. 755–768, 1995.

[3] F. Penot, "Numerical calculations of two-dimensional natural convection in isothermal open," *Numerical Heat Transfer*, vol. 5, no. 4, pp. 421–437, 1982.

[4] P. Le Quere, J. A. C. Humphrey, and F. S. Sherman, "Numerical calulations of thermally driven two-dimensional unsteady laminar flow in cavities of a rectangular cross section," *Numerical Heat Transfer*, vol. 4, no. 3, pp. 249–283, 1981.

[5] Y. L. Chan and C. L. Tien, "A numerical study of two-dimensional natural convection in rectangular cavities," *Numerical Heat Transfer*, vol. 8, no. 1, pp. 65–80, 1985.

[6] J. L. Xia and Z. W. Zhou, "Natural convection in an externally heated partially open enclosure with a heated protrusion," *Measurement and Modeling of Environmental Flows*, vol. 232, pp. 201–208, 1992.

[7] S. A. Lai and C. Reji, "Numerical prediction of natural convection in vented cavities using restricted domain approach," *International Journal of Heat and Mass Transfer*, vol. 52, no. 3-4, pp. 724–734, 2009.

[8] A. Muftuoglu and E. Bilgen, "Conjugate heat transfer in open cavities with a discrete heater at its optimized position," *International Journal of Heat and Mass Transfer*, vol. 51, no. 3-4, pp. 779–788, 2008.

[9] A. Muftuoglu and E. Bilgen, "Natural convection in an open square cavity with discrete heaters at their optimized positions," *International Journal of Thermal Sciences*, vol. 47, no. 4, pp. 369–377, 2008.

[10] Y. Liu and N. Phan-Thien, "An optimum spacing problem for three chips mounted on a vertical substrate in an enclosure," *Numerical Heat Transfer A*, vol. 37, no. 6, pp. 613–630, 2000.

[11] A. K. da Silva, S. Lorente, and A. Bejan, "Optimal distribution of discrete heat sources on a wall with natural convection," *International Journal of Heat and Mass Transfer*, vol. 47, no. 2, pp. 203–214, 2004.

[12] A. K. da Silva, S. Lorente, and A. Bejan, "Optimal distribution of discrete heat sources on a plate with laminar forced convection," *International Journal of Heat and Mass Transfer*, vol. 47, no. 10-11, pp. 2139–2148, 2004.

[13] A. K. da Silva, G. Lorenzini, and A. Bejan, "Distribution of heat sources in vertical open channels with natural convection," *International Journal of Heat and Mass Transfer*, vol. 48, no. 8, pp. 1462–1469, 2005.

[14] A. K. da Silva and L. Gosselin, "Optimal geometry of L and C-shaped channels for maximum heat transfer rate in natural convection," *International Journal of Heat and Mass Transfer*, vol. 48, no. 3-4, pp. 609–620, 2005.

A Diversity-Enhanced Constrained Particle Swarm Optimizer for Mixed Integer-Discrete-Continuous Engineering Design Problems

Semin Chun, Young-Tark Kim, and Tae-Hyoung Kim

School of Mechanical Engineering, College of Engineering, Chung-Ang University, 221 Heukseok-dong, Dongjak-gu, Seoul 156-756, Republic of Korea

Correspondence should be addressed to Tae-Hyoung Kim; kimth@cau.ac.kr

Academic Editor: Jiin Yuh Jang

Engineering optimization problems usually contain various constraints and mixed integer-discrete-continuous types of design variables. We propose an efficient particle swarm optimization (PSO) algorithm for such problems. First, we transform the constrained optimization problem into an unconstrained problem without introducing problem-dependent or user-defined parameters such as penalty factors or Lagrange multipliers (such parameters are usually required in general optimization algorithms). Then, we extend the above PSO method to handle integer, discrete, and continuous design variables in a simple manner with a high degree of precision. The proposed PSO scheme is fairly simple and therefore easy to implement. To demonstrate the effectiveness of our method, several mechanical design optimization problems are solved, and the numerical results are compared with results reported in the literature.

1. Introduction

Most engineering optimal design problems require optimization techniques that are capable of handling the problem of equality and inequality constraints. In recent years, the particle swarm optimization (PSO) algorithm [1], whose mechanism was inspired by the social and cooperative behavior of swarms such as a flock of birds or a group of people, has been applied to various optimization problems with some promising results. It is a population-based search algorithm where a randomly initialized particle of the swarm updates its state (position) in a search space based on the information of its own and other agents' historical behavior until an optimal state has been reached (see, e.g., [2–4] and the references therein). However, the original PSO is naturally a stochastic unconstrained optimization method, and thus it cannot be directly applied to constrained optimization problems which is also the case for most swarm-inspired metaheuristic search technologies. This has motivated many researchers to investigate techniques for handling constrained optimization problems, and some methods have been proposed for example, (i) the modified feasible-based mechanism [5, 6], (ii) the penalty approach [7, 8], (iii) the augmented Lagrange multiplier approach [9, 10], (iv) the violated design points redirection [11, 12], (v) the fly-back mechanism method [13], (vi) the problem reformulation into an unconstrained multi-objective formulation [14, 15], and (vii) a series of sequential quadratic programming problems [16]. The distinctive features of conventional constraint handling schemes can be found in Aziz et al. [17], Kim et al. [18], and Jansen and Perez [10]. It is worth mentioning that many of these methods could not ensure a realization of strict constraint feasibility which would limit its possible use to practical applications [10]. Further, some of them require the adoption of problem-dependent or user-defined parameters such as penalty factors or Lagrange multipliers.

In this line of researches, the author recently proposed a considerably simple and efficient way to handle the constraint conditions in PSO when solving engineering optimization problems [18]. The key feature of the proposed constrained PSO method is that the given constrained optimization problem is transformed into an unconstrained problem without introducing problem-dependent or user-defined parameters such as penalty factors or Lagrange multipliers, although such parameters are usually required in general optimization algorithms. In this approach, particles of the swarm evolve as follows (see Kim et al. [18] for details). At the first stage of evolution, all particles try to move into a feasible region where constraint satisfaction is guaranteed. Thereafter, the particles try to explore the feasible region to find an optimal solution that satisfies the given constraint conditions. This constraint handling PSO scheme has been shown to outperform conventional methods for various continuous engineering optimization problems.

On the other hand, many types of engineering optimal design problems usually contain mixed integer-discrete-continuous- (MIDC-) type design variables together with the design constraints. They involve a high degree of nonlinearity due to the combinatorial aspect of the related integer-discrete variables. The development of global stochastic optimization techniques suitable for such optimization problems is a challenging and active research area in many engineering fields. However, a rigorous implementation of integer/discrete variables, even in the framework of unconstrained PSO schemes, still represents a challenge [29]. The authors discussed in Kim et al. [18] how the aforementioned constrained PSO scheme can be combined with conventional rounding-off techniques found in many studies (see, e.g., [2, 27, 38–40]). The numerical example in this study shows that the proposed method could solve constrained optimization problems with integer, discrete, and continuous design variables in a simple manner. However, the constrained-PSO-based approach with rounding-off technique tested in Kim et al. [18] may not provide sufficient optimization reliability, in the sense that the probability of obtaining an optimal design variable is not sufficiently high. In fact, such a shortcoming may chiefly be due to the poor particle diversification characteristics which frequently causes the potential problem of premature convergence of the swarm to local optima in a practical implementation. Further, this unfavorable phenomenon becomes more persistent in the case of constrained optimizations with MIDC variables. This means that the particles' diversity is undoubtedly an indispensable issue in improving the performance of population-based search algorithms. Furthermore, without adding an additional efficient diversity enhancing mechanism to the constrained PSO, it may be very difficult to reach the true global optimal solution of such mixed variable optimization problems.

To make up for the aforementioned deficiency, we propose a novel modified constrained PSO algorithm using a simple diversity enhancing mechanism with high applicability and reliability. Many diversity-guided PSO methods have recently been proposed (see, e.g., [41–46] and the references therein). However, their effectiveness was examined only via continuous optimization problems with no constraint conditions; that is, their methods cannot be directly applied to solve constrained MIDC optimization problems. The proposed approach is designed to have particles of the swarm evolve as follows. At the beginning stage of evolution, all particles try to move into a feasible region where constraint satisfaction is guaranteed. Here, the individual particle first freely evolves in a continuous search space regardless of the variable type based on the diversity-guided velocity update law with three phases of attraction, repulsion, and a combination of attraction and repulsion (the so-called in-between phase) [43, 47]. It is worth mentioning that the PSO scheme proposed by Pant et al. [43, 47] originally targets the optimization problems with only continuous design variables and no constraint conditions. Then, for integer or discrete design variables, the corresponding continuous values of particles are rounded off to the nearest integer or to discrete values. Thereafter, the particles continue to evolve in the aforementioned manner and try to explore such a feasible region to find an optimal solution that satisfies the given constraint conditions. Note that although it may not be considered the best possible approach, the rounded-off technique has been used with promising results as demonstrated in many studies [2, 18, 24, 27, 28, 38–40, 48, 49]. The introduced three-phase velocity control method has the following features (see Pant et al. [43, 47] for details):

(i) if the population diversity is above an upper threshold, the particles attract each other,

(ii) if the diversity is below a lower threshold, the particles repel each other,

(iii) if the diversity is between upper and lower thresholds, then the individual particle is attracted by its own previous best position and is simultaneously repelled by the best known particle position.

This efficiently improves the population diversity in the framework of constrained PSO [18] and can reduce the probability of a premature convergence phenomenon (particles trapped in a local optimum due to the nature of mixed variables). Note that the procedure that transforms the given constrained optimization problem into an unconstrained problem introduces no problem-dependent extra parameters, such as penalty factors [23, 50] or Lagrange multipliers [51], which usually require time-consuming fine-tuning tasks. The proposed constrained PSO scheme, combined with the general integer and discrete variable handling technique and three phases of particle velocity control, can yield satisfactory optimal solution accuracy. We now demonstrate the scheme's distinctive features through several numerical examples.

2. Particle Swarm Optimizer for Constrained MIDC Optimization Problems

A novel PSO algorithm for constrained MIDC optimization problems with high practicality, simplicity, and implementability is proposed. Some of the advantages of the proposed method over conventional methods are discussed in detail.

2.1. Optimization Problem Description. In this study, we are interested in general constrained optimization problems that are mathematically formulated as follows:

$$\min_{\mathbf{x}} f(\mathbf{x}), \quad \mathbf{x} \in \mathbb{D} \subseteq \mathbb{R}^{n_p} \tag{1}$$

subject to

$$h_\ell(\mathbf{x}) \le 0, \quad h_\ell : \mathbb{R}^{n_p} \longrightarrow \mathbb{R}, \quad \ell = 1, 2, \ldots, m, \tag{2}$$

where n_p is the number of independent design variables, m is the number of constraint conditions, $\mathbf{x} \in \mathbb{R}^{n_p}$ is the design variable vector, $\mathbb{D}$ is the search space, $f(\mathbf{x}) \in \mathbb{R}$ is the linear/nonlinear objective function, and $h_\ell(\mathbf{x}) \in \mathbb{R}$ is the linear/nonlinear constraint function. It is assumed in general optimization problems that the design variable $\mathbf{x} \in \mathbb{R}^{n_p}$ is continuously variable. However, some engineering optimization problems require treating the design variables as not only continuously varying quantities but also integer/discrete quantities. In such cases, the design variable vector $\mathbf{x}$ can be defined as follows:

$$\mathbf{x} := \left(\mathbf{x}^{\mathrm{I}}, \mathbf{x}^{\mathrm{D}}, \mathbf{x}^{\mathrm{C}}\right)^T, \tag{3}$$

where the n_{I}-dimensional vector $\mathbf{x}^{\mathrm{I}}$ includes integer variables, the n_{D}-dimensional vector $\mathbf{x}^{\mathrm{D}}$ includes discrete variables, and the n_{C}-dimensional vector $\mathbf{x}^{\mathrm{C}}$ includes continuous variables; that is,

$$\begin{aligned} \mathbf{x}^{\mathrm{I}} &:= \left(x_1^{\mathrm{I}}, x_2^{\mathrm{I}}, \ldots, x_{n_{\mathrm{I}}}^{\mathrm{I}}\right)^T \in \mathbb{R}^{n_{\mathrm{I}}} \quad : \text{integer variables}, \\ \mathbf{x}^{\mathrm{D}} &:= \left(x_1^{\mathrm{D}}, x_2^{\mathrm{D}}, \ldots, x_{n_{\mathrm{D}}}^{\mathrm{D}}\right)^T \in \mathbb{R}^{n_{\mathrm{D}}} \quad : \text{discrete variables}, \\ \mathbf{x}^{\mathrm{C}} &:= \left(x_1^{\mathrm{C}}, x_2^{\mathrm{C}}, \ldots, x_{n_{\mathrm{C}}}^{\mathrm{C}}\right)^T \in \mathbb{R}^{n_{\mathrm{C}}} \quad : \text{continuous variables}, \end{aligned} \tag{4}$$

where $\mathbb{R}^{n_{\mathrm{I}}}$, $\mathbb{R}^{n_{\mathrm{D}}}$, and $\mathbb{R}^{n_{\mathrm{C}}}$ denote feasible subsets of integer, discrete, and continuous variables, respectively. The total number of design variables is $n_p := n_{\mathrm{I}} + n_{\mathrm{D}} + n_{\mathrm{C}}$. Note that in order to guarantee particle evolution within the given search space, it is sometimes required that the search space information be incorporated into a form of boundary constraints such that $\underline{\mathbf{x}} \le \mathbf{x} \le \overline{\mathbf{x}}$, where $\underline{\mathbf{x}} \in \mathbb{R}^{n_p}$ and $\overline{\mathbf{x}} \in \mathbb{R}^{n_p}$ denote, respectively, the vectors of the lower and upper bounds of design variables and $\le$ denotes the element-wise inequality. Thus, the considered general linear/nonlinear constrained MIDC optimization problem is defined as follows:

$$\text{find } \mathbf{x} := \left(\mathbf{x}^{\mathrm{I}}, \mathbf{x}^{\mathrm{D}}, \mathbf{x}^{\mathrm{C}}\right)^T \in \mathbb{D} \subseteq \mathbb{R}^{n_p} \tag{5}$$

to minimize $f(\mathbf{x})$ subject to (2) and the following boundary constraints:

$$h_{m+1}(\mathbf{x}) := \underline{\mathbf{x}} - \mathbf{x} \le \mathbf{0}, \qquad h_{m+2}(\mathbf{x}) := \mathbf{x} - \overline{\mathbf{x}} \le \mathbf{0}, \tag{6}$$

where $\mathbf{0}$ denotes a zero vector of an appropriate dimension. In Section 2.2, a novel PSO algorithm to find an optimal solution of the constrained MIDC problem defined in (5) is developed.

2.2. The Improved PSO for Constrained MIDC Problems. To solve the constrained MIDC optimization problem described by (5), we used the diversity enhanced particle evolution law that decreases the possibility of particles being trapped in a local optimum due to the mixed variable nature. Consider a swarm consisting of m_p particles: $\mathbf{x}_1, \mathbf{x}_2, \ldots, \mathbf{x}_{m_p}$. Each particle $\mathbf{x}_i$ is, in effect, an n_p-dimensional vector:

$$\begin{aligned} \mathbf{x}_i &:= \left(\mathbf{x}_i^{\mathrm{I}}, \mathbf{x}_i^{\mathrm{D}}, \mathbf{x}_i^{\mathrm{C}}\right)^T \\ &= \left(x_{i,1}^{\mathrm{I}}, \ldots, x_{i,n_{\mathrm{I}}}^{\mathrm{I}}, x_{i,1}^{\mathrm{D}}, \ldots, x_{i,n_{\mathrm{D}}}^{\mathrm{D}}, x_{i,1}^{\mathrm{C}}, \ldots, x_{i,n_{\mathrm{C}}}^{\mathrm{C}}\right)^T, \end{aligned} \tag{7}$$

and its velocity $\mathbf{v}_i$ is also an n_p-dimensional vector denoted by

$$\mathbf{v}_i := \left(v_{i,1}, v_{i,2}, \ldots, v_{i,n_p}\right)^T \in \mathbb{R}^{n_p}, \tag{8}$$

where $i \in [1, 2, \ldots, m_p]$. Then, to find an optimal solution for the constrained MIDC optimization problem in (5), each particle $\mathbf{x}_i \in \mathbb{R}^{n_p}$ should evolve in the search space $\mathbb{D}$. In this case, the evolution law of particles should efficiently improve the population diversity in the framework of constrained PSO [18] in order to reduce the probability of a premature convergence phenomenon (the particle being trapped in a local optimum due to the mixed variable nature). Therefore, we introduce a particle position updating algorithm using a combination of the constriction factor χ and the diversity-guided velocity update law with an attraction phase, repulsion phase, and in-between phase:

$$\mathbf{x}_i^{k+1} = \mathbf{x}_i^k + \mathbf{v}_i^{k+1}, \tag{9}$$

where $\mathbf{v}_i^{k+1}$ evolves as

$$\mathbf{v}_i^{k+1} = \begin{cases} \chi\left[\mathbf{v}_i^k + c_1\zeta_i^k\left(\mathbf{x}_i^{\mathrm{best},k} - \mathbf{x}_i^k\right) + c_2\xi_i^k\left(\mathbf{x}_{\mathrm{swarm}}^{\mathrm{best},k} - \mathbf{x}_i^k\right)\right], & \text{if } \mathrm{DIV}(k) > \mathrm{D_{high}}; \\ \chi\left[\mathbf{v}_i^k - c_1\zeta_i^k\left(\mathbf{x}_i^{\mathrm{best},k} - \mathbf{x}_i^k\right) - c_2\xi_i^k\left(\mathbf{x}_{\mathrm{swarm}}^{\mathrm{best},k} - \mathbf{x}_i^k\right)\right], & \text{if } \mathrm{DIV}(k) < \mathrm{D_{low}}; \\ \chi\left[\mathbf{v}_i^k + c_1\zeta_i^k\left(\mathbf{x}_i^{\mathrm{best},k} - \mathbf{x}_i^k\right) - c_2\xi_i^k\left(\mathbf{x}_{\mathrm{swarm}}^{\mathrm{best},k} - \mathbf{x}_i^k\right)\right], & \text{otherwise}, \end{cases} \tag{10}$$

where c_1 is the cognitive scaling factor, c_2 is the social scaling factor, and χ is the constriction factor defined as

$$\chi = \frac{2}{\left|2 - \varphi - \sqrt{\varphi^2 - 4\varphi}\right|}, \quad \varphi := c_1 + c_2 > 4, \tag{11}$$

where c_1 and c_2 are based on the authors' experience and were set at 2.02 for the example constrained MIDC optimization problems (c_1 and c_2 are usually set to 2.05 for optimization problems with only continuous design variables [18, 52]). The random numbers ζ_i^k and ξ_i^k are uniformly distributed in $[0, 1]$ and represent the stochastic behavior of the PSO.

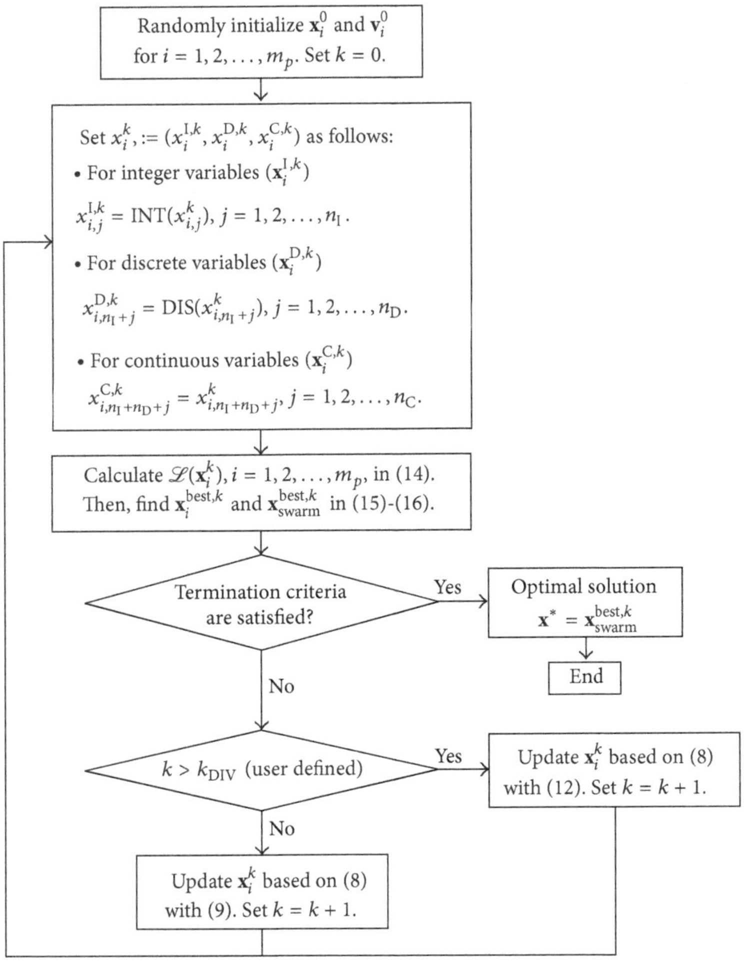

FIGURE 1: Flowchart of the proposed PSO scheme for constrained MIDC optimization problems.

$\mathbf{x}_i^{\mathrm{best},k}$ denotes the best previously obtained position of the ith particle, and $\mathbf{x}_{\mathrm{swarm}}^{\mathrm{best},k}$ denotes the best position in the entire swarm at the current iteration k, whose mathematical formulation will be given subsequently. The term $c_1\zeta_i^k(\mathbf{x}_i^{\mathrm{best},k} - \mathbf{x}_i^k)$ is associated with cognition, since it takes into account only the best position of the particle's own experience. On the other hand, $c_2\xi_i^k(\mathbf{x}_{\mathrm{swarm}}^{\mathrm{best},k} - \mathbf{x}_i^k)$ represents the social interaction of all particles [51]. In (10), the first velocity update law means the attraction phase of particles, the second law means the repulsion phase, and the final law means the in-between phase, where diversity measure DIV(k) [43, 53] is defined as

$$\mathrm{DIV}(k) = \frac{1}{m_p}\sum_{i=1}^{m_p}\sqrt{\sum_{j=1}^{n_p}\left(x_{i,j}^k - \bar{x}_j^k\right)^2}, \tag{12}$$

where

$$\bar{x}_j^k := \frac{\left(\sum_{i=1}^{m_p} x_{i,j}^k\right)}{m_p}. \tag{13}$$

Note that the constriction factor χ in PSO inherently enables particles to search different regions efficiently by avoiding premature convergence to local optima [38, 54]. Therefore, the introduced three-phase evolution method in (10) could be a better way to solve MIDC optimization problems than the approach in [43]. On the other hand, although the evolution law of particles described in (9) and (10) has an effectiveness of improving the population diversity that prevents the premature convergence phenomenon in PSO, its main drawback is the lack of fine-searching ability in the current search area. Note that the convergence of the particle swarm is necessary to refine the optimal solution [55]. Therefore, to overcome such a problem, we applied the three-phase velocity update law described in (10) and the following law in consecutive order (see Figure 1):

$$\mathbf{v}_i^{k+1} = \chi\left[\mathbf{v}_i^k + c_1\zeta_i^k\left(\mathbf{x}_i^{\mathrm{best},k} - \mathbf{x}_i^k\right) + c_2\xi_i^k\left(\mathbf{x}_{\mathrm{swarm}}^{\mathrm{best},k} - \mathbf{x}_i^k\right)\right]; \tag{14}$$

that is, the three-phase velocity update procedure from (10) was applied to an early stage of evolution of particles, and after a predetermined number of iterations (k_{DIV}), the conventional scheme described in (14) is introduced. We verified its effectiveness through several numerical examples, as discussed in Section 3.

Next, we considered how to handle the mixed integer-discrete-continuous design variables $\mathbf{x}_i := (\mathbf{x}_i^{\mathrm{I}}, \mathbf{x}_i^{\mathrm{D}}, \mathbf{x}_i^{\mathrm{C}})^T$, $i = 1, 2, \ldots, m_p$, within the framework of the PSO update scheme. To achieve this objective, we combined the particle evolution law with rounding-off techniques. We note that although the preceding scheme seems a bit rough and led to unsatisfactory optimal solution accuracy, promising results were obtained in PSO frameworks as demonstrated in previous studies (see, e.g., [2, 18, 24, 27, 28, 38–40, 48, 49]). The particle evolution procedure, which is a combination of the three-phase position update law described in (9) and (10) and rounding-off techniques, is described as follows.

(i) When particles in the swarm are initialized, each particle first freely evolves in a continuous search space regardless of the variable type based on the position and velocity updating law in (9) and (10). Therefore, all design variables take real values (i.e., $\mathbf{x}_i^k \in \mathbb{R}^{n_p}$) in this stage.

(ii) Next, for an integer variable $x_{i,j}^{\mathrm{I},k} \in \mathbf{x}_i^k$ $(j = 1, 2, \ldots, n_{\mathrm{I}})$, we exploit the rounding-off method; namely, $x_{i,j}^{\mathrm{I},k} \leftarrow \mathrm{INT}(x_{i,j}^k)$, where $\mathrm{INT}(x_{i,j}^k)$ denotes that $x_{i,j}^k \in \mathbb{R}$ is rounded off to the nearest integer. Similarly, for a discrete design variable $x_{i,n_{\mathrm{I}}+j}^{\mathrm{D},k} \in \mathbf{x}_i^k$ $(j = 1, 2, \ldots, n_{\mathrm{D}})$, $x_{i,n_{\mathrm{I}}+j}^{\mathrm{D},k} \leftarrow \mathrm{DIS}(x_{i,n_{\mathrm{I}}+j}^k)$, where $\mathrm{DIS}(x_{i,n_{\mathrm{I}}+j}^k)$ denotes that $x_{i,n_{\mathrm{I}}+j}^k \in \mathbb{R}$ takes the nearest discrete value in a given set of equally and unequally spaced discrete values.

Once all particles in the swarm have moved to their new positions according to the aforementioned procedure, then the given objective function $f(\mathbf{x})$, subject to various forms of constraints $h_\ell(\mathbf{x})$, $\ell = 1, 2, \ldots, m, m+1, m+2$, should be taken into account in any form to calculate the fitness value of each individual and to find $\mathbf{x}_i^{\mathrm{best},k}$ and $\mathbf{x}_{\mathrm{swarm}}^{\mathrm{best},k}$ in (10). Thus, we incorporated the simple constraint handling method [56] using a virtual objective function $f_v(\mathbf{x})$ with the particle evolution law. Note that $f_v(\mathbf{x}_i)$ could be a function that simultaneously satisfies the following two properties: (i) $f_v(\mathbf{x}_i) < 0$ holds for any $\mathbf{x}_i$ satisfying $h_\ell(\mathbf{x}_i) \le 0$, and (ii) $f_v(\mathbf{x}_i) < f_v(\mathbf{x}_j)$ holds whenever $f(\mathbf{x}_i) < f(\mathbf{x}_j)$ is satisfied. One possible candidate for $f_v(\mathbf{x}_i)$ is $f_v(\mathbf{x}_i) := \arctan\{f(\mathbf{x}_i)\} - \pi/2$ which was used to solve the numerical examples given in the next section (see [18, 56] for details and its distinctive features). Then, the original constrained MIDC optimization problem in (5) was modified into an unconstrained problem as follows.

For $i = 1, 2, \ldots, m_p$,

$$\min_{\mathbf{x}_i := (\mathbf{x}_i^{\mathrm{I}}, \mathbf{x}_i^{\mathrm{D}}, \mathbf{x}_i^{\mathrm{C}})^T} \mathscr{L}(\mathbf{x}_i), \tag{15}$$

where $h_{\max}(\mathbf{x}_i) := \max[h_1(\mathbf{x}_i), h_2(\mathbf{x}_i), \ldots, h_{m+2}(\mathbf{x}_i)]$ and

$$\mathscr{L}(\mathbf{x}_i) := \begin{cases} h_{\max}(\mathbf{x}_i), & \text{if } h_{\max}(\mathbf{x}_i) \ge 0, \\ f_v(\mathbf{x}_i), & \text{otherwise.} \end{cases} \tag{16}$$

Note that the procedure transforming the given constrained optimization problem into an unconstrained problem introduces no problem-dependent extra parameter such as penalty factors [23, 50] or Lagrange multipliers [51] that usually require time-consuming fine-tuning tasks. Based on the definition of a modified objective function $\mathscr{L}(\mathbf{x}_i)$ in (16), rigorous mathematical formulations of $\mathbf{x}_i^{\mathrm{best},k}$ and $\mathbf{x}_{\mathrm{swarm}}^{\mathrm{best},k}$ were obtained as follows:

$$\begin{aligned} \mathbf{x}_i^{\mathrm{best},k} &:= \arg\min_{\mathbf{x}_i^j} \left\{ \mathscr{L}\left(\mathbf{x}_i^j\right), 0 \le j \le k \right\}, \\ \mathbf{x}_{\mathrm{swarm}}^{\mathrm{best},k} &:= \arg\min_{\mathbf{x}_i^k} \left\{ \mathscr{L}\left(\mathbf{x}_i^k\right), i = 1, 2, \ldots, m_p \right\}. \end{aligned} \tag{17}$$

From the preceding observation, the overall computational steps of the proposed PSO algorithm for constrained MIDC optimization problems are described as shown in Figure 1.

In the following sections, the effectiveness of the proposed PSO scheme for constrained MIDC optimization problems is demonstrated via several engineering optimal design example problems (it may be difficult to guarantee the performance of our method theoretically due to the probabilistic nature of PSO).

3. Numerical Experimentation

Several engineering optimal design problems were investigated to evaluate the effectiveness of the proposed PSO technique described in Section 2. The best results obtained by the proposed method are compared to problems reported in the literature that were solved by various evolutionary computation and other optimization algorithms. In the following examples, all computations were performed on the MATLAB platform.

3.1. Engineering Optimization Problems

Problem 1 (design of a pressure vessel described in [19–22]). A cylindrical vessel is capped at both ends by hemispherical heads as shown in Figure 2. The objective is to minimize the total cost including the cost of the material, forming, and welding. There are four design variables: T_s (the thickness of the shell), T_h (the thickness of the head), R (the inner radius), and L (the length of the cylindrical section of the vessel not including the head). T_s and T_h are integer multiples of 0.0625 (the available thickness of rolled steel plates), and R and L are continuous. These design variables are specified as

$$\mathbf{x} = (x_1, x_2, x_3, x_4)^T := (T_s, T_h, R, L)^T. \tag{18}$$

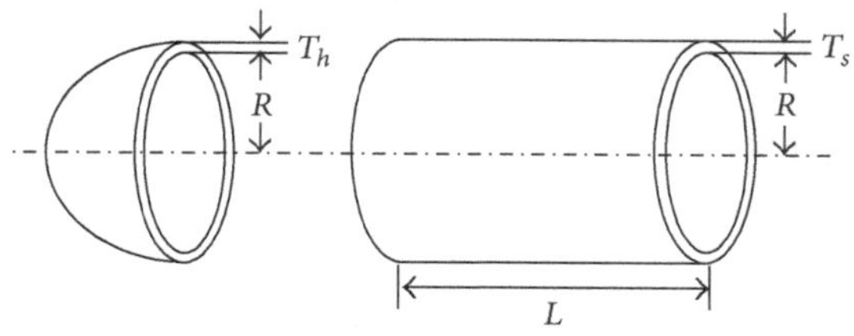

FIGURE 2: Configuration of a pressure vessel.

The optimization problem is formulated as follows:

$$\min_{\mathbf{x}} f(\mathbf{x}) := 0.6224x_1x_3x_4 + 1.7781x_2x_3^2 + 3.1661x_1^2x_4 + 19.84x_1^2x_3 \tag{19}$$

subject to

$$h_1(\mathbf{x}) := 0.0193x_3 - x_1 \le 0, \tag{20}$$

$$h_2(\mathbf{x}) := 0.00954x_3 - x_2 \le 0, \tag{21}$$

$$h_3(\mathbf{x}) := 1296000 - \pi x_3^2x_4 - \frac{4}{3}\pi x_3^3 \le 0, \tag{22}$$

$$h_4(\mathbf{x}) := x_4 - 240 \le 0, \tag{23}$$

$$h_5(\mathbf{x}) := 1.1 - x_1 \le 0, \tag{24}$$

$$h_6(\mathbf{x}) := 0.6 - x_2 \le 0, \tag{25}$$

where the search space was set as $\mathbb{D} := \{\mathbf{x} \in \mathbb{R}^4 : (1.1, 0.6, 10, 10)^T \le \mathbf{x} \le (10, 10, 150, 150)^T\}$.

Problem 2 (design of a pressure vessel described in [23–27]). For the cylindrical pressure vessel shown in Figure 2, the constrained optimization problem formulated as (19) subject to (20)–(23) was considered in [23–27]. Note that the constraint conditions described in (24) and (25) are not included in this case. Design variables $x_1(:= T_s)$ and $x_2(:= T_h)$ are integer multiples of 0.0625, and $x_3(:= R)$ and $x_4(:= L)$ are continuous. The search space was set as $\mathbb{D} := \{\mathbf{x} \in \mathbb{R}^4 : (0, 0, 10, 10)^T \le \mathbf{x} \le (99, 99, 200, 200)^T\}$.

Problem 3 (design of a speed reducer described in [24, 26]). This example deals with a design problem of the speed reducer shown in Figure 3. The design parameters are as follows: b (the face width), m (the module of teeth), n (the number of teeth on the pinion), l_1 (the length of shaft 1 between bearings), l_2 (the length of shaft 2 between bearings), d_1 (the diameter of shaft 1), and d_2 (the diameter of shaft 2). The design variables are specified as

$$\mathbf{x} = (x_1, x_2, x_3, x_4, x_5, x_6, x_7)^T := (b, m, n, l_1, l_2, d_1, d_2)^T, \tag{26}$$

where $x_3(:= n)$ is integer and the rest of $\mathbf{x}$ are continuous. The objective is to minimize the total weight of a speed reducer. The constraints include limitations on the bending stress and surface stress of the gear teeth, transverse deflections of shafts 1 and 2 due to the transmitted force, and stresses in shaft 1 and 2. The problem is formulated as follows:

$$\min_{\mathbf{x}} f(\mathbf{x}) := 0.7854x_1x_2^2\left(3.3333x_3^2 + 14.9334x_3 - 43.0934\right) - 1.508x_1\left(x_6^2 + x_7^2\right) + 7.4777\left(x_6^3 + x_7^3\right) + 0.7854\left(x_4x_6^2 + x_5x_7^2\right) \tag{27}$$

subject to

$$h_1(\mathbf{x}) := 27x_1^{-1}x_2^{-2}x_3^{-1} - 1 \le 0,$$

$$h_2(\mathbf{x}) := 397.5x_1^{-1}x_2^{-2}x_3^{-2} - 1 \le 0,$$

$$h_3(\mathbf{x}) := 1.93x_2^{-1}x_3^{-1}x_4^3x_6^{-4} - 1 \le 0,$$

$$h_4(\mathbf{x}) := 1.93x_2^{-1}x_3^{-1}x_5^3x_7^{-4} - 1 \le 0,$$

$$h_5(\mathbf{x}) := \frac{\sqrt{(745x_4/x_2x_3)^2 + 16.9 \times 10^6}}{0.1x_6^3} - 1100 \le 0,$$

$$h_6(\mathbf{x}) := \frac{\sqrt{(745x_5/x_2x_3)^2 + 157.5 \times 10^6}}{0.1x_7^3} - 850 \le 0,$$

$$h_7(\mathbf{x}) := x_2x_3 - 40 \le 0,$$

$$h_8(\mathbf{x}) := 5 - \left(\frac{x_1}{x_2}\right) \le 0,$$

$$h_9(\mathbf{x}) := \frac{x_1}{x_2} - 12 \le 0,$$

$$h_{10}(\mathbf{x}) := \left(1.5x_6 + 1.9\right)x_4^{-1} - 1 \le 0,$$

$$h_{11}(\mathbf{x}) := \left(1.1x_7 + 1.9\right)x_5^{-1} - 1 \le 0, \tag{28}$$

where $2.6 \le x_1 \le 3.6$, $0.7 \le x_2 \le 0.8$, $17 \le x_3 \le 28$, $7.3 \le x_4 \le 8.3$, $7.8 \le x_5 \le 8.3$, $2.9 \le x_6 \le 3.9$, $5.0 \le x_7 \le 5.5$.

Problem 4 (design of a speed reducer described in [22, 32–34]). It is assumed that for the same optimal design problem of the speed reducer presented in Problem 5, x_3 is an integer variable; x_1, x_2, x_4, and x_5 are integer multiples of 0.1; and x_6 and x_7 are integer multiples of 0.01. In the cited studies, the minimization problem given in (27) was modified to be

$$\min_{\mathbf{x}} f(\mathbf{x}) := 0.7854x_1x_2^2\left(3.3333x_3^2 + 14.9334x_3 - 43.0934\right) - 1.508x_1\left(x_6^2 + x_7^2\right) + 7.477\left(x_6^3 + x_7^3\right) + 0.7854\left(x_4x_6^2 + x_5x_7^2\right). \tag{29}$$

Note also that $7.3 \le x_5 \le 8.3$ was considered in this case.

Problem 5 (design of a tension/compression coil spring described in [21, 22, 27, 35, 36, 57]). In this optimal design problem for tension/compression coil spring shown in Figure 4, $x_1(:= N)$, $x_2(:= D)$, and $x_3(:= d)$ were considered

TABLE 1: Optimal design of the pressure vessel—Problem 1.

Quantity	x_1 (discrete)	x_2 (discrete)	x_3 (continuous)	x_4 (continuous)	$f(x)$
Shih and Lai [19]	1.125	0.625	47.448	119.98	8160.80
Fu et al. [20]	1.125	0.625	48.3807	111.7449	8048.619
Sandgren [21]	1.123	0.625	48.97	106.72	7982.5
Rao and Xiong [22]	1.1875	0.625	61.4483	27.4037	7284.02
Proposed method	1.125	0.625	58.29015544	43.69265624	7198.00542037

TABLE 2: Optimal design of the pressure vessel—Problem 2.

Quantity (N/A: not available)	x_1 (discrete)	x_2 (discrete)	x_3 (continuous)	x_4 (continuous)	$f(x)$
Parsopoulos and Vrahatis [23]	N/A	N/A	N/A	N/A	6154.7
Akhtar et al. [24]	0.8125	0.4375	41.9768	182.2845	6171.0
Coello and Mezura-Montes [25]	0.8125	0.4375	40.097398	176.654047	6059.946341
Mezura-Montes et al. [26]	0.8125	0.4375	42.098370	176.637146	6059.714355
He et al. [27]	0.8125	0.4375	42.0984456	176.63659584	6059.714335
Cagnina et al. [28]	0.8125	0.4375	42.098445	176.636595	6059.714335
Tomassetti [29]	0.8125	0.4375	42.098446	176.636596	6059.714337
Proposed method	0.8125	0.4375	42.0984456	176.63659584	6059.714335

TABLE 3: Statistical comparison for the pressure vessel design problem—Problem 2.

Quantity $f(x)$	Best	Mean	Worst	Standard deviation
Parsopoulos and Vrahatis [23]	6154.70	8016.37	9387.77	745.869
Mezura-Montes et al. [26]	6059.714355	6355.343115	6846.628418	256.0437950
Mezura-Montes and Coello [30]	6059.7143	6379.9380	N/A	210.0
Cagnina et al. [28]	6059.714355	6092.0498	N/A	12.1725
Aguirre et al. [31]	6059.714355	6071.0133	N/A	15.1011
Tomassetti [29]	6059.714337	6086.9	6411.4	39.2722
Kim et al. [18]	6059.714335	6059.727721	6060.074434	0.065870503
Proposed method	6059.714335	6060.33057699	6090.52620169	4.35745530

TABLE 4: Optimal design of the speed reducer—Problem 3.

Quantity	Akhtar et al. [24]	Mezura-Montes et al. [26]	Cagnina et al. [28]	Tomassetti [29]	Proposed method
x_1 (continuous)	3.503122	3.506163	3.5	3.5	3.499993
x_2 (continuous)	0.700006	0.700831	0.7	0.7	0.7
x_3 (integer)	17	17	17	17	17
x_4 (continuous)	7.549126	7.460181	7.3	7.3	7.3
x_5 (continuous)	7.859330	7.962143	7.8	7.8	7.8
x_6 (continuous)	3.365576	3.362900	3.350214	3.350215	3.35021467
x_7 (continuous)	5.289773	5.308949	5.286683	5.286683	5.28668323
$f(x)$	3008.08	3025.00513	2996.348165	2996.348165	2996.34541555

TABLE 5: Statistical comparison for the speed reducer design problem—Problem 3.

Quantity $f(x)$	Best	Mean	Worst	Standard deviation
Akhtar et al. [24]	3008.08	3012.12	3028.28	N/A
Mezura-Montes et al. [26]	3025.005127	3088.777816	3226.248291	47.361890
Cagnina et al. [28]	2996.348165	2296.3482	N/A	0.0000
Aguirre et al. [31]	2996.372448	2996.4085	N/A	0.0286
Tomassetti [29]	2996.348165	$2.9965E+03$	$3.0020E+03$	0.9121
Proposed method	2996.34541555	2996.34541555	2996.34541556	$2.46561597E-09$

TABLE 6: Optimal design of the speed reducer—Problem 4.

Quantity	Li and Azarm [32]	Li and Papalambros [33]	Kuang et al. [34]	Rao and Xiong [22]	Proposed method
x_1 (discrete)	3.5	3.5	3.6	3.5	3.5
x_2 (discrete)	0.7	0.7	0.7	0.7	0.7
x_3 (integer)	17	20	17	17	17
x_4 (discrete)	7.3	7.3	7.3	7.3	7.3
x_5 (discrete)	7.71	7.3	7.8	7.8	7.8
x_6 (discrete)	3.35	3.35	3.4	3.36	3.36
x_7 (discrete)	5.29	5.29	5.0	5.29	5.29
$f(x)$	2996.3 (infeasible)	2985.22 (infeasible)	2876.12 (infeasible)	3000.83	3000.82953716

TABLE 7: Optimal design of the tension/compression coil spring—Problem 5.

Quantity	x_1 (integer)	x_2 (continuous)	x_3 (discrete)	$f(x)$
Deb and Goyal [35]	9	1.226	0.283	2.665
Sandgren [21]	10	1.180701	0.283	2.7995
Rao and Xiong [22]	9	1.225280	0.283	2.66342
Lampinen and Zelinka [36]	9	1.223041	0.283	2.65856
He et al. [27]	9	1.223041	0.283	2.65856
Proposed method	9	1.22304101	0.283	2.65855917

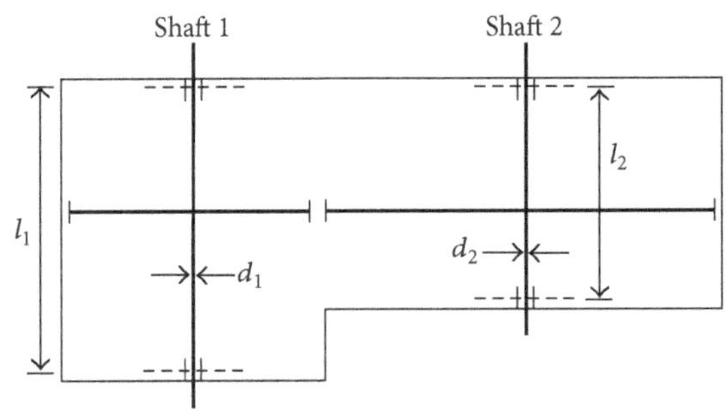

FIGURE 3: Configuration of a speed reducer.

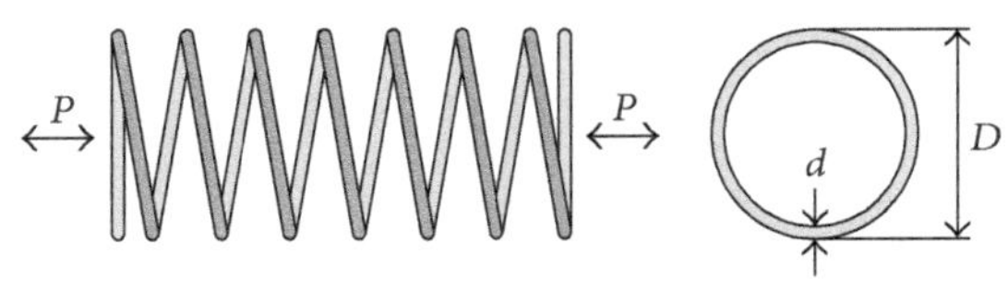

FIGURE 4: Configuration of a tension/compression coil spring.

as integer, continuous, and discrete variables, respectively, as follows:

$$x_1 = 5 + \kappa, \quad \text{where } \kappa = 0, 1, 2, \ldots, 15,$$
$$1.0 \le x_2 \le 3.0,$$
$$x_3 \in \{0.207, 0.225, 0.244, 0.263, 0.283, 0.307, 0.331, 0.362, 0.394, 0.4375, 0.50\}. \tag{30}$$

The following optimization problem was investigated:

$$\min_{\mathbf{x}} f(\mathbf{x}) := \frac{\pi^2 x_2 x_3^2 (x_1 + 2)}{4} \tag{31}$$

subject to

$$\begin{aligned} h_1(\mathbf{x}) &:= S - \frac{8K_s P_{\max} x_2}{\pi x_3^3} \ge 0, \\ h_2(\mathbf{x}) &:= l_{\max} - (\delta + 1.05(x_1 + 2)x_3) \ge 0, \\ h_3(\mathbf{x}) &:= x_3 - d_{\min} \ge 0, \\ h_4(\mathbf{x}) &:= D_{\max} - x_2 \ge 0, \\ h_5(\mathbf{x}) &:= C - 3 \ge 0, \\ h_6(\mathbf{x}) &:= \delta_{pm} - \delta \ge 0, \\ h_7(\mathbf{x}) &:= l_f - \delta - \frac{P_{\max} - P_L}{K} - 1.05(x_1 + 2)x_3 \ge 0, \\ h_8(\mathbf{x}) &:= \frac{P_{\max} - P_L}{K} - \delta_w \ge 0, \end{aligned} \tag{32}$$

with

$$C = \frac{x_2}{x_3}, \qquad K_s = \frac{4C - 1}{4C - 4} + \frac{0.615}{C},$$
$$\delta = \frac{8P_{\max} x_2^3 x_1}{G x_3^4}, \qquad K = \frac{G x_3^4}{8 x_2^3 x_1}, \tag{33}$$

where the maximum work load $P_{\max} = 1000.0$ lb, the maximum free length $l_{\max} = 14.0$ in, the minimum wire diameter $d_{\min} = 0.2$ in, the allowable maximum shear stress $S = 189000.0$ psi, the maximum outside diameter of the spring $D_{\max} = 3.0$ in, the preload compression force $P_L = 300.0$ lb, the allowable maximum deflection under preload $\delta_{pm} = 6.0$ in, the deflection from preload position to maximum load position $\delta_w = 1.25$ in, and the shear modulus of the material $G = 11.5 \times 10^6$ psi.

TABLE 8: Statistical comparison for the tension/compression coil spring—Problem 5.

Quantity $f(x)$	Best	Mean	Worst	Standard deviation
He et al. [27]	2.65856	2.738024	N/A	0.107061
Gandomi et al. [37]	2.658575665	4.3835958	7.8162919	4.6076313
Proposed method	2.65855917	2.68803204	2.69949371	0.01856616

3.2. Numerical Experimental Results and Discussion. The performance of the proposed PSO algorithm was investigated using the five MIDC optimization problems. For brevity, the swarm size m_p is 100, the maximum number of iterations is 3000, the number of experiments is 50, and $c_1 = c_2 = 2.02$ for all the examples. The maximum number of iterations for applying the three phases velocity update procedure described in (10) was set to $k_{\text{DIV}} = 2000$. Fifty runs were performed for each of the five optimization problems, and the best solutions are summarized in Tables 1 to 7. The optimal design variables and the corresponding optimal objective function values are indicated in these tables. To further assess the effectiveness of the proposed technique, the results are compared with the best optima results available in the published literature. A comparison of the optimal objective function values in Table 1 (Problem 1) and Table 4 (Problem 3) demonstrates the superiority of the proposed PSO algorithm over conventional techniques. Also, Table 2 (Problem 2), Table 6 (Problem 4), and Table 7 (Problem 5) prove our method's ability to obtain results that coincide with results found in the literature. A statistical comparison of the solution quality analysis given in Table 3 (Problem 2), Table 5 (Problem 3), and Table 8 (Problem 5) demonstrates the remarkable reliability of the proposed PSO method. Note that the statistical analysis results for Problems 1, 4, and 5 show a nearly 100% confidence level (i.e., the standard deviations are all approximately zero); these are omitted here for the sake of brevity. In addition, the presented optimal design variables are all feasible solutions (i.e., they guarantee $h_\ell(\mathbf{x}) \le 0, \forall \ell$). In Table 3, the worst value of the objective function is not particularly good, but the obtained mean value indicates that our PSO scheme is comparable to the method by Kim et al. [18] or is more robust than any conventional technique. The worst value appears only one time in the results set of 50 experiments which is confirmed in Figure 5. Overall, our experimental results demonstrate that the proposed PSO strategy for handling mixed variables in optimization problems is more likely to find the global optimum solution with greater accuracy compared to existing methods.

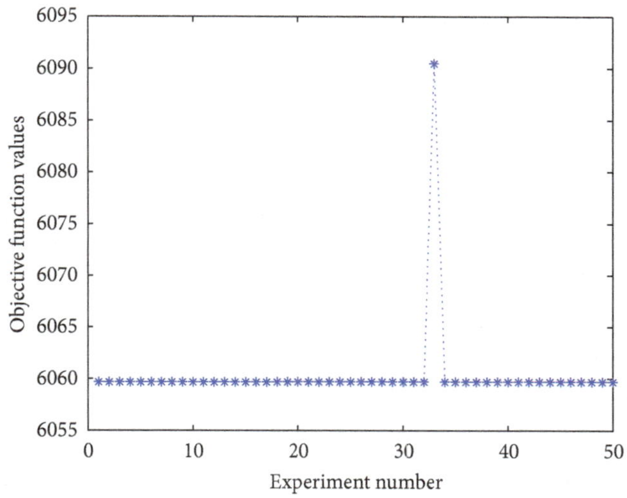

FIGURE 5: Objective function values of Problem 2 in 50 experiments.

4. Conclusion

We presented a novel modified constrained PSO algorithm using a simple diversity enhancing mechanism with high applicability and reliability for finding optimal mixed integer-discrete-continuous design variables. The proposed approach causes the particles of the swarm to evolve as follows. At the beginning stage of evolution, all particles try to move into a feasible region where constraint satisfaction is guaranteed. Here, individual particles first freely evolve in a continuous search space regardless of the variable type based on the diversity-guided velocity update law that includes three phases: attraction, repulsion, and a combination of attraction and repulsion. The introduced three-phase velocity control method improves the population diversity in the framework of constrained PSO efficiently and thus can reduce the probability of a premature convergence phenomenon, particles being trapped in a local optimum due to the mixed variable nature. For integer or discrete design variables, the corresponding continuous values of particles are rounded off to the nearest integer or discrete values. Thereafter, the particles continue to evolve in the aforementioned manner and try to explore the feasible region to find an optimal solution that satisfies the given constraint conditions. Note that the procedure transforming the given constrained optimization problem into an unconstrained problem introduces no problem-dependent extra parameters such as penalty factors or Lagrange multipliers which usually require time-consuming fine-tuning tasks. Our numerical experiment verified that the proposed constrained PSO scheme, combined with a general integer and discrete variable handling technique and three phases of particle velocity control law, can yield satisfactory optimal solution accuracy.

Acknowledgments

This research was supported by Basic Science Research Program through the National Research Foundation of Korea (NRF) funded by the Ministry of Education, Science and

Technology (no. 2012-012295) and the Chung-Ang University excellent freshman scholarship grants in 2013.

References

[1] R. C. Eberhart and J. Kennedy, "A new optimizer using particle swarm theory," in *Proceedings of the 6th International Symposium on Micro Machine and Human Science (MHS '95)*, pp. 39–43, Nagoya, Japan, 1995.

[2] K. E. Parsopoulos and M. N. Vrahatisy, "Recent approaches to global optimization problems through particle swarm optimization," *Natural Computing*, vol. 1, no. 2-3, pp. 235–306, 2002.

[3] F. van den Bergh and A. P. Engelbrecht, "A cooperative approach to particle swarm optimization," *IEEE Transactions on Evolutionary Computation*, vol. 8, no. 3, pp. 225–239, 2004.

[4] V. Kadirkamanathan, K. Selvarajah, and P. J. Fleming, "Stability analysis of the particle dynamics in particle swarm optimizer," *IEEE Transactions on Evolutionary Computation*, vol. 10, no. 3, pp. 245–255, 2006.

[5] S. O. Degertekin and M. S. Hayalioglu, "Sizing truss structures using teaching-learning-based optimization," *Computers and Structures*, vol. 119, pp. 177–188, 2013.

[6] S. O. Degertekin, "Improved harmony search algorithms for sizing optimization of truss structures," *Computers and Structures*, vol. 92-93, pp. 229–241, 2012.

[7] A. Kaveh and M. Khayatazad, "A new meta-heuristic method: ray optimization," *Computers and Structures*, vol. 112-113, pp. 283–294, 2012.

[8] P. C. Fourie and A. A. Groenwold, "The particle swarm optimization algorithm in size and shape optimization," *Structural and Multidisciplinary Optimization*, vol. 23, no. 4, pp. 259–267, 2002.

[9] K. Sedlaczek and P. Eberhard, "Using augmented Lagrangian particle swarm optimization for constrained problems in engineering," *Structural and Multidisciplinary Optimization*, vol. 32, no. 4, pp. 277–286, 2006.

[10] P. W. Jansen and R. E. Perez, "Constrained structural design optimization via a parallel augmented Lagrangian particle swarm optimization approach," *Computers and Structures*, vol. 89, no. 13-14, pp. 1352–1366, 2011.

[11] R. E. Perez and K. Behdinan, "Particle swarm approach for structural design optimization," *Computers and Structures*, vol. 85, no. 19-20, pp. 1579–1588, 2007.

[12] G. Venter and J. Sobieszczanski-Sobieski, "Particle swarm optimization," *AIAA Journal*, vol. 41, no. 8, pp. 1583–1589, 2003.

[13] L. J. Li, Z. B. Huang, F. Liu, and Q. H. Wu, "A heuristic particle swarm optimizer for optimization of pin connected structures," *Computers and Structures*, vol. 85, no. 7-8, pp. 340–349, 2007.

[14] A. I. Vaz and E. M. Fernandes, "Optimization of nonlinear constrained particle swarm," *Technological and Economic Development of Economy*, vol. 12, no. 1, pp. 30–36, 2006.

[15] G. Venter and R. Haftka, "Constrained particle swarm optimization using a bi-objective formulation," *Structural and Multidis-Ciplinary Optimization*, vol. 40, pp. 65–76, 2010.

[16] Z. Richards, "Constrained particle swarm optimisation for sequential quadratic programming," *International Journal of Modelling, Identification and Control*, vol. 8, no. 4, pp. 361–367, 2009.

[17] N. A. Aziz, M. Y. Alias, M. W. Mohemmed, and K. A. Aziz, "Particle swarm optimization for constrained and multiobjective problems: a brief review," *International Proceedings of Economics Development & Research*, vol. 6, pp. 146–150, 2011.

[18] T.-H. Kim, I. Maruta, and T. Sugie, "A simple and efficient constrained particle swarm optimization and its application to engineering design problems," *Proceedings of the Institution of Mechanical Engineers C*, vol. 224, no. 2, pp. 389–400, 2010.

[19] C. J. Shih and T. K. Lai, "Mixed-discrete fuzzy programming for nonlinear engineering optimization," *Engineering Optimization*, vol. 23, no. 3, pp. 187–199, 1995.

[20] J. F. Fu, R. G. Fenton, and W. L. Cleghorn, "A mixed integer-discrete-continuous programming method and its application to engineering design optimization," *Engineering Optimization*, vol. 17, pp. 263–280, 1991.

[21] E. Sandgren, "Nonlinear integer and discrete programming in mechnical design optimization," *ASME Journal of Mechanisms, Transmissions, and Automation in Design*, vol. 112, no. 2, pp. 223–229, 1990.

[22] S. S. Rao and Y. Xiong, "A hybrid genetic algorithm for mixed-discrete design optimization," *ASME Journal of Mechanical Design*, vol. 127, no. 6, pp. 1100–1112, 2005.

[23] K. E. Parsopoulos and M. N. Vrahatis, "Unified particle swarm optimization for solving constrained engineering optimization problems," in *Lecture Notes in Computer Science*, pp. 582–591, Springer, Berlin, Germany, 2005.

[24] S. Akhtar, K. Tai, and T. Ray, "A socio-behavioural simulation model for engineering design optimization," *Engineering Optimization*, vol. 34, pp. 341–354, 2002.

[25] C. A. C. Coello and E. Mezura-Montes, "Use of dominance-based tournament selection to handle constraints in genetic algorithms," in *Intelligent Engineering Systems Through Artificial Neural Networks*, vol. 11, pp. 177–182, ASME Press, St. Louis, Mo, USA, 2001.

[26] E. Mezura-Montes, C. A. C. Coello, and R. Landa-Becerra, "Engineering optimization using a simple evolutionary algorithm," in *Proceedings of the 15th IEEE International Conference on Tools with Artificial Intelligence*, pp. 149–156, November 2003.

[27] S. He, E. Prempain, and Q. H. Wu, "An improved particle swarm optimizer for mechanical design optimization problems," *Engineering Optimization*, vol. 36, no. 5, pp. 585–605, 2004.

[28] L. C. Cagnina, S. C. Esquivel, and C. A. C. Coello, "Solving engineering optimization problems with the simple constrained particle swarm optimizer," *Informatica*, vol. 32, no. 3, pp. 319–326, 2008.

[29] G. Tomassetti, "A cost-effective algorithm for the solution of engineering problems with particle swarm optimization," *Engineering Optimization*, vol. 42, no. 5, pp. 471–495, 2010.

[30] E. Mezura-Montes and C. A. C. Coello, "Useful infeasible solutions in engineering optimization with evolutionary algorithms," *Lecture Notes in Computer Science*, vol. 3789, pp. 652–662, 2005.

[31] A. H. Aguirre, A. E. M. Zavala, E. V. Diharce, and S. B. Rionda, "COPSO: constrained optimization via PSO algorithm," Tech. Rep. I-07-04/22-02-2007, Center for Research in Mathematics (CIMAT), 2007.

[32] W. C. Li and S. Azarm, "Optimality and constrained derivatives in two-level design optimization," *ASME Journal of Mechanical Design*, vol. 112, pp. 563–568, 1990.

[33] H. L. Li and P. Papalambros, "A production system for use of global optimization knowledge," *ASME Journal of Mechanical Design*, vol. 107, pp. 277–284, 1985.

[34] J. K. Kuang, S. S. Rao, and L. Chen, "Taguchi-aided search method for design optimization of engineering systems," *Engineering Optimization*, vol. 30, no. 1, pp. 1–23, 1998.

[35] K. Deb and M. Goyal, "Optimizing engineering designs using a combined genetic search," in *Proceedings of the 7th International Conference on Genetic Algorithms*, pp. 512–528, 1997.

[36] J. Lampinen and I. Zelinka, "Mixed integer-discrete-continuous optimization by differential evolution," in *Proceedings of the 5th International Conference on Soft Computing*, pp. 71–76, 1999.

[37] A. H. Gandomi, X.-S. Yang, and A. H. Alavi, "Mixed variable structural optimization using firefly algorithm," *Computers and Structures*, vol. 89, no. 23-24, pp. 2325–2336, 2011.

[38] L. D. S. Coelho, "An efficient particle swarm approach for mixed-integer programming in reliability-redundancy optimization applications," *Reliability Engineering and System Safety*, vol. 94, no. 4, pp. 830–837, 2009.

[39] G. Venter and J. Sobieszczanski-Sobieski, "Particle swarm optimization," *AIAA Journal*, vol. 41, no. 8, pp. 1583–1589, 2003.

[40] G. Venter and J. Sobieszczanski-Sobieski, "Multidisciplinary optimization of a transport aircraft wing using particle swarm optimization," *Structural and Multidisciplinary Optimization*, vol. 26, no. 1-2, pp. 121–131, 2004.

[41] J. Riget and J. S. Vesterstrom, "A diversity-guided particle swarm optimizer-the ARPSO," Tech. Rep. 200202, Department of Computer Science, University of Aarhus, 2002.

[42] J. Jie and J. Zeng, "Particle swarm optimization with diversity-controlled acceleration coefficients," in *Proceedings of the 3rd International Conference on Natural Computation (ICNC '07)*, pp. 150–154, Haikou, China, August 2007.

[43] M. Pant, T. Radha, and V. P. Singh, "A simple diversity guided particle swarm optimization," in *Proceedings of the IEEE Congress on Evolutionary Computation (CEC '07)*, pp. 3294–3299, Singapore, September 2007.

[44] S. Z. Zhao and P. N. Suganthan, "Diversity enhanced particle swarm optimizer for global optimization of multimodal problems," in *Proceedings of the IEEE Congress on Evolutionary Computation (CEC '09)*, pp. 590–597, Trondheim, Norway, May 2009.

[45] R. Thangaraj, M. Pant, and A. Abraham, "A new diversity guided particle swarm optimization with mutation," in *Proceedings of the World Congress on Nature and Biologically Inspired Computing (NABIC '09)*, pp. 294–299, Coimbatore, India, December 2009.

[46] T. Y. Gu, S. G. Ju, and F. Han, "A PSO algorithm with the improved diversity for feedforward neural networks," in *Proceedings of the 2nd International Symposium on Intelligent Information Technology and Security Informatics (IITSI '09)*, pp. 123–127, Moscow, Russia, January 2009.

[47] M. Pant, R. Thangaraj, and A. Abraham, *Foundations of Computational Intelligence, Global Optimization, Studies in Computational Intelligence*, vol. 3, Springer, Berlin, Germany, 2009.

[48] C. A. C. Coello, "Use of a self-adaptive penalty approach for engineering optimization problems," *Computers in Industry*, vol. 41, no. 2, pp. 113–127, 2000.

[49] X. Hu, R. C. Eberhart, and Y. Shi, "Engineering optimization with particle swarm," in *Proceedings of the IEEE Swarm Intelligence Symposium*, pp. 53–57, April 2003.

[50] K. E. Parsopoulos and M. N. Vrahatis, "Particle swarm optimization method for constrained optimization problems," in *Intelligent Technologies-Theory and Application: New Trends in Intelligent Technologies*, vol. 76 of *Frontiers in Artificial Intelligence and Applications*, pp. 214–220, IOS Press, Germany, 2002.

[51] K. Sedlaczek and P. Eberhard, "Using augmented Lagrangian particle swarm optimization for constrained problems in engineering," *Structural and Multidisciplinary Optimization*, vol. 32, no. 4, pp. 277–286, 2006.

[52] M. Clerc and J. Kennedy, "The particle swarm: explosion, stability, and convergence in a multi-dimensional complex space," *IEEE Transactions on Evolutionary Computation*, vol. 6, no. 1, pp. 58–73, 2002.

[53] A. P. Engelbrecht, *Fundamentals of Computational Swarm Intelligence*, John Wiley & Sons, New York, NY, USA, 2005.

[54] I. C. Trelea, "The particle swarm optimization algorithm: convergence analysis and parameter selection," *Information Processing Letters*, vol. 85, no. 6, pp. 317–325, 2003.

[55] A. E. M. Zavala, A. H. Aguirre, E. R. V. Diharce, and S. B. Rionda, "Constrained optimization with an improved particle swarm optimization algorithm," *International Journal of Intelligent Computing and Cybernetics*, vol. 1, no. 3, pp. 425–453, 2008.

[56] I. Maruta, T.-H. Kim, and T. Sugie, "Fixed-structure H∞ controller synthesis: a meta-heuristic approach using simple constrained particle swarm optimization," *Automatica*, vol. 45, no. 2, pp. 553–559, 2009.

[57] E. Salajegheh and G. N. Vanderplaats, "Optimum design of trusses with discrete sizing and shape variables," *Structural Optimization*, vol. 6, no. 2, pp. 79–85, 1993.

7

Numerical Investigation on Mixed Convection in Triangular Cross-Section Ducts with Nanofluids

Oronzio Manca, Sergio Nardini, Daniele Ricci, and Salvatore Tamburrino

Dipartimento di Ingegneria Aerospaziale e Meccanica, Seconda Università degli Studi di Napoli, Real Casa dell'Annunziata, Via Roma 29, 81031 Aversa (CE), Italy

Correspondence should be addressed to Oronzio Manca, oronzio.manca@unina2.it

Academic Editor: Kambiz Vafai

Convective heat transfer can be enhanced passively by changing flow geometry and boundary conditions or by improving the thermal conductivity of the working fluid, for example, introducing suspended small solid nanoparticles. In this paper, a numerical investigation on laminar mixed convection in a water-Al_2O_3-based nanofluid, flowing in a triangular cross-sectioned duct, is presented. The duct walls are assumed at uniform temperature, and the single-phase model has been employed in order to analyze the nanofluid behaviour. The hydraulic diameter is equal to 0.01 m. A fluid flow with different values of Richardson number and nanoparticle volume fractions has been considered. Results show the increase of average convective heat transfer coefficient and Nusselt number for increasing values of Richardson number and particle concentration. However, also wall shear stress and required pumping power profiles grow significantly.

1. Introduction

Heat transfer enhancement is a significant issue in the research and industry fields. Several thermal device applications employ ducts characterized by noncircular cross-sections, thus many investigations have concerned with this topic. Different shapes of cross-section area have been analyzed, like square, rhombic, rectangular, triangular, sinusoidal, elliptical ones, even with truncated corners [1–6]. In particular, triangular sectioned ducts are suitable for the construction of compact heat exchangers because of their compactness and cost effectiveness [7, 8]. In fact, compact heat exchangers with triangular-sectioned passages provide relative low fabrication costs because of the easy construction with thin materials and high mechanical strength [9]. Investigations have been performed in mixed and forced convection, considering mainly the flow regime laminar and also turbulent ones [10–15] and recently also microchannel have been analyzed [16]. Experimental test [3, 14] and analytical [17] and numerical methods [18, 19] are available in the literature. The use of numerical methods is nowadays widely adopted calculator powers and scheme strongly improved. Several researchers developed and validated unstructured grid meshes both in laminar and turbulent flow regime [18, 19] in order to predict the friction and heat transfer features in triangular ducts. Moreover, different fluids have been considered like gas, liquid, and non-Newtonian ones [11, 20].

Forced convection in a triangular duct is affected by several parameters, including the apex angle, the hydraulic diameter, the axial length, and the flow conditions. A comprehensive review of theoretical and experimental studies on laminar forced convection and heat transfer in ducts having noncircular sections has been performed by Shah and London [21]. In particular, they provided the flow and thermal performances of equilateral, isosceles, and right triangular ducts in developed and thermal developing flow. For the fully developed flow, a Nusselt number value equal to 2.47, 3.111, and 1.892 were reported for T, H_1 and H_2 boundary conditions, respectively. A few works are reported on natural [22] or mixed convection [10] in triangular sectioned ducts. Ali and Al-Ansary [22] analyzed the natural convection in vertical triangular ducts in laminar and transition regime, giving the critical values of the modified Rayleigh number for the transition to turbulent flow. They underlined the local axial (perimeter averaged) heat transfer coefficient decrease in the laminar region and increase in the transition one. Talukdar and Shah [10] studied the effect of Rayleigh number on bulk mean temperature and Nusselt number in triangular ducts with different apex angles. They pointed out the increase of these parameters for increasing Rayleigh numbers and the higher heat transfer rate from the bottom boundary.

According to Bergles [23] there are several enhancement techniques available. Among them, passive techniques, employing special surface geometries, rough surfaces, extended surfaces or swirl flow devices, and fluid additives for the enhancement can be adopted [24]. An innovative way to improve the fluid thermal conductivity consists of the introduction of nanosize particles suspended in the working fluids. The so-called nanofluids are composed by a base fluid and solid particles with a diameter smaller than 100 nm. They are becoming more and more popular, and there is a fast growth in the number of scientific research activities in these topics in order to evaluate solutions to develop efficient heat exchangers or cooling devices [25–27]. In fact, several investigations pointed out that nanofluid heat transfer average coefficients could increase by more than 20% in comparison with the base fluid coefficients also in the case of low nanoparticle concentrations [28]. However, a critical discussion about the lack of agreement among results from different research groups must be pointed out. There are still some difficulties in formulating efficient theoretical models for the predictions [29, 30], but this topic is still incomplete, as recently underlined by Gherasim et al. [31]. It is well known that the evaluation of properties, like thermal conductivity and viscosity, differs from different research groups because of the numerical and experimental approaches and processes adopted [32, 33].

The number of theoretical and experimental investigations on convective heat transfer in confined flows with nanofluids is growing. For example, experimental tests were performed on forced convective heat transfer in laminar and turbulent flow regime of nanofluids inside tubes in [34, 35]. Correlations for the Nusselt number, using nanofluids composed of water and Cu, TiO_2, and Al_2O_3 nanoparticles were proposed. Experimental results for the convective heat transfer of Al_2O_3 (27–56 nm)/water-based nanofluids flowing through a copper tube in laminar regime was reported in [36]. An experimental investigation was carried out to study the mixed laminar convection of Al_2O_3-water nanofluid inside an inclined copper tube submitted to a uniform wall heat flux at its outer surface by Ben Mansour et al. [37]. Results showed that the experimental heat transfer coefficient decreases slightly with an increase of particle volume concentration from 0% to 4%. They developed correlations to calculate the Nusselt number in the fully developed region for horizontal and vertical tubes, for Rayleigh number from 5×10^5 to 9.6×10^5, Reynolds number from 350 to 900 and particle volume concentrations up to 4%. Moreover, also the number of numerical studies on nanofluids is growing,

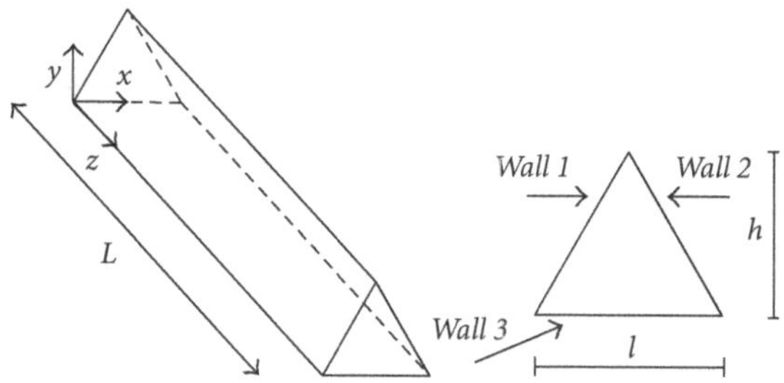

FIGURE 1: Sketch of the geometrical model.

and the analysis can be carried out by using two approaches. It is possible to consider that the continuum assumption is still valid for fluids with suspended nanosize particles at low volume fractions or to adopt two-phase models in order to describe both the fluid and the solid phase. The single-phase model with physical and thermal properties, all assumed to be constant with temperature, was employed in [38–41]. The advantages of adopting nanofluids with respect to the heat transfer mechanism were discussed in [38, 39], but it was also found that the presence of nanoparticles led to significant effects on growing wall shear stress and pumping power for heated tubes in laminar and turbulent regime. For a given Reynolds number, buoyancy force has a negative effect on the Nusselt number, while the nanoparticles concentration had a positive effect on the heat transfer enhancement and also on the skin friction reduction, as pointed out in [40], where the mixed convection in horizontal curved tubes was performed. Mirmasoumi and Behdzadmehr [41] adopted a two-phase model in order to describe the strong influence of particle concentration on the thermal parameters, above all on the tube bottom part and also at the near wall region.

In the present paper, a numerical investigation on laminar mixed convection in ducts with an equilateral triangular cross-section is presented. A constant and uniform temperature is applied on the walls. Results are given to evaluate the fluid dynamic and thermal features of the considered geometry with different Richardson number values, adopting Al_2O_3/water nanofluid as the working fluid. Different nanoparticle volume fractions are considered and the single-phase model has been assumed.

2. Governing Equations

A computational analysis of a three-dimensional model concerning with the mixed convection in a triangular cross sectioned duct is considered, and its scheme is reported in Figure 1. The aim is to evaluate its thermal and fluid-dynamic behaviours and study the temperature and velocity fields in the cases with Al_2O_3/water nanofluid as working fluid. A constant uniform temperature is applied on the walls, such as the T_1 boundary condition indicated in [21], and its value is evaluated on the basis of the chosen Richardson numbers. The inlet average velocity is set in the ranges of laminar regime. Single-phase model is employed and the fluid properties are considered constant with temperature.

Governing equations of continuity, momentum and energy, for a three-dimensional steady state laminar, incompressible flow, with temperature independent thermophysical properties and assuming the Boussinesq approximation, in rectangular coordinates are given in the following:

continuity:

$$\frac{\partial u}{\partial x}+\frac{\partial v}{\partial y}+\frac{\partial w}{\partial z}=0, \tag{1}$$

momentum:

$$\left[u\frac{\partial u}{\partial x}+v\frac{\partial u}{\partial y}+w\frac{\partial u}{\partial z}\right]=-\frac{1}{\rho}\frac{\partial P}{\partial x}+\nu\left[\frac{\partial^2 u}{\partial x^2}+\frac{\partial^2 u}{\partial y^2}+\frac{\partial^2 u}{\partial z^2}\right], \tag{2}$$

$$\left[u\frac{\partial v}{\partial x}+v\frac{\partial v}{\partial y}+w\frac{\partial v}{\partial z}\right]=-\frac{1}{\rho}\frac{\partial P}{\partial y}+\nu\left[\frac{\partial^2 v}{\partial x^2}+\frac{\partial^2 v}{\partial y^2}+\frac{\partial^2 v}{\partial z^2}\right]+\beta g(T-T_{\text{in}}), \tag{3}$$

$$\left[u\frac{\partial w}{\partial x}+v\frac{\partial w}{\partial y}+w\frac{\partial w}{\partial z}\right]=-\frac{1}{\rho}\frac{\partial P}{\partial x}+\nu\left[\frac{\partial^2 w}{\partial x^2}+\frac{\partial^2 w}{\partial y^2}+\frac{\partial^2 w}{\partial z^2}\right], \tag{4}$$

energy:

$$\left[u\frac{\partial T}{\partial x}+v\frac{\partial T}{\partial y}+w\frac{\partial T}{\partial z}\right]=\lambda\left[\frac{\partial^2 T}{\partial x^2}+\frac{\partial^2 T}{\partial x^2}+\frac{\partial^2 T}{\partial x^2}\right]. \tag{5}$$

The assigned boundary conditions are the following:

(i) inlet section: uniform velocity and temperature profiles,

(ii) outlet section: outflow condition with velocity components and temperature derivatives equal to zero,

(iii) duct surfaces: velocity components equal to zero and assigned temperature value.

The use of the single-phase model does not exclude the possibility to take into account for phenomena like the Brownian motion. In fact, the relations adopted to evaluate the thermal conductivity and viscosity consider the component linked to the Brownian motion as well as the particle diameter effects, and they were adopted by several authors, also in the case of nanofluid mixed convection [42, 43].

3. Physical Properties of Nanofluids

The working fluid is pure water or a water/Al_2O_3-based nanofluid. The nanoparticle diameter of 30 nm and different volume fractions equal to 1%, 3%, and 5% have been considered. In Table 1, the values of density, specific heat, dynamic viscosity, thermal conductivity, and thermal expansion coefficient, given by Rohsenow et al. [44], are reported

TABLE 1: Properties of pure water and Al_2O_3 particles at T = 293 K.

Material	ρ (kg/m^3)	c_p (J/kgK)	μ (Pas)	λ (W/mK)	β (1/K)
Al_2O_3	3880	773	//	36	//
Water	998.2	4182	993×10^{-6}	0.597	2.1×10^{-4}

TABLE 2: Thermophysical properties of the working fluids.

ϕ	ρ (kg/m^3)	c_p (J/kgK)	μ (Pas)	λ (W/Mk)	β (1/K)
0%	998.2	4182	993×10^{-6}	0.597	2.100×10^{-4}
1%	1027	4053	1110×10^{-6}	0.617	2.098×10^{-4}
3%	1085	3816	1233×10^{-6}	0.643	2.093×10^{-4}
5%	1142	3603	1724×10^{-6}	0.664	2.086×10^{-4}

for pure water and Al_2O_3 particles at the reference temperature of 293 K. The concentration of nanoparticles influences the working fluid properties. The single-phase model was adopted. Thus, fluid properties must be evaluated by employing relations, available in the literature, in order to compute the thermal and physical properties of the considered nanofluids [29, 36, 40, 45, 46]. Their values are given in Table 2. Density was evaluated by using the classical formula valid for conventional solid-liquid mixtures, while the specific heat values and thermal expansion coefficient ones were calculated by assuming the thermal equilibrium between particles and surrounding fluid [36, 40, 46]

density:

$$\rho_{\text{nf}}=(1-\phi)\rho_{\text{bf}}+\phi\rho_p, \tag{6}$$

specific heat:

$$\left(\rho c_p\right)_{\text{nf}}=(1-\phi)\left(\rho c_p\right)_{\text{bf}}+\phi\left(\rho c_p\right)_p, \tag{7}$$

thermal expansion coefficient:

$$\frac{\beta_{\text{nf}}}{\beta_{\text{bf}}}=\frac{1}{\left((1-\phi/\phi)\left(\rho_{\text{bf}}/\rho_p\right)\right)}\frac{\beta_p}{\beta_{\text{bf}}}+\frac{1}{\left((\phi/1-\phi)\left(\rho_{\text{bf}}/\rho_p\right)+1\right)}. \tag{8}$$

For the viscosity as well as for thermal conductivity, formula given by [29, 40, 45] were adopted, because these relations are expressed as a function of particle volume concentration and diameter

dynamic viscosity:

$$\mu_{\text{nf}}=\mu_{\text{bf}}+\frac{\rho_p V_B d_p^2}{72C\delta}, \tag{9}$$

thermal conductivity:

$$\frac{\lambda_{\text{nf}}}{\lambda_{\text{bf}}}=1+64.7\phi^{0.746}\left(\frac{\text{d}_{\text{bf}}}{\text{d}_p}\right)^{0.3690}\left(\frac{\lambda_p}{\lambda_{\text{bf}}}\right)\text{Pr}^{0.9955}\text{Re}^{1.2321}, \tag{10}$$

where $V_B = \sqrt{18K_bT/\pi\rho_p d_p}$ and K_b is the Boltzmann constant, given by $K_b = 1.36 \times 10^{-26}$; the other terms are given by:

$$C = \mu_{bf}^{-1}\left[\left(c_1 d_p + c_2\right)\phi + \left(c_3 d_p + c_4\right)\right], \quad (11)$$

with $c_1 = -0.000001133$, $c_2 = -0.000002771$, $c_3 = 0.00000009$ and $c_4 = -0.000000393$, and the distance between nanoparticles, δ, obtained by

$$\delta = \sqrt[3]{\frac{\pi}{6\phi}}d_p, \quad (12)$$

while $\mathrm{Pr} = \mu/\rho_{bf}\alpha_{bf}$ and $\mathrm{Re} = \rho_{bf}K_bT/3\pi\mu^2L_{bf}$ with L_{bf} is the mean free path of water (0.17 nm).

4. Geometrical Configuration and Data Reduction

A numerical thermal and fluid-dynamic study on a three-dimensional duct model with an equilateral triangular section was carried out. The total length L is 2.0 m, while the internal edge one l is 0.017 m, as shown in Figure 1; the consequent hydraulic diameter, $d_h = 4A/P_h$ is equal to 0.01 m. The duct includes a developing section at the entrance region with a length of 1.5 m and a 0.4 m long test section, followed by the outlet one. The working fluid is water or a mixture of water and Al_2O_3 nanoparticles with a diameter of 30 nm, at different volume fractions equal to 1%, 3% and 5%.

The dimensionless parameters of Reynolds number, Grashof number, Richardson number, Nusselt number, and friction factor are considered for the data reduction and they are expressed by:

$$\mathrm{Re} = \frac{Vd_h}{\nu}, \quad (13)$$

$$\mathrm{Gr} = \frac{g\beta(T_w - T_{in})d_h^3}{\nu^2}, \quad (14)$$

$$\mathrm{Ri} = \frac{\mathrm{Gr}}{\mathrm{Re}^2}, \quad (15)$$

$$\mathrm{Nu}_{av} = \frac{\dot{q}d_h}{(T_w - T_m)\lambda_f}, \quad (16)$$

$$f = 2\Delta P\frac{d_h}{L}\frac{1}{\rho V^2}, \quad (17)$$

where V is the average inlet velocity, $\dot{q}$ is the heat flux, T_w and T_m represent the temperature of the surface and the fluid bulk one, respectively.

5. Numerical Model

The governing equations, reported in the previous section, are solved by means of the finite volume method by means of Fluent code [47]. A steady-state solution and a segregated method are chosen to solve the governing equations, which are linearized implicitly with respect to dependent variables of the equation. A second-order upwind scheme is chosen for energy and momentum equations. The SIMPLE coupling scheme is chosen to couple pressure and velocity. The convergence criteria of 10^{-4}, 10^{-6} and 10^{-8} are assumed for the residuals of density, velocity components, and energy ones, respectively, for all the considered simulations. It is assumed that the incoming flow is laminar at ambient temperature and pressure. Reynolds number is set equal to 100 and different values of Grashof number, ranging from 0 to 50000, were considered. On the solid walls, the no-slip condition was applied and velocity inlet and outflow ones were given for inlet and outlet sections.

Four different unstructured mesh distributions were tested on a triangular sectioned duct at Re = 100 and 250 in order to perform the grid-independence analysis. They had 152190, 302778, 563104, and 1146232 nodes, respectively. The third grid case was adopted, because it ensured a good compromise between the computational time and the accuracy requirements. In fact, comparing the third and fourth configurations, differences of 0.58% and 0.19% at most were evaluated in terms of average Nusselt number and pressure coefficient, in the case of pure water.

The validation has been performed by comparing the numerical results for Re = 100 and Ri = 0 with literature data for T_1 boundary condition. For the local and average Nusselt number, the comparison has been accomplished with Wilbulswas results [21], presented as a function of the axial coordinate for the thermal entrance region, z^*, defined as $z^* = z/(d_h\,\mathrm{Re})$. For the friction factor, data from Fleming-Sparrow, Miller-Han, and Gangal [21] have been considered. Data reduction has been carried out by means of z^+ parameter, the axial coordinate for the hydrodynamic entrance region, given by $z^* = z/(d_h\mathrm{Pe})$, with the Peclet number, Pe, defined as $\mathrm{Pe} = Vd_h/\alpha$. The present numerical results in terms of average and local Nusselt number and friction factors are in good agreement with the given correlations, as shown in Figure 2. In particular, a difference of 3% is observed for local and average Nusselt number at most, while a maximum error of 1% is evaluated in terms of friction factor.

6. Results and Discussion

A computational analysis of a three-dimensional model, regarding the nanofluid mixed convection in a triangular cross-sectioned duct is considered. A uniform temperature has been applied on the duct walls, according to the considered Richardson number. The inlet velocities ensure the steady laminar regime, and they correspond to Re = 100. The working fluids are pure water or a water/Al_2O_3-based nanofluid at different volume fractions. The single-phase model approach is adopted. The range of the considered Richardson numbers and volume fractions are given below:

(i) Reynolds number, Re: 100,

(ii) Richardson number: 0.0, 0.1, 0.5, 1, 2, 3, and 5,

(iii) particle concentrations, ϕ 0%, 1%, 3% and 5%.

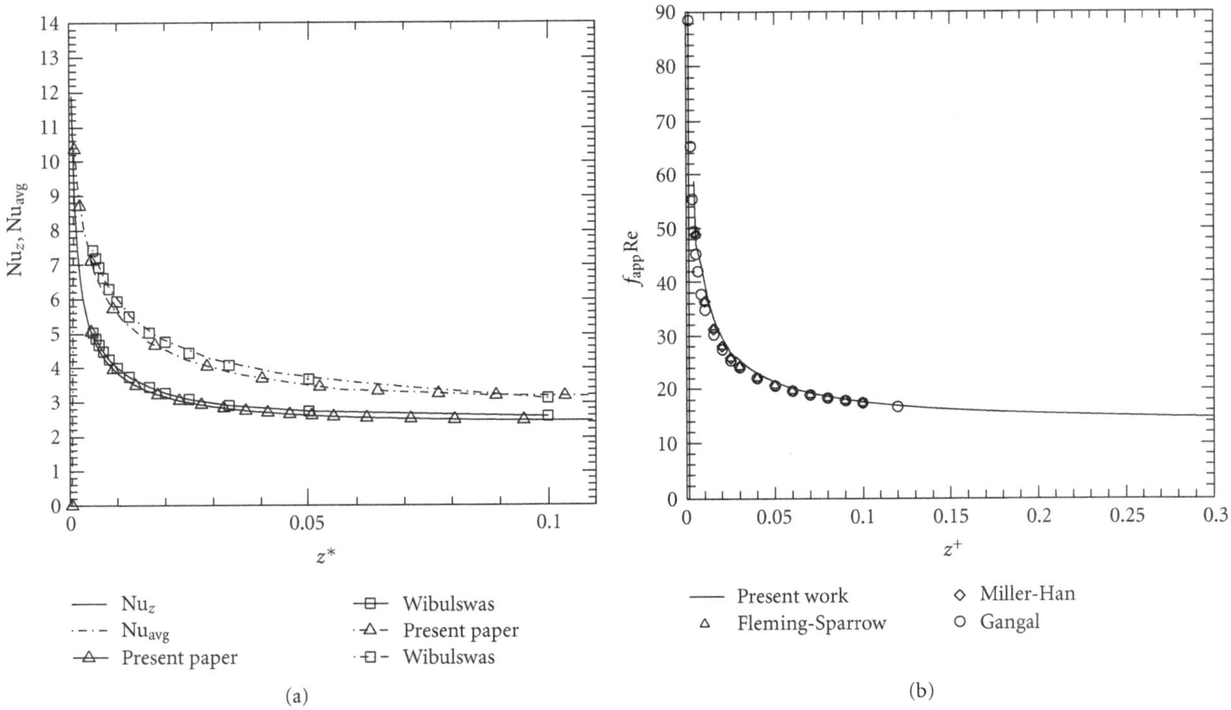

FIGURE 2: Validation of numerical results: (a) local and average Nusselt number; (b) friction factor.

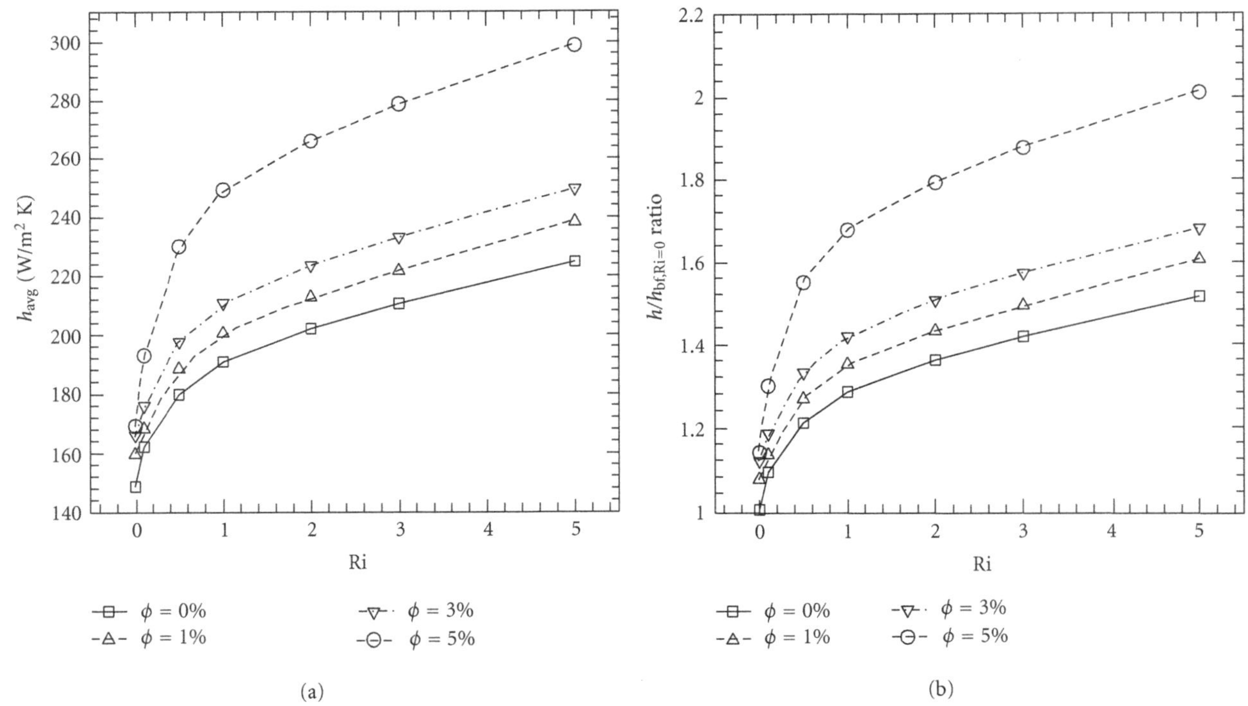

FIGURE 3: Convective heat transfer coefficient profiles as a function of Ri, ϕ = 0%, 1%, 3% and 5%: (a) average convective heat transfer coefficient; (b) average convective heat transfer coefficient enhancement, referred to pure water case at Ri = 0.

Results are presented in terms of average convective heat transfer coefficient, average Nusselt number, wall shear stress and required pumping power profiles in the test section.

Figure 3(a) depicts the average convective heat transfer coefficient profiles as a function of Ri for different values of nanoparticle volume concentration. The effect of buoyancy leads to an increase in terms of heat transfer coefficients, and a sharp growth is detected at low Richardson numbers. In fact, in the case of pure water h_{avg} is equal to about 148, 162, 178, 190, and 225 W/m^{2}K for Ri = 0, 0.1, 0.5, 1, 5,

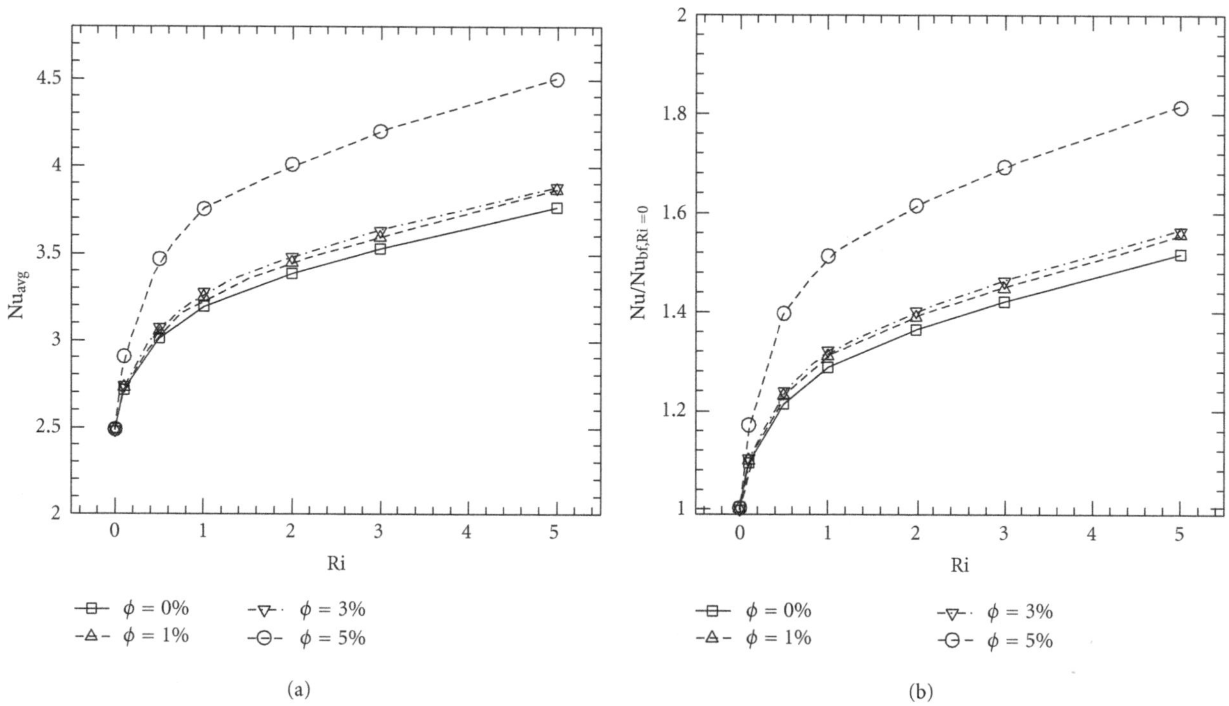

FIGURE 4: Nusselt number profiles as a function of Ri, ϕ = 0%, 1%, 3% and 5%: (a) average Nusselt number; (b) average Nusselt number enhancement, referred to pure water case at Ri = 0.

respectively. A similar behaviour is observed for the configurations with nanofluids. However, increasing values of convective heat transfer are evaluated as particle concentration increases. In fact, h_{avg} is equal to about 160, 170, 188, 199, 208, and 238 W/m^2K at $\phi = 1\%$ while at $\phi = 5\%$ h_{avg} is equal to about 168, 193, 230, 250, 266, and 298 W/m^2K for Ri = 0, 0.1, 0.5, 1, 2 and 5, respectively. Figure 3(b) describes the average convective heat transfer enhancement in comparison with the pure water case at Ri = 0. The maximum improvement is detected for the cases, characterized by $\phi = 5\%$, which show heat transfer coefficient values equal to 1.14, 1.55, 1.68, and 2.01 times greater than the reference case for Ri = 0, 0.5, 1, and 5. Smaller increases are observed for decreasing values of nanoparticle volume concentration. For example, the enhancement ratio is equal to 1.06, 1.26, 1.33, and 1.60 for $\phi = 1\%$ at Ri = 0, 0.5, 1 and 5 while for $\phi = 3\%$ it is equal to 1.10, 1.32, 1.41, and 1.68.

The average Nusselt profiles as a function of Ri for different values of nanoparticle volume concentration are presented in Figure 4(a). For fully developed laminar flow at Ri = 0, the average Nusselt number is equal to 2.47 for triangular ducts with walls at a uniform temperature. In mixed convection, the fully developed condition is reached much farther upstream than for pure forced convection, depending on the importance of buoyancy effects. However, average Nusselt number in the test section tends to increase as Ri and nanoparticle fraction increase. In fact, at Ri = 5, Nu_{avg} = 3.76, 3.87, 3.90, and 4.5 for at ϕ = 0%, 1%, 3% and 5%. The consequent enhancement in terms of Nusselt number is depicted in Figure 4(b); it is less significant if compared with Figure 3 because of the nanofluid thermal conductivity increase. In fact, for ϕ equal to 1% and 3% the average Nusselt number is 4% and 6% higher than the cases with pure water on average. A more significant enhancement is provided by the cases with $\phi = 5\%$.

The heat transfer enhancement is partly due to the increase of fluid velocity, because the simulations have been carried out at a constant Reynolds number. In fact, the increase of nanofluid viscosity is not balanced by the increase of density. Another possible reason is linked to the Brownian motion which could become very important in the case of laminar flow regime. However, a further analysis about the comparison criteria among the results in terms of heat transfer coefficients for base fluid and nanofluid is necessary as observed by Prabhat et al. [48].

The working fluid flows into a triangular cross-sectioned duct with walls at a uniform temperature; thus, the heat transfer mechanism is different if the bottom wall behaviour is compared with the inclined ones when the buoyancy effects are considered. In fact, the orientation of hot surfaces and their inclination must be taken into account even in internal flows. For Ri = 0 no differences are detected among the heated walls. Figure 5 allows to describe the different behaviour of bottom surface and upper ones. In general, convective heat transfer coefficient profiles result to be higher for *wall 1* for all the considered Ri values, as shown by Figure 5(a). Profiles tend to increase as Ri increases; in fact, for pure water h_{avg} is equal to about 210 and 370 W/m^2K for Ri = 0.1 and 5, respectively. The introduction of nanoparticles leads to a significant enhancement of heat transfer coefficients even at low Ri numbers and particle concentration. For ϕ = 5%, the highest heat transfer coefficient value is detected for

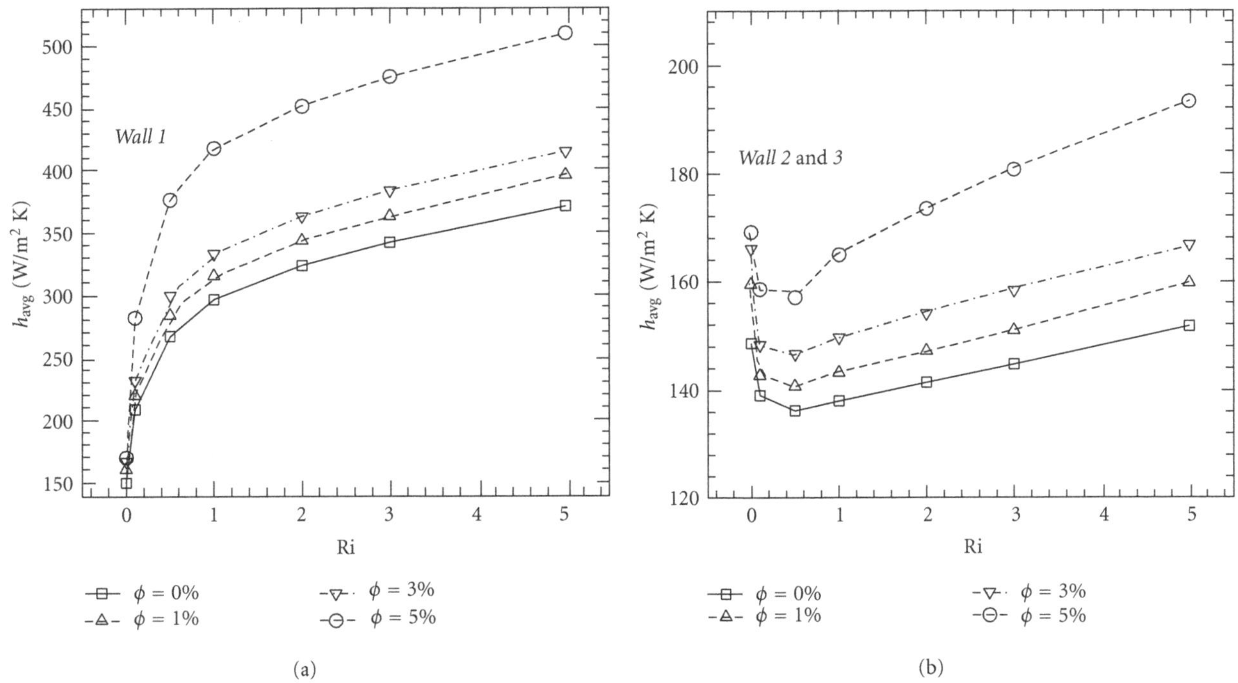

FIGURE 5: Average convective heat transfer coefficient profiles as a function of Ri, ϕ = 0%, 1%, 3% and 5%: (a) *wall 1*; (b) *wall* 2 and 3.

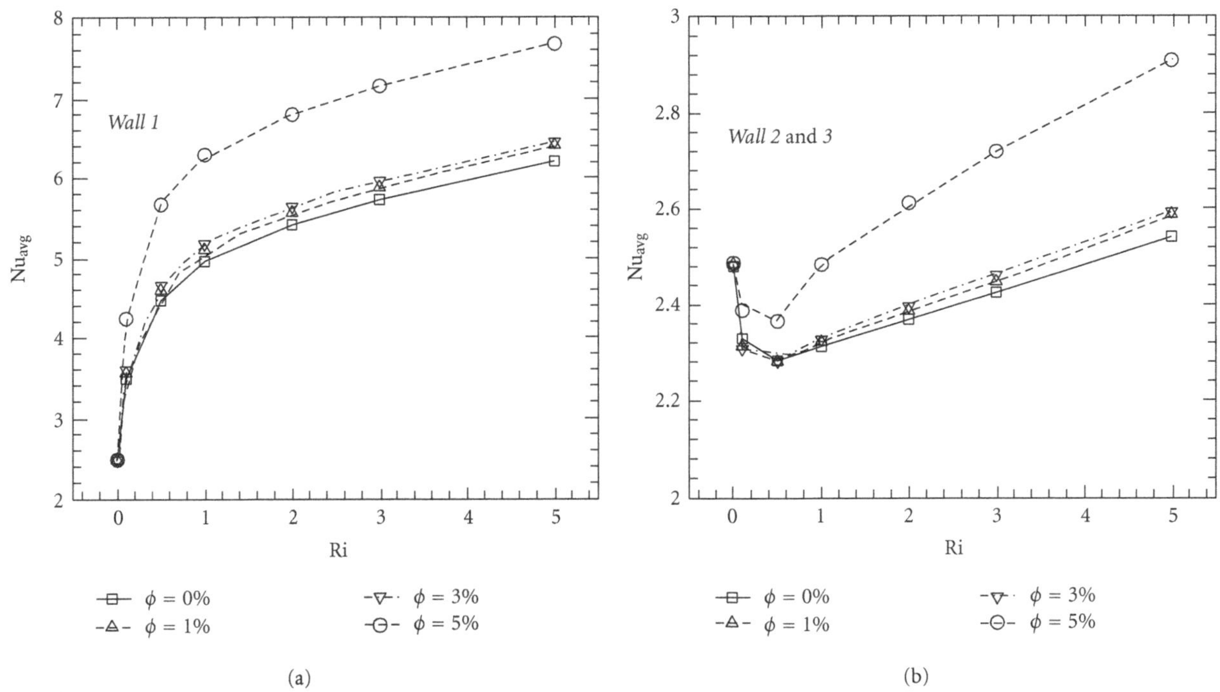

FIGURE 6: Average Nusselt number profiles as a function of Ri, ϕ = 0%, 1%, 3% and 5%: (a) *wall 1*; (b) *wall 2* and 3.

Ri = 5, and it is equal to 510 W/m^{2}K. *Wall 2* and *3* present smaller heat transfer coefficient values, as reported in Figure 5(b). In fact, at Ri = 5h_{avg} is equal to 152, 160, 170, and 194 W/m^{2}K for ϕ = 0%, 1%, 3%, and 5%, respectively. Moreover, for Ri < 2, buoyancy determines negative effects on the heat transfer mechanism if results are compared with ones obtained in the case of forced convection. The results in terms of average Nusselt number are shown in Figure 6. The highest Nusselt numbers for *wall 1* are equal to 6.2, 6.4, 6.5, and 7.7 for ϕ = 0%, 1%, 3%, and 5%, respectively; the profiles for *wall 2* and *wall 3* tends to decrease as Ri increases until Ri is equal to 1 for ϕ = 0%, 1% and 3%, and 1 for ϕ = 5%, then they increase.

The use of nanofluids as working fluids instead of pure water provides a heat transfer enhancement, according to the particle volume fraction, but it leads to increasing values of

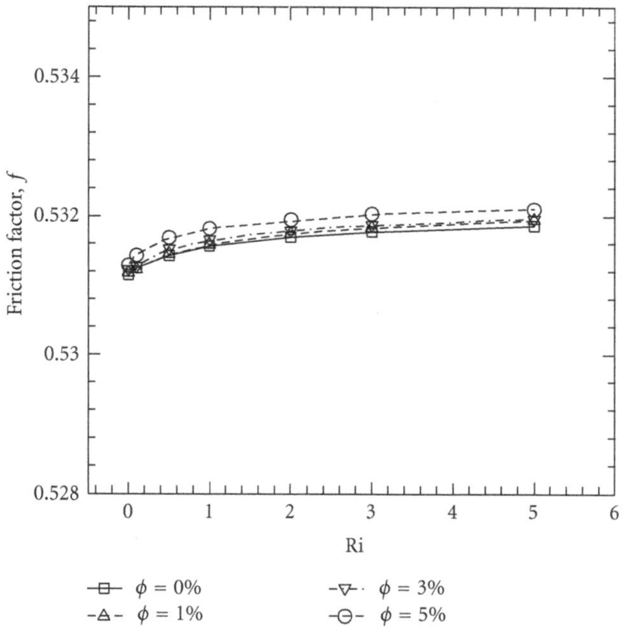

FIGURE 7: Friction factor profiles as a function of Ri, ϕ = 0%, 1%, 3% and 5%.

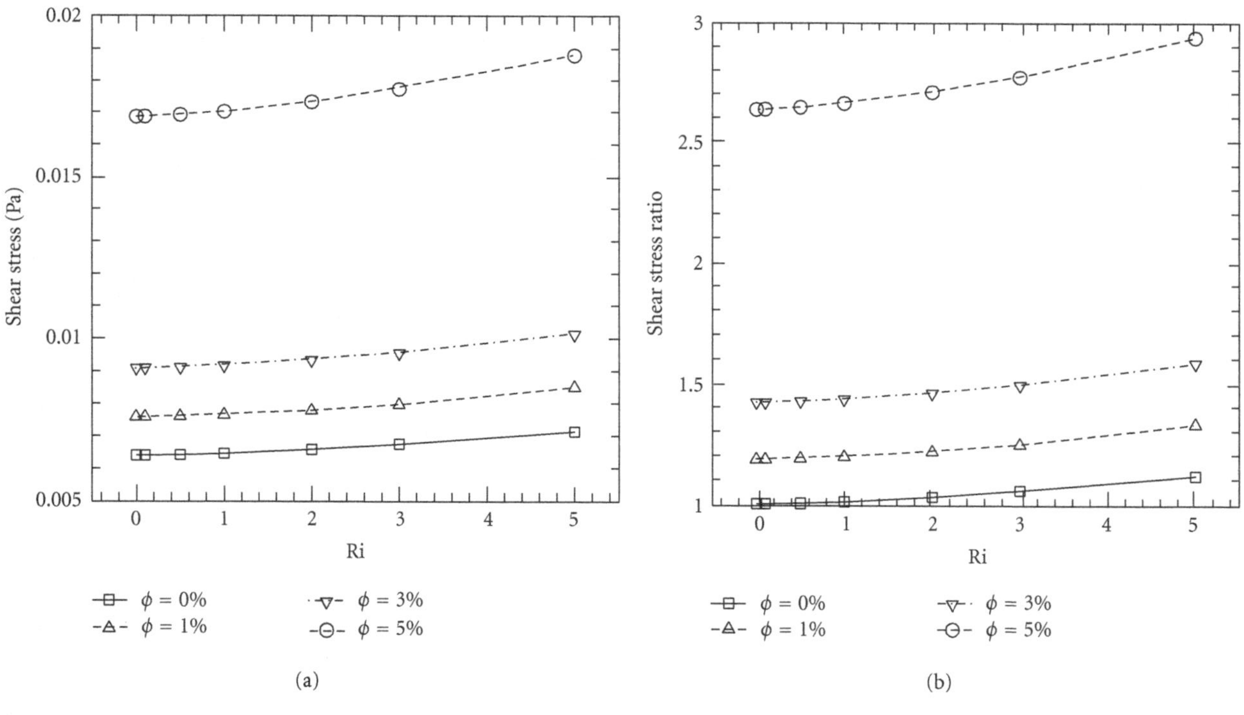

FIGURE 8: Wall shear stress profiles as a function of Ri, ϕ = 0%, 1%, 3% and 5%: (a) average wall shear stress; (b) average wall shear stress, referred to the pure water case at Ri = 0.

wall shear stress, while the friction factor keeps substantially constant as reported in Figure 7, because the flow is fully developed. The average wall shear stress profiles as a function of Ri are depicted in Figure 8(a). Values grow as the buoyancy effects become more significant. In fact, the highest values are observed at Ri = 5; moreover, wall shear stress increases as particle concentration increases and the maximum values are evaluated for ϕ = 5%. In fact, for ϕ = 5% wall shear stress is 2.63, 2.65, 2.68, 2.75, and 2.94 times higher than the value calculated at Ri = 0 with pure water, as depicted in Figure 8(b). For lower particle concentrations, wall shear stress ratio is smaller than the ones calculated at ϕ = 5% and, for example, it is equal to 1.18, 1.20, and 1.32 for ϕ = 1% and Ri = 0.1, 1, and 5, respectively, while it is equal to 1.40, 1.42, and 1.58 for ϕ = 3%. *Wall 1* is featured by higher wall shear stress values than the ones calculated for *wall 2* and

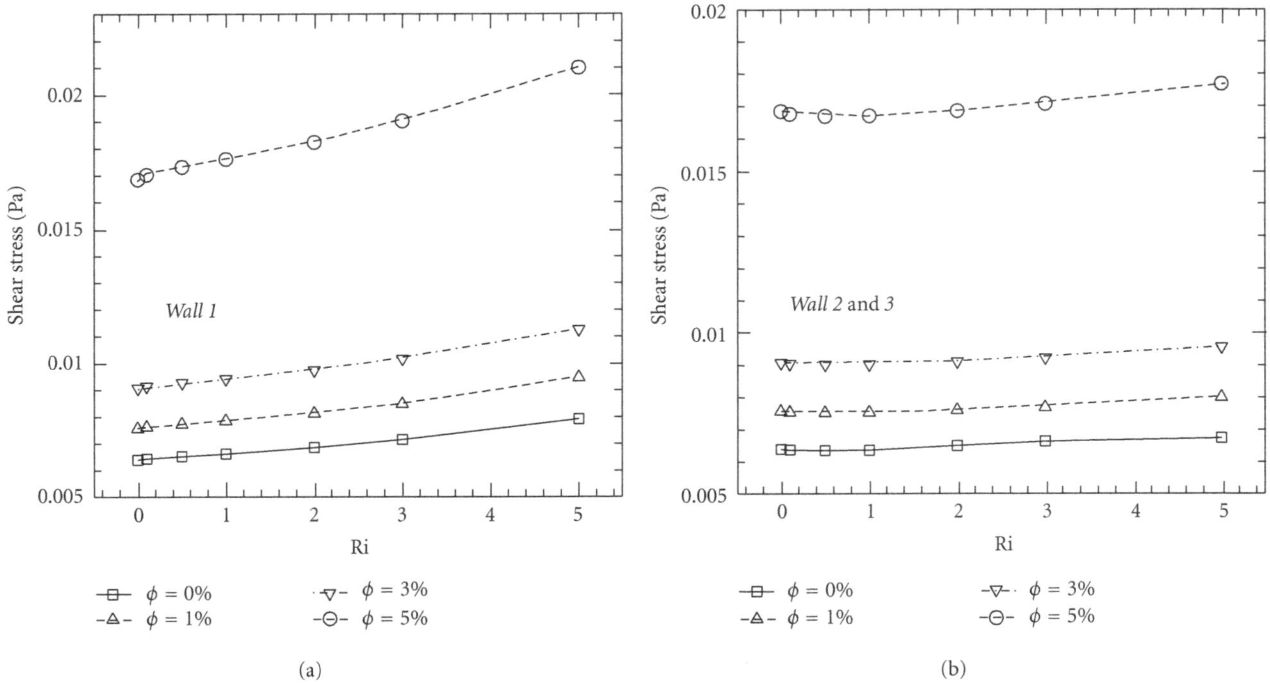

FIGURE 9: Average wall shear stress profiles as a function of Ri, ϕ = 0%, 1%, 3% and 5%: (a) *wall 1*; (b) *wall 2* and 3.

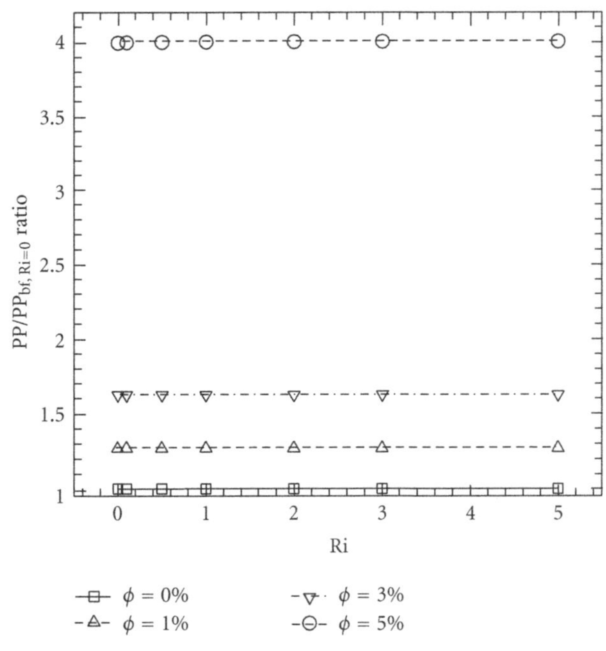

FIGURE 10: Required pumping power profiles, referred to the pure water case at Ri = 0, as a function of Ri, ϕ = 0%, 1%, 3% and 5%.

wall 3, as observed in Figure 9. Profiles increase as Ri, and nanoparticle concentration increases if *wall 1* is considered, while they keep substantially constant in the case of *wall 2* and *wall 3*. The friction factor keeps substantially constant as reported in Figure 7, because the flow is fully developed.

The required pumping power ratio profiles, referred to the water cases at Ri = 0, are carried out in Figure 10. Pumping power is defined as $PP = \dot{V}\Delta P$, and its profiles tend to increase as ϕ grows, while a very little dependence on Ri is observed. The pumping power ratio is equal to 1.26, 1.61, and 4.0 for ϕ = 1%, 3%, and 5%, respectively.

Figures 11 and 12 describe the fully developed flow regime in terms of velocity and dimensionless temperature profiles for the vertical symmetry axis of the channel at z/d_h = 125 and 150. Figure 11 shows that u/u_{max} profiles substantially overlap each other for different values of particle concentration and Richardson number. The highest values are detected at y/h equal to about 0.30, while

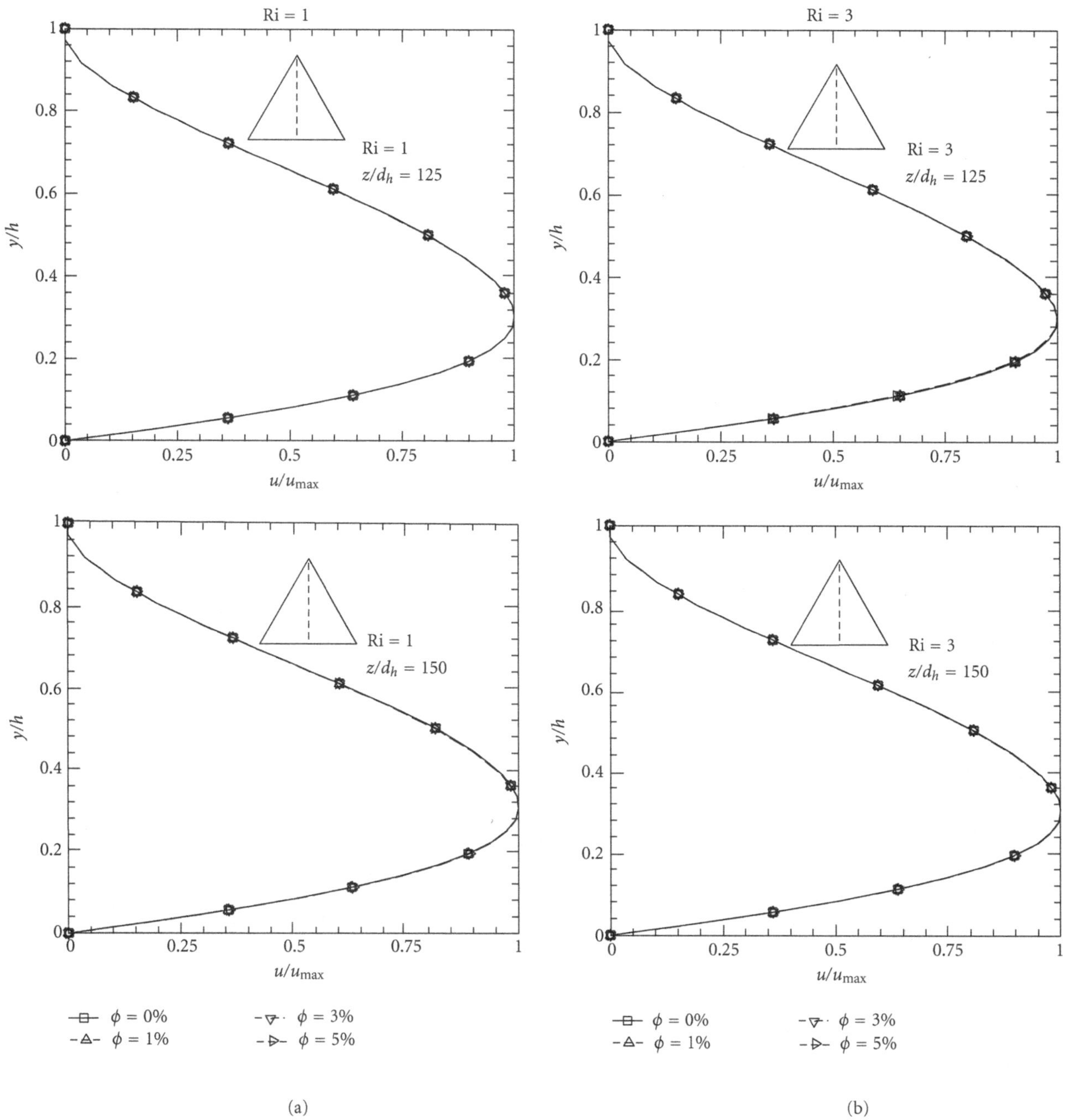

FIGURE 11: Velocity profiles for different values of particle concentration and Ri = 1 and 3: (a) z/d_h = 125; (b) z/d_h = 150.

the maximum values of dimensionless temperature, defined as $T^* = (T - T_w)/(T_m - T_w)$, is evaluated at y/h equal to about 0.20. The temperature profiles keeps similar if they are compared at the same values of Ri and volume particle concentration, as shown in Figure 12.

7. Conclusions

In this paper, a numerical investigation about the laminar mixed convection in water/Al_2O_3-based nanofluids flowing into a triangular sectioned duct is carried out. The laminar flow regime was considered at Re = 100, and Ri numbers ranging from 0 to 5 were assumed. A constant and uniform temperature is applied on the duct walls, depending on the Ri number. The single-phase model was adopted in order to analyze the behaviour in the case of nanofluids as working fluid. Thus, the considered nanoparticle volume concentrations were equal to 0%, 1%, 3%, and 5%.

The introduction of nanoparticles significantly raises the convective heat transfer coefficients as particle concentration grows as well as Ri number. This effect is very significant at low Ri numbers. In fact, the maximum improvement is detected for the cases, characterized by ϕ = 5%, which show heat transfer coefficient values equal to 1.14, 1.55, 1.68, and 2.01 times greater than the pure water case at Ri = 0. However, the wall shear stress and the required pumping

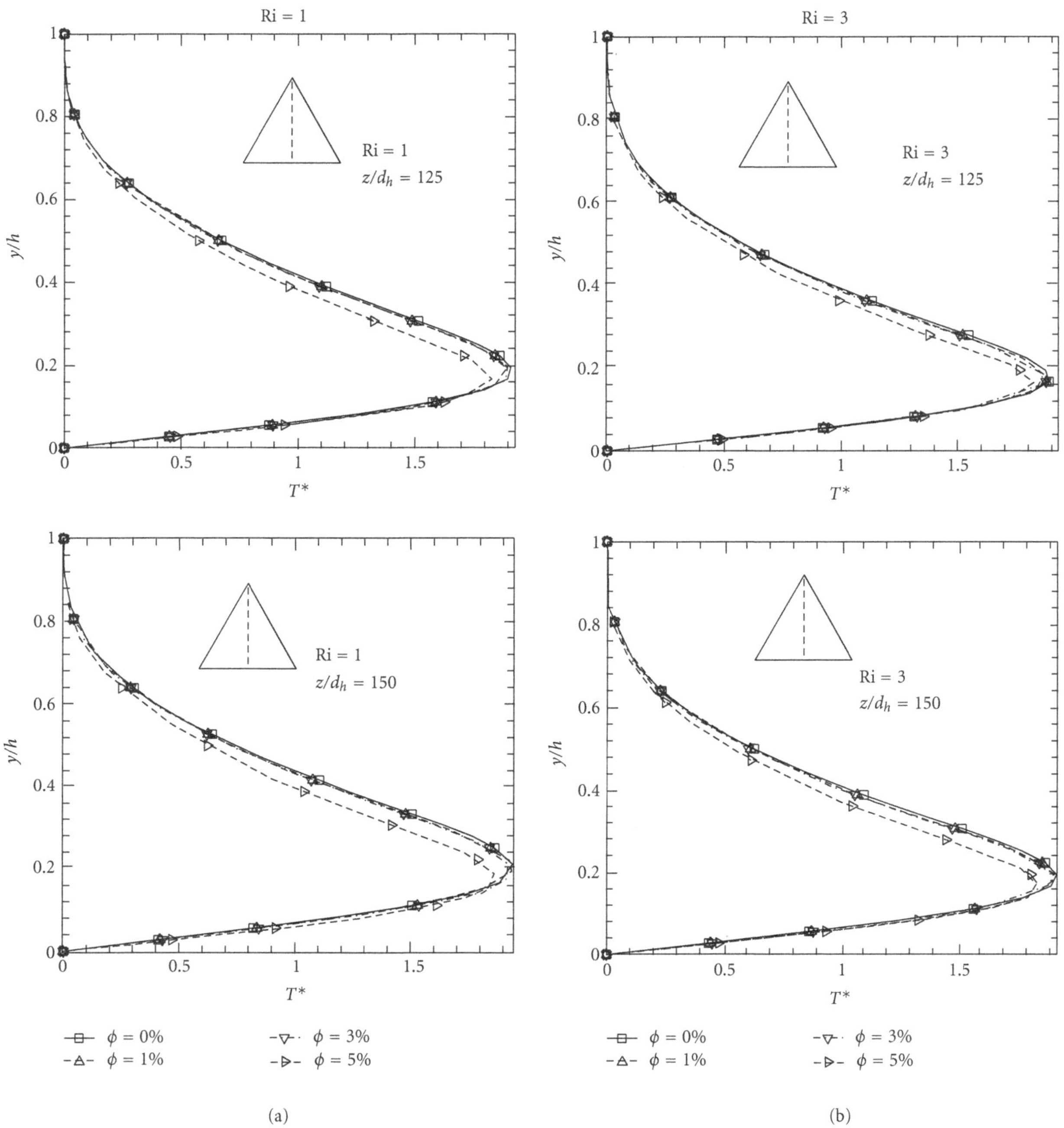

FIGURE 12: Dimensionless temperature profiles for different values of particle concentration and Ri = 1 and 3: (a) $z/d_h = 125$; (b) $z/d_h = 150$.

power increase, and their values become very high at high concentrations. This effect is amplified at low Ri numbers, and an increase of about 30% in terms of wall shear stress is evaluated at Ri = 0 and 0.1 in comparison with the pure water results. Moreover, the pumping power ratio, referred to the pure water cases at Ri = 0, is equal to 1.26, 1.61, and 4.0 for $\phi = 1\%$, 3%, and 5%, respectively.

It should be remarked that further investigations need to be accomplished in order to understand the main physical reasons of the heat enhancement using the nanofluids in laminar regime, particularly, in mixed convection, as underlined in [48].

Nomenclature

A: Cross-section area (m^2)
c_p: Specific heat (J/kgK)
d: Diameter (m)
f: Friction factor (17)
g: Gravitational acceleration (m/s^2)
Gr: Grashof number (14)
H: Duct height (m)
h: Heat transfer coefficient (W/m^2K)
l: Duct internal edge length (m)
L: Duct length (m)

Nu: Nusselt number(16)
P: Pressure (Pa)
Pe: Peclet number
PP: Required pumping power (W)
q: Target surface heat flux (W/m^2)
Re: Reynolds number(13)
Ri: Richardson number(15)
T: Temperature (K)
u, v, w: Velocity components (m/s)
V: Average velocity (m/s)
$\dot{V}$: Volume flow rate (m^3/s)
x,y,z: Spatial coordinates (m)
α: Thermal diffusivity (m^2/s)
β: Volumetric expansion coefficient (1/K)
δ: Distance between nanoparticles (m)
λ: Thermal conductivity (W/mK)
μ: Dynamic viscosity (Pas)
ν: Kinematic viscosity (m^2/s)
ρ: Density (kg/m^3)
ϕ: Nanoparticle volumetric concentration
avg: Average
bf: Base fluid
f: Fluid
h: Hydraulic
m: Mass
nf: Nanofluid
p: Particle
w: Wall.

Acknowledgment

This work was supported by SUN with a 2010 grant and by MIUR with Articolo D.M. 593/2000 Grandi Laboratori "EliosLab."

References

[1] I. Uzun and M. Unsal, "A numerical study of laminar heat convection in ducts of irregular cross-sections," *International Communications in Heat and Mass Transfer*, vol. 24, no. 6, pp. 835–848, 1997.

[2] Y. Asako and M. Faghri, "Three-dimensional laminar heat transfer and fluid flow characteristics in the entrance region of a rhombic duct," *Journal of Heat Transfer*, vol. 110, no. 4, pp. 855–861, 1988.

[3] R. K. Shah, "Laminar flow friction and forced convection heat transfer in ducts of arbitrary geometry," *International Journal of Heat and Mass Transfer*, vol. 18, no. 7-8, pp. 849–862, 1975.

[4] S. R. Montgomery and P. Wibulswas, "Laminar flow heat transfer ducts of rectangular cross-section," in *Proceedings of the 3rd International Heat Transfer Conference*, pp. 85–96, New York, NY, USA, 1966.

[5] K. Velusamy, V. K. Garg, and G. Vaidyanathan, "Fully developed flow and heat transfer in semi-elliptical ducts," *International Journal of Heat and Fluid Flow*, vol. 16, no. 2, pp. 145–152, 1995.

[6] R. Gupta, P. E. Geyer, D. F. Fletcher, and B. S. Haynes, "Thermohydraulic performance of a periodic trapezoidal channel with a triangular cross-section," *International Journal of Heat and Mass Transfer*, vol. 51, no. 11-12, pp. 2925–2929, 2008.

[7] C. W. Leung and S. D. Probert, "Forced-convective internal cooling of a horizontal equilateral-triangle cross-sectioned duct," *Applied Energy*, vol. 50, no. 4, pp. 313–321, 1995.

[8] W. M. Kays and A. L. London, *Compact Heat Exchangers*, McGraw-Hill, New York, NY, USA, 3rd edition, 1984.

[9] L. Zhang, "Laminar flow and heat transfer in plate-fin triangular ducts in thermally developing entry region," *International Journal of Heat and Mass Transfer*, vol. 50, no. 7-8, pp. 1637–1640, 2007.

[10] P. Talukdar and M. Shah, "Analysis of laminar mixed convective heat transfer in horizontal triangular ducts," *Numerical Heat Transfer, Part A*, vol. 54, no. 12, pp. 1148–1168, 2008.

[11] A. Lawal, "Mixed convection heat transfer to power law fluids in arbitrary cross-sectional ducts," *Journal of Heat Transfer*, vol. 111, no. 2, pp. 399–405, 1989.

[12] G. E. Schneider and B. L. LeDain, "Fully developed Laminar heat transfer in triangular passages," *Journal of Energy*, vol. 5, no. 1, pp. 15–21, 1981.

[13] R. W. Hanks and R. C. Cope, "Laminar-turbulent transitional flow phenomena in isosceles triangular cross-section ducts," *American Institute of Chemical Engineers Journal*, vol. 16, no. 4, pp. 528–535, 1970.

[14] C. A. C. Altemani and E. M. Sparrow, "Turbulent heat transfer and fluid flow in an unsymmetrically heated triangular duct," *Journal of Heat Transfer*, vol. 102, no. 4, pp. 590–597, 1980.

[15] C. W. Leung, T. T. Wong, and H. J. Kang, "Forced convection of turbulent flow in triangular ducts with different angles and surface roughnesses," *Heat and Mass Transfer/Waerme- und Stoffuebertragung*, vol. 34, no. 1, pp. 63–68, 1998.

[16] M. Bahrami, M. Michael Yovanovich, and J. Richard Culham, "A novel solution for pressure drop in singly connected microchannels of arbitrary cross-section," *International Journal of Heat and Mass Transfer*, vol. 50, no. 13-14, pp. 2492–2502, 2007.

[17] K. Rehme, "Simple method of predicting friction factors of turbulent flow in non-circular channels," *International Journal of Heat and Mass Transfer*, vol. 16, no. 5, pp. 933–950, 1973.

[18] Y. G. Lai, "An unstructured grid method for a pressure-based flow and heat transfer solver," *Numerical Heat Transfer, Part B*, vol. 32, no. 3, pp. 267–281, 1997.

[19] S. Chen, T. L. Chan, C. W. Leung, and B. Yu, "Numerical prediction of laminar forced convection in triangular ducts with unstructured triangular grid method," *Numerical Heat Transfer, Part A*, vol. 38, no. 2, pp. 209–224, 2000.

[20] S. Syrjala, "Accurate prediction of friction factor and nusselt number for some duct flows of power-law non-newtonian fluids," *Numerical Heat Transfer, Part A*, vol. 41, no. 1, pp. 89–100, 2002.

[21] R. K. Shah and A. L. London, *Laminar Flow Forced Convection in Ducts*, Academic Press, New York, NY, USA, 1978.

[22] M. E. Ali and H. Al-Ansary, "Natural convection heat transfer from vertical triangular ducts," in *Proceedings of the 2009 ASME Summer Heat Transfer Conference, (HT '09)*, vol. 2, pp. 421–428, San Francisco, Calif, USA, July 2009.

[23] A. E. Bergles, "Some perspective on enhanced heat transfer, second-generation heat transfer technology," *Journal of Heat Transfer*, vol. 110, no. 4, pp. 1082–1096, 1988.

[24] R. L. Webb and N. H. Kim, *Principles of Enhanced Heat Transfer*, Taylor & Francis Group, New York, NY, USA, 2nd edition, 2005.

[25] J. Buongiorno, "Convective transport in nanofluids," *Journal of Heat Transfer*, vol. 128, no. 3, pp. 240–250, 2006.

[26] S. U. S. Choi, "Nanofluids: from vision to reality through research," *Journal of Heat Transfer*, vol. 131, no. 3, pp. 1–9, 2009.

[27] O. Manca, P. Mesolella, S. Nardini, and D. Ricci, "Numerical study of a confined slot impinging jet with nanofluids," *Nanoscale Research Letters*, vol. 6, pp. 1–16, 2011.

[28] Y. Xuan and Q. Li, "Heat transfer enhancement of nanofluids," *International Journal of Heat and Fluid Flow*, vol. 21, no. 1, pp. 58–64, 2000.

[29] C. H. Chon, K. D. Kihm, S. P. Lee, and S. U. S. Choi, "Empirical correlation finding the role of temperature and particle size for nanofluid (Al2O3) thermal conductivity enhancement," *Applied Physics Letters*, vol. 87, no. 15, Article ID 153107, pp. 1–3, 2005.

[30] B. C. Pak and Y. I. Cho, "Hydrodynamic and heat transfer study of dispersed fluids with submicron metallic oxide particles," *Experimental Heat Transfer*, vol. 11, no. 2, pp. 151–170, 1998.

[31] I. Gherasim, G. Roy, C. T. Nguyen, and D. Vo-Ngoc, "Heat transfer enhancement and pumping power in confined radial flows using nanoparticle suspensions (Nanofluids)," *International Journal of Thermal Sciences*, vol. 50, pp. 369–377, 2010.

[32] P. Keblinski, R. Prasher, and J. Eapen, "Thermal conductance of nanofluids: is the controversy over?" *Journal of Nanoparticle Research*, vol. 10, no. 7, pp. 1089–1097, 2008.

[33] J. Buongiorno, D. C. Venerus, N. Prabhat et al., "A benchmark study on the thermal conductivity of nanofluids," *Journal of Applied Physics*, vol. 106, no. 9, Article ID 094312, 2009.

[34] Y. Xuan and Q. Li, "Investigation on convective heat transfer and flow features of nanofluids," *Journal of Heat Transfer*, vol. 125, no. 1, pp. 151–155, 2003.

[35] D. Wen and Y. Ding, "Experimental investigation into convective heat transfer of nanofluids at the entrance region under laminar flow conditions," *International Journal of Heat and Mass Transfer*, vol. 47, no. 24, pp. 5181–5188, 2004.

[36] S. E. B. Maiga, C. T. Nguyen, N. Galanis, and G. Roy, "Heat transfer behaviours of nanofluids in a uniformly heated tube," *Superlattices and Microstructures*, vol. 35, no. 3–6, pp. 543–557, 2004.

[37] R. Ben Mansour, N. Galanis, and C. T. Nguyen, "Experimental study of mixed convection with water-Al2O3 nanofluid in inclined tube with uniform wall heat flux," *International Journal of Thermal Sciences*, vol. 50, pp. 403–410, 2010.

[38] S. Maiga, S. J. Palm, C. T. Nguyen, G. Roy, and N. Galanis, "Heat transfer enhancement by using nanofluids in forced convection flows," *International Journal of Heat and Fluid Flow*, vol. 26, no. 4, pp. 530–546, 2005.

[39] S. E. B. Maiga, C. T. Nguyen, N. Galanis, G. Roy, T. Maré, and M. Coqueux, "Heat transfer enhancement in turbulent tube flow using Al2O3 nanoparticle suspension," *International Journal of Numerical Methods for Heat and Fluid Flow*, vol. 16, no. 3, pp. 275–292, 2006.

[40] M. Akbari and A. Behzadmehr, "Developing mixed convection of a nanofluid in a horizontal tube with uniform heat flux," *International Journal of Numerical Methods for Heat and Fluid Flow*, vol. 17, no. 6, pp. 566–576, 2007.

[41] S. Mirmasoumi and A. Behzadmehr, "Numerical study of laminar mixed convection of a nanofluid in a horizontal tube using two-phase mixture model," *Applied Thermal Engineering*, vol. 28, no. 7, pp. 717–727, 2008.

[42] S. Allahyari, A. Behzadmehr, and S. M. H. Sarvari, "Conjugate heat transfer of laminar mixed convection of a nanofluid through an inclined tube with circumferentially non-uniform heating," *Nanoscale Research Letters*, vol. 6, no. 1, pp. 1–13, 2011.

[43] S. Allahyari, A. Behzadmehr, and S. M. H. Sarvari, "Conjugate heat transfer of laminar mixed convection of a nanofluid through a horizontal tube with circumferentially non-uniform heating," *International Journal of Thermal Sciences*, vol. 50, pp. 1963–1972, 2011.

[44] W. M. Rohsenow, J. P. Hartnett, and Y. I. Cho, *Handbook of Heat Transfer*, McGraw-Hill, New York, NY, USA, 3rd edition, 1998.

[45] N. Masoumi, N. Sohrabi, A. Behzadmehr, and S. M. H. Sarvari, "A new model for calculating the effective viscosity of nanofluids," *Journal of Physics D*, vol. 42, no. 5, Article ID 055501, 2009.

[46] K. Khanafer, K. Vafai, and M. Lightstone, "Buoyancy driven heat transfer enhancement in a two-dimensional enclosure utilizing nanofluids," *International Journal of Heat and Mass Transfer*, vol. 46, no. 19, pp. 3639–3653, 2003.

[47] Fluent Inc., "FLUENT computational fluid dynamic code version 6.3 user guide," 2006, http://www.fluent.com/.

[48] N. Prabhat, J. Buongiorno, and L. Hu, "Convective heat transfer enhancement in nanofluids: real anomaly or analysis artifact?" in *Proceedings of the ASME/JSME 8th Thermal Engineering Joint Conference, (AJTEC '11)*, Honolulu, Hawaii, USA, March 2011.

8

On Full-Tensor Permeabilities of Porous Media from Numerical Solutions of the Navier-Stokes Equation

Yi Wang,[1,2] Shuyu Sun,[2] and Bo Yu[1]

[1] *National Engineering Laboratory for Pipeline Safety, Beijing Key Laboratory of Urban Oil and Gas Distribution Technology, China University of Petroleum, Beijing 102249, China*
[2] *Computational Transport Phenomena Laboratory, Division of Physical Science and Engineering, King Abdullah University of Science and Technology, Thuwal 23955-6900, Saudi Arabia*

Correspondence should be addressed to Shuyu Sun; shuyu.sun@kaust.edu.sa

Academic Editor: Toshio Tagawa

A numerical method is proposed to compute full-tensor permeability of porous media without artificial simplification. Navier-Stokes (N-S) equation and Darcy's law are combined to design these numerical experiments. This method can successfully detect the permeability values in principle directions of the porous media and the anisotropic degrees. It is found that the same configuration of porous media may possess isotropic features at lower Reynolds numbers while manifesting anisotropic features at higher Reynolds numbers due to the nonlinearity from convection. Anisotropy becomes pronounced especially when convection is dominant.

1. Introduction

Permeability is a key parameter of porous media because it relates with many parameters, such as the infiltration, dielectric strength and thermophysics. It may strongly affect the design and production process of fibrous reinforcements, cement paste and so forth and related scientific researches. The permeability may vary over several orders of magnitude, and it plays an important role in petroleum migration and reservoir performance [1]. Therefore, the accurate measurement of permeability is always a hot topic in both industrial and academic fields. Many of these researches adopt experimental approaches, such as free-space methods, open-ended coaxial probe techniques, cavity resonators, transmission-line techniques, and gas-dynamic methods [2, 3].

Jensen and Heriot-Watt [4] proposed a statistical model for small-scale permeability using minipermeameter and core plug measurements. They suggested the minipermeameter measurement as a better choice. Wang et al. [5] discussed two experimental methods to determine the absolute values of in-plane permeability. They concluded that both the radial flow measurement method and the unidirectional flow measurement method were recommended to obtain reliable permeability data. Ferland et al. [6] proposed a concurrent method to estimate permeability at low cost. However, the experimental procedures can introduce uncertainty significantly on the estimated permeability. They suggested some special treatments to increase the reliability of experimental data. Some other researchers [7–10] also realized that permeability is difficult to measure although its definition is simple. Many factors, such as flow rate, pressure, fluid properties, handling process by human factors, edge effect, wall effect, single-equipment reproducibility, and between-equipment repeatability, could strongly influence the results of measurements. Therefore, the ability to obtain consistent permeability data depends on skilled and experienced experimental design, reproducible preparation of specimens, operation of equipment, and evaluation of measured raw data. This is the reason many published data observed by different persons often have significant differences for the same material. Weitzenböck et al. [11] pointed out that numerous practical problems caused three-dimensional (3D) permeability measurements to be very difficult. Then, they proposed an approach to measure the permeability by a two-dimensional (2D) radial flow method, which allows the experimental axes not to align with the principal direction of permeability [12, 13].

Some researchers made efforts on measuring transport properties (including permeability) by numerical simulation. Keehm et al. obtained good results in estimating permeability and electric conductivity of complicated pore geometries using Lattice-Boltzmann method (LBM) [14]. Kameda et al. applied LBM to estimate 3D permeability through 2D images of small fragments of rocks. They obtained a valid permeability-porosity trend by using a significant number of such small fragments in statistical sense [15]. Saenger et al. estimated permeability through LBM flow simulation and compared mechanical and transport properties for the same digital rocks [16].

Since all physical parameters can be easily and exactly controlled in numerical simulations, the shortcomings of laboratory experiments mentioned previously can be avoided. In particular, numerical methods have the advantage of separately studying different physical processes coexisting in nature, which are uneasily separated in laboratory. It is easier to form standardization of measurement methods and obtain unchangeable results. It will be a very efficient and economical way for measuring properties of porous media combining with the digital rock physics [17, 18]. In the present paper, we studied how to establish this digital laboratory method for measuring permeability of porous media through directly solving the N-S equation, other than reconstructing it by LBM method. To the best knowledge of the authors, all studies of permeability prediction have been concerned only with diagonal permeability tensor actually. Full tensor of permeability has not been studied extensively since it has more general meanings for practical applications. Thus, we expand our research object from the simple diagonal permeability tensor to more general full permeability tensor. We select gravity in periodic domain as the driven force, instead of pressure widely used in previous studies [1–18]. This can avoid the edge effect encountered in experimental measurements. As a first trial, we assume incompressible single-phase flow to pass through the porous media in rectangular geometry, but the methodology can be extended to more complex applications, for example two-phase flow using diffuse interface models, which will be pursued in our future researches. The basic principles and proposed methods in this study will be presented next.

2. Principles and Methods

2.1. General Principles. The basic law that is used by experimental and numerical studies for measuring permeability is the Darcy's law:

$$\mathbf{u}_D = -\frac{\mathbf{k}}{\mu}\left(\nabla p - \rho \mathbf{g}\right), \qquad (1)$$

where $\mathbf{u}_D$ is Darcy velocity (m/s), $\mathbf{k}$ is the permeability tensor (m^2), $\mathbf{g}$ is the gravitational acceleration ($\mathrm{m/s^2}$), ∇p is the pressure gradient (Pa/m), ρ is the density of fluid ($\mathrm{kg/m^3}$), and μ is the dynamic viscosity of fluid ($\mathrm{Pa \cdot s}$).

The N-S equation (2) and the continuity equation (3) that describe fluid flow in the continuum sense are as follows (Δ is the Laplace operator):

$$\frac{\partial \mathbf{u}}{\partial t} + (\mathbf{u} \cdot \nabla)\,\mathbf{u} = -\frac{1}{\rho}\nabla p + \frac{\mu}{\rho}\Delta \mathbf{u} + \mathbf{g}, \qquad (2)$$

$$\nabla \cdot \mathbf{u} = 0. \qquad (3)$$

Conventional methods for measuring permeability neglect gravity and drive the fluid flow by pressure because it is easier to control, especially for laboratory experiments. Here, we select to use the full expressions of (1)~(3) to determine the permeability. Periodic boundary condition is adopted so that boundary pressure needs not to be considered. It is pretty easier for numerical implementation and more sensible in nature because subsurface formation is in fact a periodic system or statistically periodic one. The general procedure is stated below.

Step 1. Solve (2) and (3) for incompressible single-phase flow in porous media and obtain several velocity fields as samples.

Step 2. Obtain volumetric velocity of the whole domain according to the samples obtained in Step 1.

Step 3. Solve permeability for the whole domain by directly setting the volumetric velocity as the Darcy velocity in (1).

The detailed numerical methods used in these three steps will be introduced in the next section.

2.2. Numerical Methods. As stated previously, we only consider simple 2D rectangular cases in the present study. The incompressible single-phase flow in porous media is modeled as the flow passing through a square barrier in a square domain with periodic boundary conditions. The numerical model is shown in Figure 1(a). L is the domain side length while a is the side length of the barrier. Γ_P means periodic boundary.

On this kind of domain, the vectors and tensors in (1)~(3) have the following forms:

$$\mathbf{u} = \begin{bmatrix} u_x \\ u_y \end{bmatrix}, \qquad \mathbf{g} = \begin{bmatrix} g_x \\ g_y \end{bmatrix},$$
$$\mathbf{k} = \begin{bmatrix} k_{xx} & k_{xy} \\ k_{yx} & k_{yy} \end{bmatrix}, \qquad \nabla p = \begin{bmatrix} \dfrac{\partial p}{\partial x} \\ \dfrac{\partial p}{\partial y} \end{bmatrix}. \qquad (4)$$

Here $g_x = g\cos\theta$, $g_y = g\sin\theta$, $g = 9.807\,\mathrm{m/s^2}$, k_{xx}, and k_{yy} are the diagonal components of the permeability tensor while k_{xy} and k_{yx} are the off-diagonal components of the permeability tensor. These four variables are all independent of each other. Three typical directions of gravity are chosen to calculate the full-tensor permeability. They are represented by $\theta = 3\pi/2, \pi$, and $\pi/4$, respectively, and named "Sample a," and "Sample b," and "Sample c" as shown in Figure 1(b).

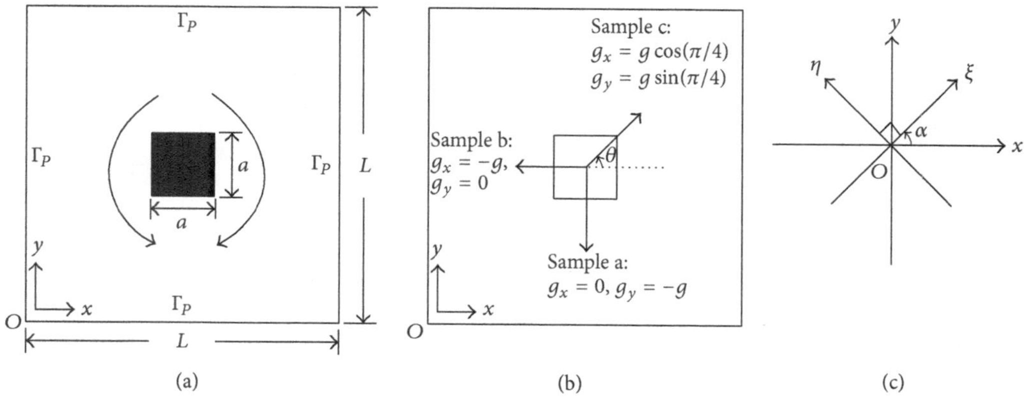

FIGURE 1: Calculation methodology: (a) numerical model; (b) sampling; and (c) coordinate rotation.

The numerical algorithm for solving (2)~(3) in "Step 1" is projection method with fully explicit spatial discretization (but solve pressure implicitly) using second-order finite difference central type scheme. Cyclic tridiagonal matrix algorithm (CTDMA) is applied to accelerate the convergence of pressure iteration. The mesh size is set as 80×80. Flow is considered to reach the steady state as the temporal truncation $|\partial \mathbf{u}/\partial t|$ averaged over the whole domain is lower than 10^{-6} m/s^2.

The volumetric velocity in "Step 2" is defined as follows:

$$\mathbf{U} = \frac{1}{V_d}\int_{V_d} \mathbf{u} dV_d, \tag{5}$$

where $\mathbf{U} = \begin{bmatrix} U_x \\ U_y \end{bmatrix}$ is the volumetric velocity with the components U_x and U_y in the x and y directions. V_d is the volume of the whole domain. Equation (5) has a discrete expression in this study:

$$U_x = \frac{(\sum u_x)}{N_x}, \qquad U_y = \frac{(\sum u_y)}{N_y}, \tag{6}$$

where N_x and N_y are the total number of u_x and u_y on grid cells, respectively. The values of u_x and u_y can be determined by solving the N-S equation so that U_x and U_y can be determined.

The volumetric velocity in (5) is actually the Darcy velocity in (1); that is, $\mathbf{u}_D = \mathbf{U}$. In the Darcy scale averaged in the whole domain, pressure gradient should be zero because of the periodic boundary condition; that is, $\nabla p = 0$. Thus, (1) can be rewritten as

$$\mathbf{U} = \frac{\mathbf{k}}{\mu}\rho \mathbf{g}. \tag{7}$$

For Sample a ($(g_y)_a = -g$), (7) becomes

$$(U_x)_a = \frac{k_{xy}}{\mu}\rho (g_y)_a, \tag{8a}$$

$$(U_y)_a = \frac{k_{yy}}{\mu}\rho (g_y)_a. \tag{8b}$$

For Sample b ($(g_x)_b = -g$), (7) becomes

$$(U_x)_b = \frac{k_{xx}}{\mu}\rho (g_x)_b, \tag{9a}$$

$$(U_y)_b = \frac{k_{yx}}{\mu}\rho (g_x)_b. \tag{9b}$$

For Sample c ($(g_x)_c = g\cos(\pi/4)$, $(g_y)_c = g\sin(\pi/4)$), (7) becomes

$$(U_x)_c = \frac{k_{xx}}{\mu}\rho (g_x)_c + \frac{k_{xy}}{\mu}\rho (g_y)_c, \tag{10a}$$

$$(U_y)_c = \frac{k_{yx}}{\mu}\rho (g_x)_c + \frac{k_{yy}}{\mu}\rho (g_y)_c. \tag{10b}$$

Theoretically, (8a), (8b), (9a), and (9b) are adequate to determine the four components of the full-tensor permeability. However, $(U_x)_a$ and $(U_y)_b$ are always very close to zeros since they are orthogonal to the bulk flow directions and large numerical errors may exist. Thus, k_{xy} and k_{yx} determined by (8a) and (9b) are always very close to zeros. This apparently violates the fact that k_{xy} and k_{yx} may be far from zeros for some cases. The application of (8a) and (9b) actually makes the full-tensor permeability decay to a diagonal permeability so that they should be canceled. To predict the off-diagonal components accurately, (10a) and (10b) are utilized since the bulk flow in Sample c has equivalent projections in the x and y directions so that $(U_x)_c$ and $(U_y)_c$ are both accurate. Therefore, the final equations to determine the four components of the permeability tensor should be the four equations (8b), (9a), (10a), and (10b), which are used in "Step 3." The procedure is calculating k_{xx} by (9a) and k_{yy} by (8b) and substituting them to (10a) and (10b) to obtain k_{xy} and k_{yx}, respectively.

In practice, it is of interest to detect the maximum permeability, the minimum permeability, principle direction, and anisotropy so that essential features of reservoir, which are independent of coordinate systems and samplings, can be described. Therefore, the original permeability tensor can be transformed by rotating the original xOy coordinate

TABLE 1: Characteristic parameters.

Case	V (m/s)	R_h (m)	Re	Da	$\text{ReDa}^{1/2}$
Case 1	1.16×10^{-6}	2.50×10^{-8}	2.90×10^{-8}	6.25×10^{-4}	7.25×10^{-10}
Case 2	6.62×10^{-7}	5.00×10^{-8}	3.31×10^{-8}	2.50×10^{-3}	1.66×10^{-9}
Case 3	3.96×10^{-7}	7.50×10^{-8}	2.97×10^{-8}	5.63×10^{-3}	2.23×10^{-9}
Case 4	2.32×10^{-7}	1.00×10^{-7}	2.32×10^{-8}	1.00×10^{-2}	2.32×10^{-9}
Case 5	1.28×10^{-7}	1.25×10^{-7}	1.60×10^{-8}	1.56×10^{-2}	2.00×10^{-9}
Case 6	6.24×10^{-8}	1.50×10^{-7}	9.36×10^{-9}	2.25×10^{-2}	1.40×10^{-9}
Case 7	2.51×10^{-8}	1.75×10^{-7}	4.39×10^{-9}	3.06×10^{-2}	7.69×10^{-10}
Case 8	7.12×10^{-9}	2.00×10^{-7}	1.42×10^{-9}	4.00×10^{-2}	2.85×10^{-10}
Case 9	8.58×10^{-10}	2.25×10^{-7}	1.93×10^{-10}	5.06×10^{-2}	4.34×10^{-11}

to a new $\xi O \eta$ coordinate according to tensors' characteristics [19]. The angle between the $\xi O \eta$ coordinate system and the xOy coordinate system is represented by the angle α (Figure 1(c)). Therefore, a series of transformed permeability tensors $\mathbf{k}_{\text{trans}} = \begin{bmatrix} k_{\xi\xi} & k_{\xi\eta} \\ k_{\eta\xi} & k_{\eta\eta} \end{bmatrix}$ can be obtained by continuously rotating the $\xi O \eta$ coordinate system, that is, continuously changing the value of angle α. Once $k_{\xi\eta} \approx 0$, $k_{\eta\xi} \approx 0$ is achieved at a certain angle α_p; it represents the case that fluid flows no longer in any tangential directions but only in the two normal directions ξ and η so that maximum and minimum components of permeability among all transformed permeability tensors can be examined ($k_{\max} = \max(k_{\xi\xi}, k_{\eta\eta})$, $k_{\min} = \min(k_{\xi\xi}, k_{\eta\eta})$). In this case, α_p represents the principle direction. To examine the principle direction precisely, $\alpha = 0$~$360°$ with interval $1°$ is adopted. Another important parameter is the anisotropy of the permeability in the porous media. Here, we define the ratio A_k to determine the anisotropic degree of permeability: $A_k = k_{\max}/k_{\min}$. The tensor transformation formula [19] is shown as follows:

$$k_{\xi\xi} = k_{xx} \cos\alpha \cos\alpha + k_{yy} \cos\beta \cos\beta + k_{xy} \cos\alpha \cos\beta + k_{yx} \cos\beta \cos\alpha, \tag{11}$$

$$k_{\eta\eta} = k_{xx} \cos\gamma \cos\gamma + k_{yy} \cos\alpha \cos\alpha + k_{xy} \cos\gamma \cos\alpha + k_{yx} \cos\alpha \cos\gamma, \tag{12}$$

$$k_{\xi\eta} = k_{xx} \cos\alpha \cos\gamma + k_{yy} \cos\beta \cos\alpha + k_{xy} \cos\alpha \cos\alpha + k_{yx} \cos\beta \cos\gamma, \tag{13}$$

$$k_{\eta\xi} = k_{xx} \cos\gamma \cos\alpha + k_{yy} \cos\alpha \cos\beta + k_{xy} \cos\gamma \cos\beta + k_{yx} \cos\alpha \cos\alpha, \tag{14}$$

where $\beta = \pi/2 - \alpha$, $\gamma = \pi/2 + \alpha$.

3. Results and Discussion

Two important factors are discussed here: porosity and Reynolds number. To study them separately, two groups of numerical cases are designed. One is changing porosity by changing the size of barrier but keeping constant length of domain. The other one is changing the length of domain but keeping constant porosity. These two factors are discussed in the following two sections. Fluid passing through the porous media is water with density $\rho = 1000\,\text{kg/m}^3$ and dynamic viscosity $\mu = 0.001\,\text{Pa} \cdot \text{s}$.

3.1. Discussion on Porosity. The domain size is set as $L = 10^{-6}$ m. The size of barrier is set as $a = (1/10)L \sim (9/10)L$ with interval $(1/10)L$ so that 9 cases named Case 1~Case 9 are generated. According to the size of the computational domain and spatial element sizes on the fixed mesh, time step is set as its maximum value available: $\Delta t = 10^{-11}$ s.

The flow simulation results by directly solving the Navier-Stokes equation are shown in Figure 2. The contours represent the magnitudes of velocity (i.e., the characteristic speed to be defined later), while the vectors represent the directions of velocity. It can be seen that the flow fields are well represented in the 9 cases. The square barriers are clearly identified by the vectors. We define characteristic speed as $V = \sqrt{U_x^2 + U_y^2}$, characteristic length as the hydraulic radius $R_h = a^2/(4a) = a/4$, Reynolds number as $\text{Re} = \rho V R_h/\mu$, and Darcy number as $\text{Da} = (R_h/L)^2$. The characteristic parameters are listed in Table 1 (results for Samples a, b, and c are the same so that the data in Table 1 are any of them). The very low Reynolds number (order of magnitude 10^{-10}~10^{-8}) indicates that the flows are very slow laminar flows. It can be seen that the dimensionless number $\text{ReDa}^{1/2}$ has the order of magnitude 10^{-11}~10^{-9}, which satisfies the suggested restriction of Darcy's law $\text{ReDa}^{1/2} \ll 1$ by Bear and Cheng [20].

Porosity is defined as the ratio of void space volume to the total volume as follows:

$$\phi = 1 - \frac{a^2}{L^2}. \tag{15}$$

Thus, the 9 cases relate to 9 different values of porosity. Full permeability tensors obtained from flow simulation results of the 9 cases are listed in Table 2. It can be seen that the diagonal components of the permeability tensor all decrease with decreasing porosity. Their values show that the studied porous media fall within the ranges of oil rocks ($10^{-11}\,\text{m}^2$~$10^{-14}\,\text{m}^2$), sandstone ($10^{-14}\,\text{m}^2$~$10^{-16}\,\text{m}^2$), and good limestone ($10^{-16}\,\text{m}^2$~$10^{-18}\,\text{m}^2$) [20]. The diagonal

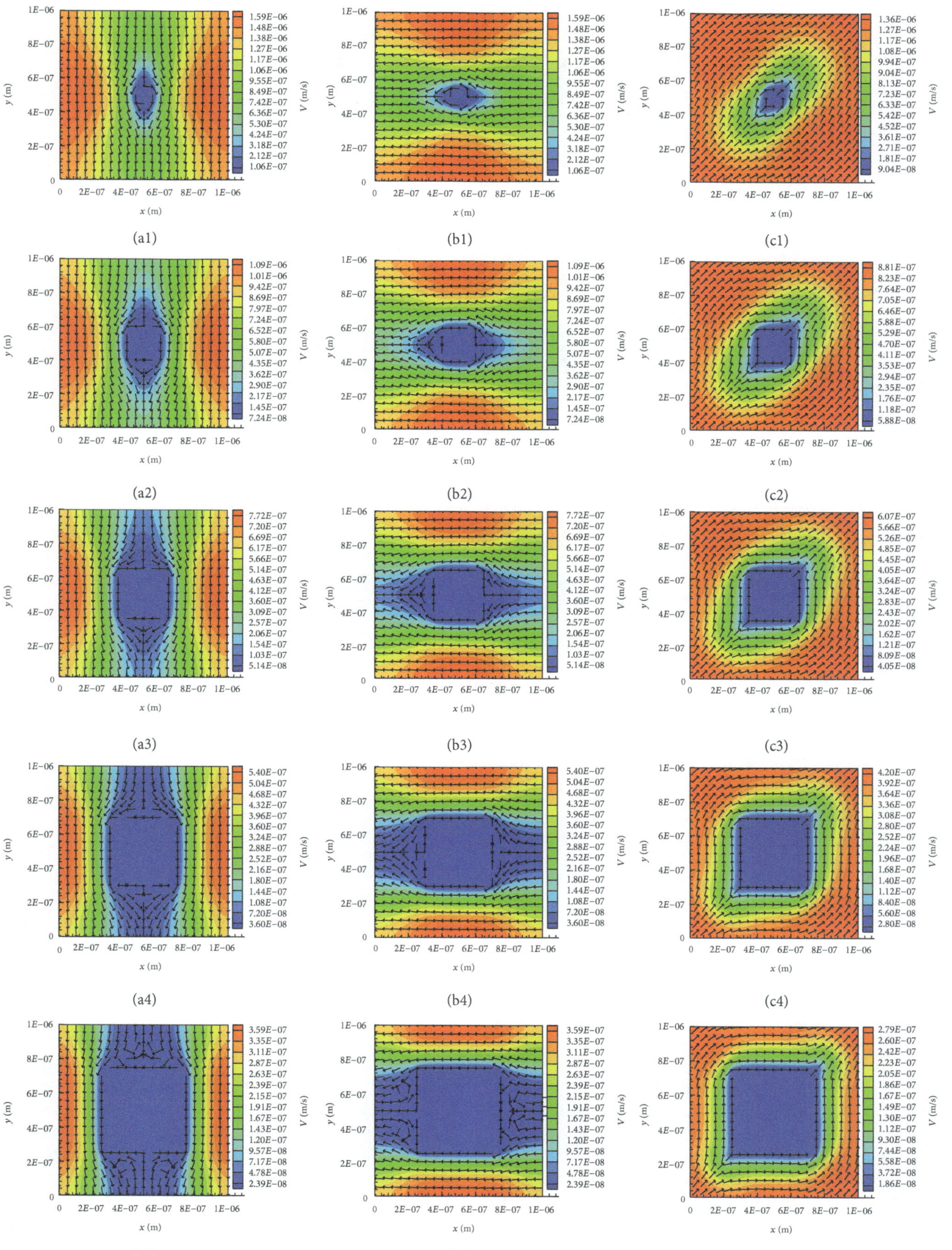

(a)

FIGURE 2: Continued.

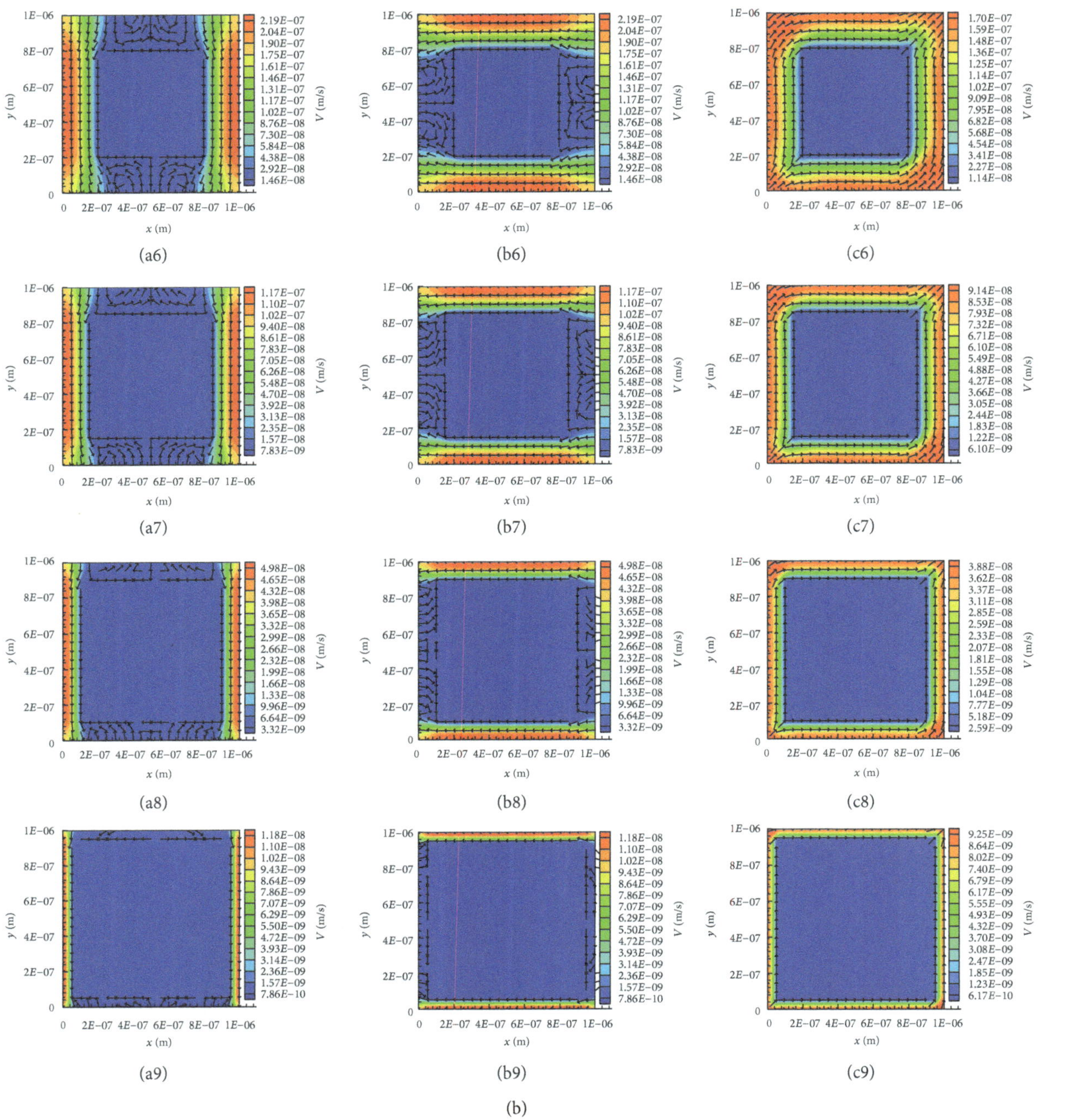

FIGURE 2: Flow fields for different porosities from Navier-Stokes simulation: columns a, b, and c represent Sample a, Sample b, and Sample c; numbers 1~9 represent Case 1~Case 9; the contour shows total speed while the vector shows directions of total velocity.

components of transformed permeability tensor are shown in Figure 3 for the 9 porosities; there always exist $k_{\xi\xi} = k_{\eta\eta}$ and $k_{\xi\xi} = k_{xx}$, $k_{\eta\eta} = k_{yy}$. These mean that the two diagonal components of the full permeability tensor are the same and are always equal to each other along with the rotating coordinate. This indicates that the permeable properties are all the same for any direction so that principle direction does not exist. Therefore, the full-tensor permeabilities are always diagonal tensors with equivalent components for the 9 porosities. The off-diagonal components, which are 3~6 orders of magnitudes lower than the diagonal components (Table 2), should be considered as numerical errors. In this case, the off-diagonal components should be considered as zeros. Revealed from the mathematical meaning, physical meaning is that the porous media in the 9 cases are all isotropic. Corresponding characteristic parameters are listed in Table 3.

3.2. Discussion on Reynolds Number. Besides the porosity, Reynolds number is also a key factor which may strongly affect velocity fields so that the results of permeability may be affected. Thus, it is important to check whether

TABLE 2: Full-tensor permeabilities at different porosities.

Case	ϕ	k_{xx} (m^2)	k_{yy} (m^2)	k_{xy} (m^2)	k_{yx} (m^2)
Case 1	0.99	1.1808×10^{-13}	1.1808×10^{-13}	-5.6848×10^{-16}	-5.6363×10^{-16}
Case 2	0.96	6.7481×10^{-14}	6.7481×10^{-14}	-3.1129×10^{-17}	-2.8981×10^{-17}
Case 3	0.91	4.0362×10^{-14}	4.0362×10^{-14}	-6.2013×10^{-19}	2.6772×10^{-19}
Case 4	0.84	2.3702×10^{-14}	2.3702×10^{-14}	-2.7576×10^{-19}	3.5447×10^{-20}
Case 5	0.75	1.3026×10^{-14}	1.3026×10^{-14}	-8.3818×10^{-20}	1.6370×10^{-20}
Case 6	0.64	6.3596×10^{-15}	6.3596×10^{-15}	7.1193×10^{-21}	-2.1080×10^{-20}
Case 7	0.51	2.5614×10^{-15}	2.5614×10^{-15}	5.8343×10^{-21}	-3.6183×10^{-20}
Case 8	0.36	7.2643×10^{-16}	7.2643×10^{-16}	1.0640×10^{-21}	-1.0052×10^{-20}
Case 9	0.19	8.7493×10^{-17}	8.7493×10^{-17}	6.4756×10^{-23}	-1.9648×10^{-22}

TABLE 3: Characteristic permeabilities and anisotropies at different porosities.

Case	ϕ	k_{max} (m^2)	k_{min} (m^2)	A_k (= k_{max}/k_{min})
Case 1	0.99	1.1808×10^{-13}	1.1808×10^{-13}	1.0000
Case 2	0.96	6.7481×10^{-14}	6.7481×10^{-14}	1.0000
Case 3	0.91	4.0362×10^{-14}	4.0362×10^{-14}	1.0000
Case 4	0.84	2.3702×10^{-14}	2.3702×10^{-14}	1.0000
Case 5	0.75	1.3026×10^{-14}	1.3026×10^{-14}	1.0000
Case 6	0.64	6.3596×10^{-15}	6.3596×10^{-15}	1.0000
Case 7	0.51	2.5614×10^{-15}	2.5614×10^{-15}	1.0000
Case 8	0.36	7.2643×10^{-16}	7.2643×10^{-16}	1.0000
Case 9	0.19	8.7493×10^{-17}	8.7493×10^{-17}	1.0000

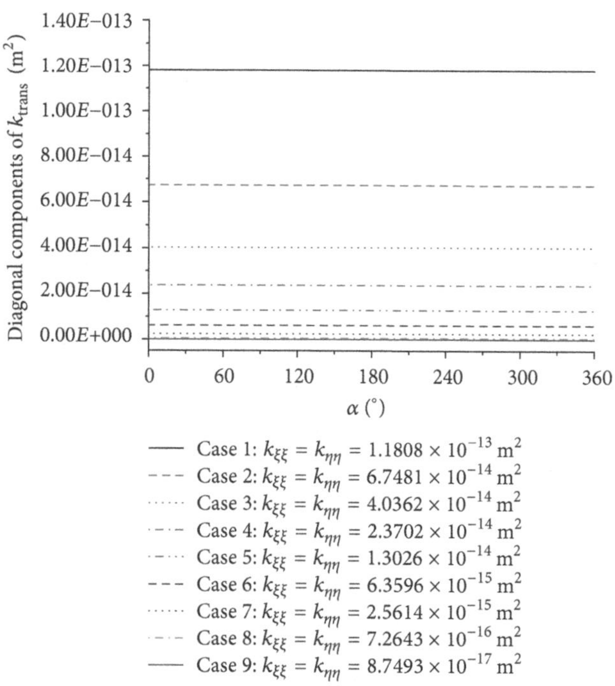

FIGURE 3: Diagonal components of transformed permeability tensor at different porosities.

the Reynolds number, actually controlled by the domain size, may change the conclusions in Section 3.1. Therefore, the domain size is gradually scaled up from $L = 10^{-6}$ m by keeping porosity at 0.64 (Case 6 in Tables 2 and 3). The results are shown in Table 4.

From 10^{-6} m to 10^{-4} m, the diagonal components of the full-tensor permeability just increase two orders of magnitudes once the domain size increases one order of magnitude. The off-diagonal components should still be considered as numerical errors and set to be zeros as stated previously. These three cases are quite similar, and their difference is only the magnitude. It can be verified by the flow fields in Figures 4(a1), 4(b1), 4(c1)~Figures 4(a3), 4(b3), 4(c3), Figures 5(a)~5(c), and Table 5 that the domain size scaling up from 10^{-6} m to 10^{-4} m just leads the velocity to increase proportionally. They are all isotropic. However, the situation is different when the domain size increases up to 10^{-3} m (see line 5 in Table 4). The off-diagonal components are no longer several orders of magnitudes smaller than the diagonal components so that the full-tensor permeability is no longer diagonal. Figures 4(a4)~4(c4) show that the fluid tends to depart the solid walls and the vortex expands. Figure 5(d) shows that the components of the transformed permeability tensor are no longer constant. Maximum and minimum values of the diagonal components (k_{max}, k_{min}) can always be obtained at the same time once the off-diagonal components become zeros. Their values (Table 5) show that the porous media at the domain length of 10^{-3} m enter the range of clean stand (10^{-9} m^2~10^{-12} m^2) which is a better aquifer than the previous ones [20]. Correspondingly, four angles (α_I, α_{II}, α_{III}, and α_{IV}) can be obtained, which are shown in Table 6. Surprisingly from Table 5, the porous media start to show anisotropic property (A_k = 1.9241) when the domain length reaches 10^{-3} m.

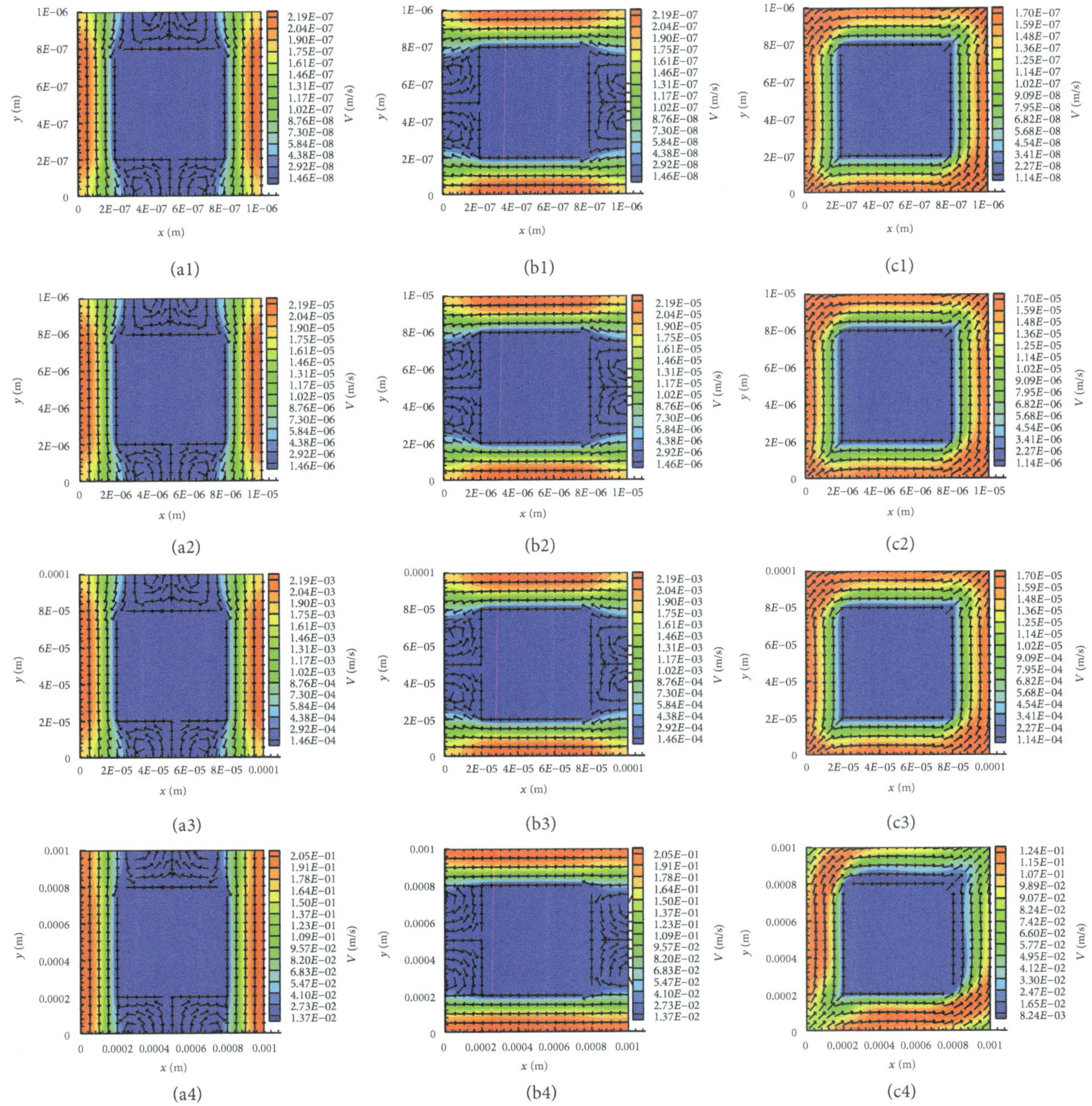

Figure 4: Flow fields for different Reynolds numbers from Navier-Stokes simulation: columns a, b, and c represent Sample a, Sample b, and Sample c; numbers 1~4 represent domain size $L = 10^{-6}$ m, 10^{-5} m, 10^{-4} m, and 10^{-3} m, respectively; the contours show total speed while the vectors show directions of total velocity.

It can be summarized from Table 6 that α_p is always equal to angle α_{II} or α_{IV} when k_{max} equals $k_{\xi\xi}$. These two angles are on the same line across the second and the fourth quadrants in the original xOy coordinate (see Figure 1(c)). This line is the direction where permeability achieves maximum value, that is, the easiest direction where fluid can pass through, so that this is the principle direction. Accordingly, the direction fluid receives the most resistance (i.e., the direction for k_{min}) is the line represented by angles α_I and α_{III}, so that this line is the direction orthogonal to the principle direction.

When the length of domain is greater than or equal to 10^{-2} m, the flow simulation is divergent. This is probably because the mesh size 80 × 80 is not dense enough for larger domain. Larger domains with finer grid, which are very time consuming, will be studied in the future.

To discuss the previous phenomenon that different Reynolds numbers may generate essentially different permeability tensors, or in other words predict different types of porous media for the same configuration, four characteristic parameters are shown in Table 7. R_u and R_v are defined as

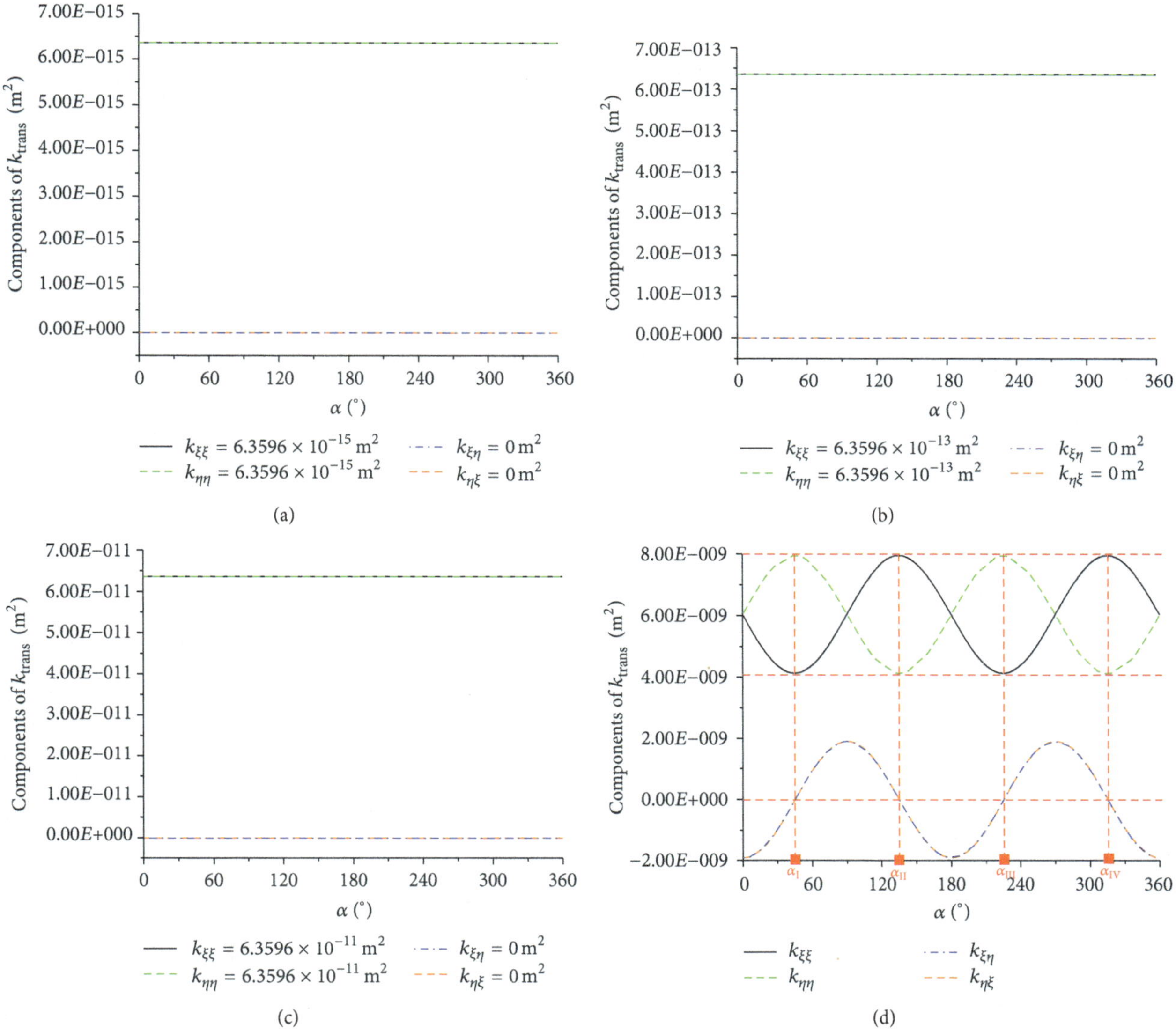

FIGURE 5: Components of transformed permeability tensor at different Reynolds numbers: (a) $L = 10^{-6}$ m, (b) $L = 10^{-5}$ m, (c) $L = 10^{-4}$ m, and (d) $L = 10^{-3}$ m.

TABLE 4: Full-tensor permeabilities at different Reynolds numbers.

L (m)	k_{xx} (m^2)	k_{yy} (m^2)	k_{xy} (m^2)	k_{yx} (m^2)
10^{-6}	6.3596×10^{-15}	6.3596×10^{-15}	7.1193×10^{-21}	-2.1080×10^{-20}
10^{-5}	6.3596×10^{-13}	6.3596×10^{-13}	7.5167×10^{-19}	-2.1064×10^{-18}
10^{-4}	6.3596×10^{-11}	6.3596×10^{-11}	-7.4280×10^{-17}	-3.6338×10^{-16}
10^{-3}	6.0366×10^{-9}	6.0366×10^{-9}	-1.9077×10^{-9}	-1.9077×10^{-9}
$\geq 10^{-2}$	Divergence	Divergence	Divergence	Divergence

TABLE 5: Characteristic permeabilities and anisotropies at different Reynolds numbers.

L (m)	ϕ	k_{max} (m^2)	k_{min} (m^2)	A_k ($= k_{max}/k_{min}$)
10^{-6}	0.64	6.3596×10^{-15}	6.3596×10^{-15}	1.0000
10^{-5}	0.64	6.3596×10^{-13}	6.3596×10^{-13}	1.0000
10^{-4}	0.64	6.3596×10^{-11}	6.3596×10^{-11}	1.0000
10^{-3}	0.64	7.9443×10^{-9}	4.1289×10^{-9}	1.9241

Table 6: Angles at zero off-diagonal components of transformed permeability for $L = 10^{-3}$ m.

L (m)	α_I	α_{II}	α_{III}	α_{IV}	α_p
10^{-3}	45°	135°	225°	315°	135°, 315°
Summary	$k_{max} = k_{\eta\eta}$ $k_{min} = k_{\xi\xi}$	$k_{max} = k_{\xi\xi}$ $k_{min} = k_{\eta\eta}$	$k_{max} = k_{\eta\eta}$ $k_{min} = k_{\xi\xi}$	$k_{max} = k_{\xi\xi}$ $k_{min} = k_{\eta\eta}$	$\alpha_p = \alpha_{II}$ or α_{IV}

Table 7: Characteristic parameters at different Reynolds numbers.

L (m)	Re	$\text{ReDa}^{1/2}$	R_u	R_v
10^{-6}	9.36×10^{-9}	1.40×10^{-9}	2.79×10^{-9}	2.79×10^{-9}
10^{-5}	9.36×10^{-6}	1.40×10^{-6}	2.79×10^{-6}	2.79×10^{-6}
10^{-4}	9.36×10^{-3}	1.40×10^{-3}	2.79×10^{-3}	2.79×10^{-3}
10^{-3}	8.88	1.33	1.35	1.35

the ratios of the mean convection effect to the mean diffusion effect in the momentum equation (2):

$$R_u = \frac{\overline{u(\partial u/\partial x) + v(\partial u/\partial y)}}{\overline{(\mu/\rho)(\partial^2 u/\partial x^2 + \partial^2 u/\partial y^2)}},$$
$$R_v = \frac{\overline{u(\partial v/\partial x) + v(\partial v/\partial y)}}{\overline{(\mu/\rho)(\partial^2 v/\partial x^2 + \partial^2 v/\partial y^2)}}, \qquad (16)$$

where the superscript "—" represents the average value over the whole domain. It is clear in Table 7 that the fluid flows are actually diffusion dominated for the domain lengths smaller than 10^{-3} m since R_u and R_v are all much smaller than 1. However, R_u and R_v are greater than 1 at the domain length of 10^{-3} m. This means that the flow is dominated by the convection effect of fluid. It is well known from their intentions that the diffusion effect tends to transport variables to all directions uniformly while the convection effect tends to transport variables along specific directions. Therefore, the existence of convection dominant flow may be the main reason for the anisotropy. It is worth to point out that Table 7 also verifies the conclusion made by Bear and Cheng [20] that Darcy's law is usually valid as long as Reynolds number is lower than 1 (occurred at the domain lengths 10^{-6} m ~ 10^{-4} m where Re = 9.36×10^{-9}~9.36×10^{-3}) but sometimes as high as 10 (occurred at the domain length 10^{-3} m where Re = 8.88 and the suggested restriction $\text{ReDa}^{1/2} \ll 1$ is violated).

4. Conclusion

A numerical method is proposed to compute the permeability in the form of full tensor. The flow simulation results show that flow fields can be well represented by solving the Navier-Stokes equation directly. Original and transformed permeability tensors are obtained so that maximum and minimum components of permeability with principle directions and anisotropies are detected successfully. With this information, the directions of largest and smallest resistances for fluid flow in porous media can be inferred easily.

Through the analyses on the porosity effect and the Reynolds number effect, it is found that porous media with the same porosity and the same configuration can manifest different levels of anisotropy at different Reynolds numbers. At the Reynolds numbers that diffusion dominates flow, isotropy is a good description. At the Reynolds number that convection dominates flow, anisotropy occurs. Thus, it is important to pay attention to Reynolds numbers for porous media applications. This is especially important for applications with large Darcy velocities, such as flow and transport in packed columns and in fractured geological media.

5. Future Issues to Be Addressed

The previous discussions show that the proposed method in the present study is an easy way to determine full-tensor permeability numerically and related characteristics. However, many issues still need to be addressed in future works. Only two of them are briefly listed below as an example.

(1) Method for solving the Navier-Stokes equation in porous media should be improved to accelerate the computation. Current speed is not acceptable for engineering applications. Multigrid method needs to be developed for the flow around solids so that the iteration of pressure can be largely accelerated. High-resolution time integration scheme is expected to largely reduce the number of time steps needed. Since only steady-state results are needed for obtaining the full-tensor permeability, a method directly solving the steady-state Navier-Stokes equation is also expected so that a large amount of time integrations can be avoided.

(2) Shape, number, position, and so forth of inner barriers should be intensively studied to know the application and limitation of this method and to improve precision. Denser meshes are also expected to study what will happen for larger Reynolds numbers.

Acknowledgments

The work presented in this paper has been supported in part by the project entitled "Simulation of Subsurface Geochemical Transport and Carbon Sequestration," funded by the GRP-AEA Program at KAUST. The work has also been supported in part by National Science Foundation of China (no. 51206186, no. 51174206) and Science Foundation of China, University of Petroleum, Beijing (no. 2462012KYJJ0403, no. 2462012KYJJ0404).

References

[1] C. L. Dinwiddie, F. J. Molz III, and J. W. Castle, "A new small drill hole minipermeameter probe for in situ permeability measurement: fluid mechanics and geometrical factors," *Water Resources Research*, vol. 39, no. 7, 2003.

[2] B. J. Wolfson and S. M. Wentworth, "Complex permeability and permeability measurement using a rectangular waveguide," *Microwave and Optical Technology Letters*, vol. 27, no. 3, pp. 180–182, 2000.

[3] A. N. Ivanov, S. N. Kozlova, and A. V. Pechenov, "Permeability measurement," *Measurement Techniques*, vol. 43, no. 12, pp. 1086–1088, 2000.

[4] J. L. Jensen and U. Heriot-Watt, "A model for small-scale permeability measurement with applications to reservoir characterization," in *Proceedings of the SPE/DOE 7th Symposium on Enhanced Oil Recovery*, pp. 891–900, April 1990.

[5] T. J. Wang, C. H. Wu, and L. J. Lee, "In-plane permeability measurement and analysis in liquid composite molding," *Polymer Composites*, vol. 15, no. 4, pp. 278–288, 1994.

[6] P. Ferland, D. Guittard, and F. Trochu, "Concurrent methods for permeability measurement in resin transfer molding," *Polymer Composites*, vol. 17, no. 1, pp. 149–158, 1996.

[7] R. Gauvin, F. Trochu, Y. Lemenn, and L. Diallo, "Permeability measurement and flow simulation through fiber reinforcement," *Polymer Composites*, vol. 17, no. 1, pp. 34–42, 1996.

[8] Y. Luo, I. Verpoest, K. Hoes, M. Vanheule, H. Sol, and A. Cardon, "Permeability measurement of textile reinforcements with several test fluids," *Composites A*, vol. 32, no. 10, pp. 1497–1504, 2001.

[9] T. F. Fwa, S. A. Tan, C. T. Chuai Y, and K. Guwe, "Expedient permeability measurement for porous pavement surface," *The International Journal of Pavement Engineering*, vol. 2, pp. 259–270, 2001.

[10] R. Arbter, J. M. Beraud, C. Binetruy et al., "Experimental determination of the permeability of textiles: a benchmark exercise," *Composites A*, vol. 42, no. 9, pp. 1157–1168, 2011.

[11] J. R. Weitzenböck, R. A. Shenoi, and P. A. Wilson, "Measurement of principal permeability with the channel flow experiment," *Polymer Composites*, vol. 20, no. 2, pp. 321–335, 1999.

[12] J. R. Weitzenböck, R. A. Shenoi, and P. A. Wilson, "Radial flow permeability measurement. Part A: theory," *Composites A*, vol. 30, no. 6, pp. 781–796, 1999.

[13] J. R. Weitzenböck, R. A. Shenoi, and P. A. Wilson, "Radial flow permeability measurement. Part B: application," *Composites A*, vol. 30, no. 6, pp. 797–813, 1999.

[14] Y. Keehm, T. Mukerji, and A. Nur, "Computational rock physics at the pore scale: transport properties and diagenesis in realistic pore geometries," *Leading Edge*, vol. 20, no. 2, pp. 180–183, 2001.

[15] A. Kameda, J. Dvorkin, Y. Keehm, A. Nur, and W. Bosl, "Permeability-porosity transforms from small sandstone fragments," *Geophysics*, vol. 71, no. 1, pp. N11–N19, 2006.

[16] E. H. Saenger, F. Enzmann, Y. Keehm, and H. Steeb, "Digital rock physics: effect of fluid viscosity on effective elastic properties," *Journal of Applied Geophysics*, vol. 74, no. 4, pp. 236–241, 2011.

[17] B. Quintal, M. Frehner, C. Madonna, N. Tisato, M. Kuteynikova, and E. H. Saenger, "Integrated numerical and laboratory rock physics applied to seismic characterization of reservoir rocks," *Leading Edge*, vol. 30, no. 12, pp. 1360–1367, 2011.

[18] Y. Keehm, T. Mukerji, and A. Nur, "Computational rock physics at the pore scale: transport properties and diagenesis in realistic pore geometries," *Leading Edge*, vol. 20, no. 2, pp. 180–183, 2001.

[19] R. L. Bishop and S. I. Goldberg, *Tensor Analysis on Manifolds*, Dover Publications, 1968.

[20] J. Bear and A. H. D. Cheng, *Modeling Groundwater Flow and Contaminant Transport*, vol. 23 of *Theory and Applications of Transport in Porous Media*, Springer Science+Business Media B.V., 2010.

Effect of Red Mud and Copper Slag Particles on Physical and Mechanical Properties of Bamboo-Fiber-Reinforced Epoxy Composites

Sandhyarani Biswas,[1] Amar Patnaik,[2] and Ritesh Kaundal[2]

[1] *Department of Mechanical Engineering, NIT, Rourkela 769008, India*
[2] *Department of Mechanical Engineering, NIT, Hamirpur 177005, India*

Correspondence should be addressed to Sandhyarani Biswas, biswas.sandhya@gmail.com

Academic Editor: Mohamed S. Aly-Hassan

In the present work, a series of bamboo-fiber-reinforced epoxy composites are fabricated by using red mud and copper slag particles as filler materials. A filler plays an important role in determining the properties and behavior of particulate composites. The effects of these two fillers on the mechanical properties of bamboo-epoxy composites are investigated. Comparative analysis shows that with the incorporation of these fillers, the tensile strength of the composites increases significantly, whereas the flexural strength and impact strength decrease with increase in filler content (red mud and copper slag fillers) in the epoxy-bamboo fiber composites. The density and hardness are also affected by the type and content of filler particles. It is found that the addition of copper slag filler improves the hardness of the bamboo-epoxy composites, whereas the addition of red mud filler reduces the hardness value of bamboo-epoxy composites. The study reveals that the addition of copper slag filler in bamboo-epoxy composites shows better physical and mechanical properties as compared to the red-mud-filled composites.

1. Introduction

In the recent years, for making the low-cost engineering materials the use of natural fibers in polymers composites has brought forth a lot of interest. The new environmental legislation as well as consumer demands has pressured manufacturing industries (especially automotive, packaging and construction) to search new materials that can replace the conventional nonrenewable reinforcing materials such as glass fibre, carbon fiber, and so forth [1]. The advantages of natural fibers over the traditional glass fibers are low density, good specific strength and modulus, economical viability, enhanced energy recovery, and good biodegradability [2]. There are also some disadvantages of natural fiber-reinforced polymer composites such as the incompatibility between the hydrophilic natural fibres and hydrophobic thermoplastic and thermoset matrices requiring appropriate use of physical and chemical treatments to enhance the adhesion between fibre and the matrix [3]. Generally the cellulosic fibers, like henequen, sisal, coconut fiber (coir), jute, palm, bamboo, wood, paper in their natural condition, and several waste cellulosic products such as shell flour, wood flour, and pulp have been used as reinforcement materials in different thermosetting and thermoplastic resins [4–11].

Being a conventional construction material since ancient times, bamboo fiber is a good campaigner for use as a natural fiber in composite materials. The reason that many studies focus on bamboo is because it is an abundant natural resource in Asia, and its overall mechanical properties are comparable to those of wood. Bamboo is a naturally occurring composite material, which grows abundantly in most of the tropical countries. This is one of the oldest building materials used by human kind. The attractive physical and mechanical properties that can be obtained with bamboo-fiber- reinforced composites, such as high specific modulus, strength, and thermal stability, have been well documented in the literature. Jain et al. [12] have compared the Bamboo fiber reinforced epoxy composites $(BFRP)_{epoxy}$

with Bamboo-fiber-reinforced unsaturated polyester composites. $(BFRP)_{USP}$ in terms of their cost and mechanical strength. Jindal [13] has observed that tensile strength of bamboo-fiber-reinforced plastic (BFRP) composite is comparatively equivalent to that of the mild steel, whereas their density is only 12% of that of the mild steel. Hence, the BFRP composites can be extremely useful in structural applications. Takagi and Ichihara [14] have studied the effect of fiber content and fiber length on mechanical properties of bamboo-fiber-reinforced green composites (BFGC). Fiber length up to 15 mm has given positive effect on tensile strength and flexural strength.

In the fiber-reinforced polymer composites, fillers play an important role. The addition of particulate fillers has shown a great promise and has been a subject of considerable interest due to the improvement in performance of polymers and their composites in industrial and structural applications due to the addition of fillers. Filler materials are used in composite materials to reduce material costs, to improve mechanical properties to some extent, and in some cases to improve processability. They also increase the properties like hardness and abrasion resistance and reduce the shrinkage. The physical and mechanical characteristics can further be modified by adding a solid filler phase to the matrix body during the composite preparation. Biswas and Satapathy [15] study the effect of red mud filler particles on the epoxy-bamboo fiber composites and epoxy-glass fiber composites. Also Biswas et al. [16] studied the effect of ceramic fillers on mechanical properties of bamboo-fiber-reinforced epoxy composites and results show that the addition of ceramic fillers in bamboo-fiber-reinforced epoxy composites enhanced the mechanical properties of the composites. Although a great deal of work has been published on the effect of various types and content of fillers on polymer composites, the potential of ceramic-rich industrial wastes for such use in polymeric matrices has rarely been explored. Industrial development over the last decades has generated huge amounts of toxic and hazardous inorganic waste like, fly ash, cement by pass dust, copper slag, red mud, and so forth, which contain appreciable amounts of hazardous elements.

Production of alumina from bauxite by the Bayer's process is associated with the generation of red mud as the major waste material in alumina industries worldwide. Depending upon the quality of bauxite, the quantity of red mud generated varies from 55 to 65% of the bauxite processed [17]. The enormous quantity of red mud discharged by these industries poses an environmental and economical problem. Red mud, as the name suggests, is brick red in colour and slimy having average particle size of about 80–100 μm. It comprises of the iron, titanium, and the silica part of the parent ore along with other minor constituents. Depending on the source, these residues have a wide range of composition: Fe_2O_3 20–60%, Al_2O_3 10–30%, SiO_2 2–20%, Na_2O_2 10%, CaO 2–8%, and TiO_2 traces 2–8%. Copper slag is produced during matte smelting and conversion steps in the pyro-metallurgical production of copper. During matte smelting, two separate liquid phases, copper-rich matte (sulphides) and slag (oxides) are formed. It has been estimated that for every tonne of refined copper produced, about 2.2 tonne of slag is generated and every year, approximately 24.6 million tonne of slag is generated in copper production worldwide. Slag containing <0.8% copper is either discarded as waste or sold as products with properties similar to those of natural basalt (crystalline) or obsidian (amorphous). Copper slag having average particle size of about 80–100 μm. The composition of the copper slag is as follows: Fe_2O_3 35.3%, SiO_2 36.6%, CaO 10%, Al_2O_3 8.1%, CuO 0.37%, MgO 4.38%, Na_2O 0.47%, K_2O 3.45%, PbO 0.12%, Zn 0.97%, and Cu 0.24%. The objectives of this study were to investigate the physical and mechanical properties of particulate (red mud and copper slag) filled epoxy-bamboo fiber hybrid composites and present the comparison between the effect of red mud and copper slag filler on bamboo-epoxy composites.

2. Preparation of Composites

The filler composites of four different compositions are prepared for this study. Bamboo fiber was collected from local supplier and is used as the common reinforcing phase in all the four compositions. The bamboo fiber has an elastic modulus 1.9 GPa and possesses a density of 1.32 g/cm^3. Epoxy resin (Elastic modulus 3.42 GPa, density 1.36 g/cm^3), manufactured by Ciba Geigy and locally supplied by Northern Polymers Ltd. New Delhi, India, is the matrix material. Here the red mud and copper slag are used as filler materials. The composites are made by conventional hand lay-up technique. The low temperature curing epoxy resin and corresponding hardener (HY951) are mixed in a ratio of 10 : 1 by weight as recommended. After that red mud and copper slag fillers are mixed in the epoxy resin before the bamboo fiber is reinforced in the matrix, the composites are prepared in two different sets. In first set, the red mud, (0, 5, 10, and 15 wt.%) is used as filler materials and mixed in the epoxy resin before the bamboo fiber is reinforced in the matrix. In second set, copper slag (0, 5, 10, and 15 wt.%) is used as a filler materials and mixed in the epoxy resin before the bamboo fiber is reinforced in the matrix. For all the composition, bamboo fiber-loading (weight fraction of bamboo fiber in the composite) is kept fixed, that is, 45 wt.% of bamboo fiber. The castings are put under load for about 24 h for proper curing at room temperature. Specimens of suitable dimension are cut using a diamond cutter for physical, mechanical characterization, and erosion testing.

3. Physical and Mechanical Characterization

3.1. Density. The theoretical density of the composite materials in terms of weight fraction can easily be calculated as in the following equation given by Agarwal and Broutman [18]:

$$\rho_{ct} = \frac{1}{\left(W_f/\rho_f\right) + \left(W_m/\rho_m\right)}. \quad (1)$$

In this equation, W and ρ represent the weight fraction and density of the fiber and matrix materials. Equation (1) is suitable only for two phase composite materials, but in the present investigation the composites consist of three

Table 1: Density and void fraction of red mud filled short epoxy-bamboo fiber composites.

Designation	Composition	Theoretical density (g/cm^3)	Experimental density (g/cm^3)	Voids fraction (%)
EBR-1	Epoxy + 45 wt.% Bamboo fiber + 0 wt.% red mud	1.17	1.15	1.71
EBR-2	Epoxy + 45 wt.% Bamboo fiber + 5 wt.% red mud	1.21	1.18	2.48
EBR-3	Epoxy + 45 wt.% Bamboo fiber + 10 wt.% red mud	1.23	1.17	4.88
EBR-4	Epoxy + 45 wt.% Bamboo fiber + 15 wt.% red mud	1.25	1.16	7.20

EBR: Red mud filled Bamboo fiber reinforced Epoxy composites.

Table 2: Density and void fraction of copper slag filled short epoxy-bamboo fiber composites.

Designation	Composition	Theoretical density (g/cm^3)	Experimental density (g/cm^3)	Voids fraction (%)
EBC-1	Epoxy + 45 wt.% Bamboo fiber + 0 wt.% copper slag	1.17	1.15	1.71
EBC-2	Epoxy + 45 wt.% Bamboo fiber + 5 wt.% copper slag	1.27	1.21	4.72
EBC-3	Epoxy + 45 wt.% Bamboo fiber + 10 wt.% copper slag	1.34	1.25	6.72
EBC-4	Epoxy + 45 wt.% Bamboo fiber + 15 wt.% copper slag	1.49	1.38	7.38

components, that is, matrix, fiber and particulate filler. Therefore, the theoretical density of the theses composites can be calculated easily by using the following modified form of (1) written as

$$\rho_{\text{ct}} = \frac{1}{\left(W_f/\rho_f\right) + \left(W_m/\rho_m\right) + \left(W_p/\rho_p\right)}, \quad (2)$$

where the suffix "ρ" indicates the particulate filler materials.

Whereas the actual density (ρ_{ce}) of the composites can be determined experimentally by using simple water immersion technique. The volume fraction of voids (V_v) in the composites is calculated using the following equation:

$$V_v = \frac{\rho_{\text{ct}} - \rho_{\text{ct}}}{\rho_{\text{ct}}}. \quad (3)$$

3.2. Micro-Hardness. Micro-hardness measurement is done using a micro-hardness tester equipped with a square-based pyramidal (angle 136° between opposite faces) diamond indenter by applying a load of 10 N.

3.3. Tensile Test. The tensile test is performed on flat specimens. During the tensile test, a uniaxial load is applied through both ends of the specimen. The ASTM standard test method for tensile properties of fiber resin composites has the designation D 3039-76. The length of the test section should be 200 mm. The tensile test is performed in the universal testing machine (UTM) Instron 1195, and results are analyzed to calculate the tensile strength of composite samples.

3.4. Flexural Strength. The flexural strength is calculated by three point bend test, and the data is recorded during the test. The flexural strength (F.S) of any composite specimen is determined using the following equation. The test is conducted as per ASTM standard (D2344-84) using the same universal testing machine (UTM)

$$\text{F.S} = \frac{3PL}{2bt^2}, \quad (4)$$

where P is maximum load, b the width of specimen and t the thickness of specimen, and L the span length of the sample.

3.5. Impact Test. Low velocity instrumented impact tests are carried out on composite specimens. The tests are done as per ASTM D 256 using an impact tester. The pendulum impact testing machine ascertains the notch impact strength of the material by shattering the V notched (45°) specimen with a pendulum hammer, measuring the spent energy, and relating it to the cross section of the specimen. The standard specimen for ASTM D 256 is 60 mm × 10 mm × 4 mm and the depth under the notch is 2 mm.

4. Results and Discussions

4.1. Effect of Filler Content on Void Faction of Particulate Filled Epoxy-Bamboo Fiber Composites. Determining the properties of the composites, density is the most important factor. Density generally depends upon the relative proportion of the matrix and the reinforcing materials. The theoretical and measured densities along with the corresponding volume fraction of voids for red mud filled and copper filled epoxy-bamboo fiber composites are presented in Tables 1 and 2. It is noticed that the density calculated theoretically using (1) is not in agreement with the experimental calculated density value. Therefore, the difference between the theoretical and experimental density value is the measure of the voids and pores present in the composites.

It is seen from Tables 1 and 2 that the volume fraction of voids increases with increasing the filler concentration in the epoxy-bamboo fiber composites. In case of 0 wt.% red mud and copper slag filled epoxy-bamboo fiber composite, the volume fraction of voids is less. However, with the addition

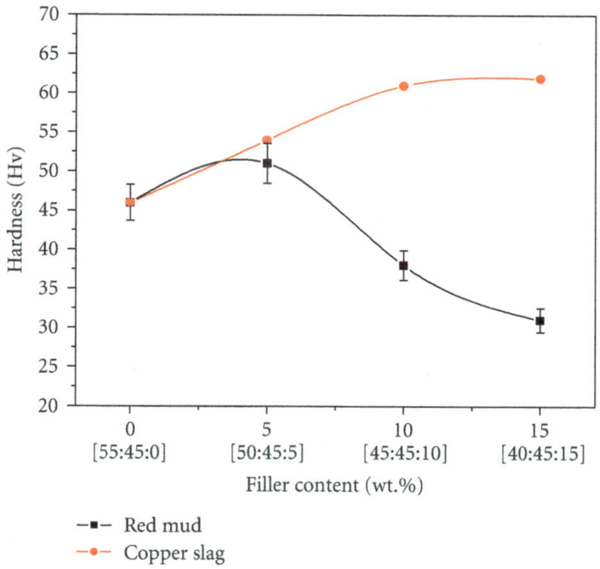

FIGURE 1: Variation of Micro-hardness with red mud and copper slag filler percentage.

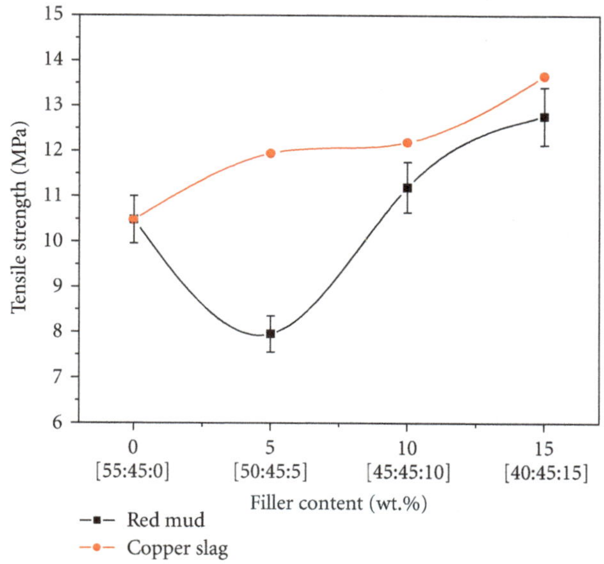

FIGURE 2: Variation of Tensile strength with red mud and copper slag filler percentage.

of filler materials, more voids are found in the composites. As the filler content increases from 5 wt.% to 15 wt.%, the volume fraction of voids is also found to be increasing proportionately. The composites filled with 15 wt.% of red mud and copper slag filled epoxy-bamboo fiber composites showd the maximum volume fraction of voids as compare to the others composites. Hence, based up on the above analysis, it is clearly demonstrated that with increase in filler concentration the void content of the composites goes on increasing. On comparison of red mud and copper slag filled epoxy-bamboo fiber composites, it is seen that the composites filled copper slag filler shows the higher values of voids as compared to the red mud filled composites. This is due to the higher density of the copper slag fillers as compared to the red mud filler.

The mechanical performance of the composites is generally influenced by the presence of voids, and the estimation of the quality of the composites is based upon the knowledge of void contents in the composites. The air-filled cavities formed inside the composite is known as porosity, and it is, often unavoidable part in all the composites during the fabrication process. The void may be originated during the mixing and integration of two or more different material parts. Higher void contents usually mean lower fatigue resistance, greater susceptibility to water penetration, and weathering [18]. It is understandable that a good composite should have fewer voids. However, presence of void is unavoidable in composite making particularly through hand-lay-up route.

4.2. Effect of Filler Content on Hardness of Particulate Filled Epoxy-Bamboo Fiber Composites. Surface hardness of the composites is considered as one of the most important factors that govern the nature and properties of the composites. The measured hardness values of both the red mud and copper slag filled and unfilled epoxy-bamboo fiber composites are shown in Figure 1. It is seen that in case of red mud filled composites, the hardness value decreases with increase in the filler contents.

Whereas in case of copper slag, the hardness value increases with increases in the fillers contents. In case of hardness test, a compression or pressing stress is in action. So the matrix phase and the solid filler phase would press together and touch each other more significantly. Thus the interface can transfer pressure more effectively although the interfacial bond may be poor.

4.3. Effect of Filler Content on Tensile Strength of Particulate Filled Epoxy-Bamboo Fiber Composites. Tensile properties of red mud and copper slag filled epoxy-bamboo fiber composites were studied. Tensile strength of the bamboo-epoxy filled with red mud and copper slag filler content ranged from 0 wt.% to 15 wt.% is shown in Figure 2. It is seen that the tensile strength of all the composites increases with increase in the both the fillers contents.

However, the composite filled with 5 wt.% of red mud filler contents shows the lower tensile strength value as compared to 0 wt.% of red mud filled epoxy-bamboo fiber composite. The decreased in tensile strength with 5 wt.% of red mud filler content is due to the poor adhesion between the fiber, filler, and the matrix, because without proper adhesion at higher loads reinforcements promote void formation. The particle loading and particle size, the particle, fiber and matrix interfacial adhesion significantly affects the strength of the composites. The effective stress transfer is the most important factor, which leads to the strength of composites materials. When the bonding between the particles and matrix is poor, the stress transfer at the particle/polymer interface is insufficient. Therefore, discontinuity in the form of debonding exists because of non-adherence of particle to polymer. Due to this (debonding),

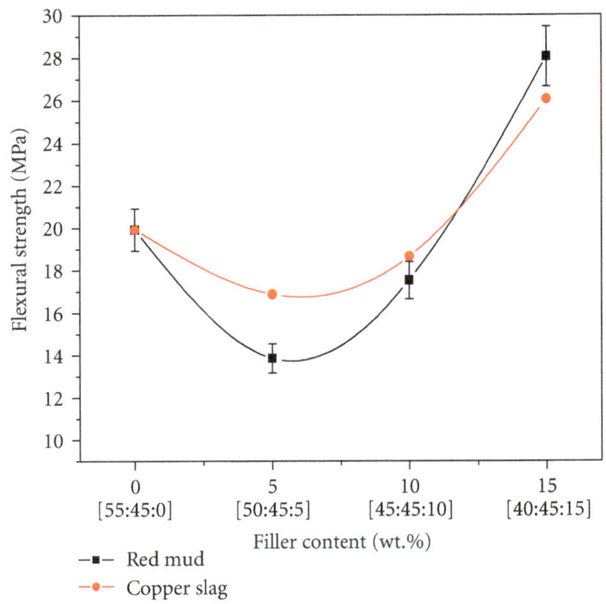

FIGURE 3: Variation of Flexural strength with red mud and copper slag filler percentage.

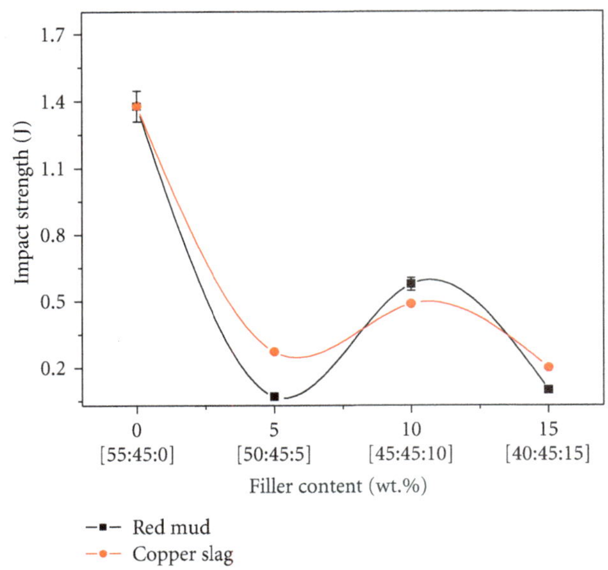

FIGURE 4: Variation of Impact Strength with red mud and copper slag filler percentage.

the particle cannot carry any load and the composite strength decreases with increasing particle loading [19]. However, for composites containing well-bonded particles, addition of particles to a polymer will lead to an increase in strength of the composites. So due to the good bonding of red mud and copper slag fillers particles with matrix, the tensile strength of all the composites increases with increases in the both the fillers contents.

4.4. Effect of Filler Content on Flexural Strength of Particulate Filled Epoxy-Bamboo Fiber Composites. Figure 3 shows the flexural strength of red mud and copper slag filled epoxy-bamboo fiber composites obtained experimentally from three bend tests.

It is observed that the epoxy-bamboo fiber composites filled with 5 wt.% and 10 wt.% of red mud and copper slag particles have the lower value of flexural strength as compared to the 0 wt.% of red mud and copper slag filled epoxy-bamboo fiber composites, whereas the composite filled with 15 wt.% of red mud and copper slag particles shows the higher value of flexural strength as compared to the 0 wt.% of red mud and copper slag filled bamboo-epoxy composites. This increase in the flexural strength may be related to the presence of red mud and copper slag particulates located at the interface of the fiber and matrix. The reduction in the flexural strengths of the composites with filler content is probably caused by an incompatibility of the particulates and the epoxy matrix, leading to poor interfacial bonding. The lower values of flexural properties may also be attributed to fiber-to-fiber interaction, voids, and dispersion problems. Also it is observed that the cooper slag filled composites show the better flexural strength as compare to the red mud filled bamboo epoxy composites.

4.5. Effect of Filler Content on Impact Strength of Particulate Filled Epoxy-Bamboo Fiber Composites. The ability of a material to resist breaking under a sudden loading or the ability of material to withstand the fracture under the action of high speed stress is known as the impact strength. In the polymer materials, the impact strength is the most widely specified mechanical property [20]. The impact performance of particulate filled fiber-reinforced composites depends on many factors like the nature of the constituent, the filler, fiber, matrix interface, the fabrication, geometry of the composite, and the test conditions. The impact failure of a composite occurs by factors like matrix fracture, and fibre/matrix debonding. The applied load transferred by shear to fibres may exceed the fibre/matrix interfacial bond strength and debonding occurs. When the stress level exceeds the fibre strength, fibre fracture occurs, which involves energy dissipation [21].

Figure 4 shows the impact strength of both the red mud and copper slag filled epoxy-bamboo fiber composites. The value of impact strength shows a reduction with the addition of both the red mud and copper slag in epoxy-bamboo fiber composites. It is seen that with the increase in the both the filler contents, the impact strength of all the composites decreases. Increased filler contents in the matrix resulted in composites becoming stiffer and harder. This will reduce the composite's resilience and toughness and lead to lower impact strength.

5. Conclusions

This study shows that successful fabrication of epoxy-bamboo fiber composites filled with red mud and copper sag particles is possible. Addition of these fillers modifies the physical (density and hardness) and mechanical properties (tensile, flexural, and impact strength) of the bamboo-epoxy composites. The tensile strength of both the red mud and copper slag filled composites increases with increase in the filler contents, whereas the flexural strength and impact strength of both the filled composites decrease with increase in the fillers filler contents. The micro-hardness and density of the composites are also greatly influenced by the type and content of fillers. In case of micro-hardness, with the addition of copper slag fillers the hardness value of epoxy-bamboo fiber composites increases, whereas with the addition of red mud filler the hardness value decreases with increase in the filler contents. Hence, while fabricating a composite of specific requirements, there is a need for the choice of appropriate filler material and for optimizing its content in the composite system.

Acknowledgment

The corresponding author is grateful to Department of Science and Technology, Govrnment of India for providing the financial support for this research work under the Research Project Ref. no. SR/FT/ET-026/2009.

References

[1] A. K. Bledzki and J. Gassan, "Composites reinforced with cellulose based fibres," *Progress in Polymer Science*, vol. 24, no. 2, pp. 221–274, 1999.

[2] J. Bolton, "The potential of plant fibres as crops for industrial use," *Outlook on Agriculture*, vol. 24, no. 2, pp. 85–89, 1995.

[3] J. Gassan and V. S. Gutowski, "Effects of corona discharge and UV treatment on the properties of jute-fibre expoxy composites," *Composites Science and Technology*, vol. 60, no. 15, pp. 2857–2863, 2000.

[4] H. Belmares, A. Barrera, E. Castillo et al., "New composite materials from natural hard fibres," *Industrial & Engineering Chemistry Product Research and Development*, vol. 20, pp. 555–561, 1981.

[5] C. A. Cruz-Ramos, E. M. Saenz, E. C. Bautista, M. D. Simposium, and N.-D Polimeros, Universidad Nacional Autonoma de Mexico, D.F., vol. 153, 1982.

[6] M. N. Casaurang-Martinez, S. R. Peraza-Sanchez, and C. A. Cruz-Ramos, "Dissolving-grade pulps from henequen fiber," *Cellulose Chemistry and Technology*, vol. 24, pp. 629–683, 1990.

[7] M. N. Cazaurang-Martinez, P. J. Herrera-Franco, P. I. Gonzalez-Chi, and M. Aguilar-Vega, "Physical and mechanical properties of henequen fibers," *Journal of Applied Polymer Science*, vol. 43, no. 4, pp. 749–756, 1991.

[8] S. Jain, R. Kumar, and U. C. Jindal, "Mechanical behaviour of bamboo and bamboo composite," *Journal of Materials Science*, vol. 27, no. 17, pp. 4598–4604, 1992.

[9] S. Varghese, B. Kuriakose, and S. Thomas, "Stress relaxation in short sisal-fiber-reinforced natural rubber composites," *Journal of Applied Polymer Science*, vol. 53, no. 8, pp. 1051–1060, 1994.

[10] G. Ahlblad, A. Kron, and B. Stenberg, "Effects of plasma treatment on mechanical properties of rubber/cellulose fibre composites," *Polymer International*, vol. 33, no. 1, pp. 103–109, 1994.

[11] V. G. Geethamma, R. Joseph, and S. Thomas, "Short coir fiber-reinforced natural rubber composites: effects of fiber length, orientation and alkali treatment," *Journal of Applied Polymer Science*, vol. 55, pp. 583–594, 1995.

[12] S. Jain, U. C. Jindal, and R. Kumar, "Development and fracture mechanism of the bamboo/polyester resin composite," *Journal of Materials Science Letters*, vol. 12, no. 8, pp. 558–560, 1993.

[13] U. C. Jindal, "Development and testing of bamboo fibers reinforced plastic Composites," *Journal of Composite Materials*, vol. 20, no. 1, pp. 19–29, 1986.

[14] H. Takagi and Y. Ichihara, "Effect of fiber length on mechanical properties of "green" composites using a starch-based resin and short bamboo fibers," *JSME International Journal A*, vol. 47, no. 4, pp. 551–555, 2004.

[15] S. Biswas and A. Satapathy, "A comparative study on erosion characteristics of red mud filled bamboo-epoxy and glass-epoxy composites," *Materials and Design*, vol. 31, no. 4, pp. 1752–1767, 2010.

[16] S. Biswas, A. Satapathy, and A. Patnaik, "Effect of ceramic fillers on mechanical properties of bamboo fiber reinforced epoxy composites: a comparative study," *Advanced Materials Research*, vol. 123–125, pp. 1031–1034, 2010.

[17] B. K. Mahapatra, M. B. S. Rao, R. B. Rao, and A. K. Paul, *Characteristics of Red Mud Generated at NALCO Refinery*, Light Metals, Damanjodi, India.

[18] B. D. Agarwal and L. J. Broutman, *Analysis and Performance of Fiber Composites*, John Wiley and Sons, 2nd edition, 1990.

[19] S. Y. Fu, X. Q. Feng, B. Lauke, and Y. W. Mai, "Effects of particle size, particle/matrix interface adhesion and particle loading on mechanical properties of particulate-polymer composites," *Composites Part B*, vol. 39, no. 6, pp. 933–961, 2008.

[20] S. J. Park and J. S. Jin, "Effect of silane coupling agent on interphase and performance of glass fibers/unsaturated polyester composites," *Journal of Colloid and Interface Science*, vol. 242, no. 1, pp. 174–179, 2001.

[21] J. L. Thomason and M. A. Vlung, "The influence of fibre length and concentration on the properties of glass fibre reinforced polypropylene: (7) interface strength and fibre strain in injection moulded long fibre PP at high fibre content," *Composites A*, vol. 28, pp. 277–288, 1997.

10

FSI Simulation of Common Carotid under Normal and High Blood Pressures

S. M. Abdul Khader,[1] Anurag Ayachit,[2] B. Raghuvir Pai,[1] V. R. K. Rao,[2] and S. Ganesh Kamath[3]

[1] *Department of Mechanical Engineering, MIT, Manipal 576104, India*
[2] *Department of Radiodiagnosis & Imaging, KMC, Manipal 576104, India*
[3] *Department of Cardiovascular & Thoracic Surgery, KMC, Manipal 576104, India*

Correspondence should be addressed to S. M. Abdul Khader, smak.quadri@gmail.com

Academic Editor: Hakan F. Oztop

There has been a tremendous progress in most of the computational domains in recent times. One of the key areas is analyzing the complex interaction of blood flow through elastic artery. Such study will aid in predicting the progression and behaviour of diseases like atherosclerosis and hypertension. In the present investigation, a simple straight common carotid artery having an eccentric blockage is analyzed for normal as well as high blood pressure condition. The clinical data has been obtained from ultrasound measurements on a volunteer patient and a model is generated, ANSYS, a commercially available Finite Element Analysis software. The ANSYS Fluid-Structure Interaction (FSI) solver is used to perform the coupled field transient analysis for three pulse cycles and the obtained results demonstrate the pulsatile flow. The simulation shows that the influence of hypertension has high deformation than normal case, severely altering the flow dynamics in downstream of the stenosis. The results obtained agree very well with available clinical observation of normal blood pressure condition.

1. Introduction

The simulation of blood flow has helped in investigating the complex interaction of flow dynamics within cardiovascular system. The altered haemodynamics as observed in development of cardiovascular disease such as atherosclerosis (stenosis) and aneurysms further influences the development of the disease and arterial deformity and changes the regional blood rheology. The study of physiologically realistic pulsatile flow through stenosis has profound implications for the diagnosis and treatment of vascular disease. Although the actual geometry of a stenotic artery may be varied and complex, it has been frequently simplified as symmetrical or asymmetrical constriction in a cylindrical tube [1]. The flow through constricted tubes is generally characterized by a high velocity jet generated from the narrowest section and flow separation distal to the stenosis. The characteristics of the prestenotic flow is usually laminar; however the flow in poststenotic is extremely turbulent depending upon the flow conditions and severity of the stenosis. The accurate simulation of realistic physiological phenomenon is possible only through FSI and using CFD alone will not be beneficial to demonstrate the interaction of blood flow through elastic artery.

There are numerous FSI studies conducted in the past assuming both the simple and complex geometry models generated from *in vivo* data. Initially, the flow dynamics in poststenotic region was demonstrated on a simple straight tube stenosed model [2]. In this study, 25%, 50%, and 75% axisymetrical and symmetrical models were analyzed. A highly complex flow is observed in the downstream side of stenosis leading to flow separation as observed from simulated results. Later, the flow interaction with elastic structure was adopting using interactive approach. This study was more realistic as experimental data was used as boundary condition for computation [3]. Similar study was conducted on stenotic arteries using both thick and thin wall model. The simulated results revealed that severity of stenosis and asymmetry in thick-wall models changed not only the wall geometry, but also the stiffness of the tube wall which in turn affected the wall deformation [4].

Since the arterial wall has a complex structure, it cannot be treated as a homogenous material. Considering this fact, FSI simulation of blood flow in carotid was conducted on a nonlinear thick-wall model [5]. The obtained results were more significant compared to analysis performed considering linear model. The asymmetric stenosis was studied to quantify the effects of stenosis severity, eccentricity, and pressure conditions on blood flow and artery compression. It is observed that, particularly in poststenotic region, the velocity field is highly dependent on the flow waveform along with the uniform distribution of flow recirculation along the circumference of the walls. Also the strong flow reversal and recirculation were found to be uniformly distributed around the circumference of the walls in the downstream of stenosis. From the available literature, it is observed that most of the studies have been performed at normal blood pressure. But the important aspect of flow dynamics in the downstream of stenosis under the influence of high blood pressure is yet to be studied as it can prove to be one of the major risk factors [6]. The flow resistance in arteries increases abruptly due to the influence of high-grade stenosis. This forces the body to raise the blood pressure to maintain the necessary blood supply. The constriction and high blood pressure causes high flow velocity, high wall shear stress, and low pressure at the throat of the stenosis, while in the distal side of stenosis, there will be low wall shear stress, flow separation, and wall compression.

In the present study, a patient specific case is investigated having eccentric stenosis using normal blood pressure. The results are compared with results of the influence of high blood pressure, so as to prognose the crucial factors. The results indicate the significant changes with hypertension. Considerable changes in the wall deformation, wall shear stress, flow behaviour in the downstream of the stenosis and flow separation zone are observed.

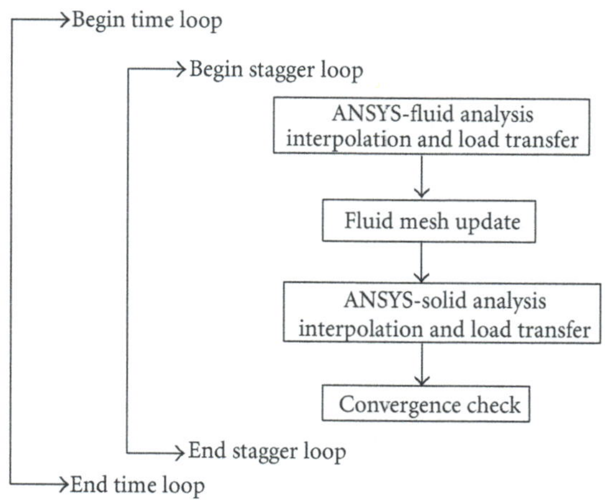

FIGURE 1: FSI algorithm.

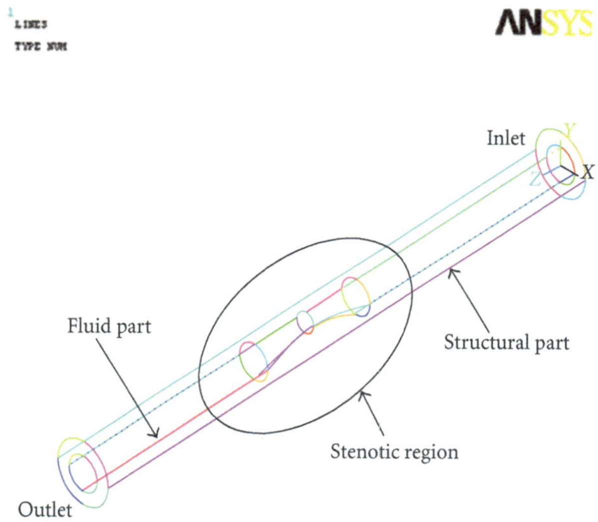

FIGURE 2: Stenosed model.

2. Methodology

The blood flow in carotid artery is assumed to be laminar, incompressible and governed by the Navier-Stokes equations of incompressible flows. The fluid domain is solved using modified momentum equation adopting moving velocity along with continuity equation as given in equation below [7, 8]:

$$\frac{\partial}{\partial t}\int_{\Omega}\rho\partial\Omega+\int_{S}\rho(v-v_b)\cdot n\partial S = \int_{S}\left(\tau_{ij}i_j-Pi_i\right)\cdot n\partial S+\int_{\Omega}b_i\partial\Omega, \tag{1}$$

where ρ is the density, τ is the stress tensor, v is the velocity vector, vb is the grid velocity, P is the pressure, and b_i is the body force at time t.

The artery wall is assumed to be elastic, isotropic, incompressible, and homogeneous. The transient dynamic structural solution is given by the following equation [4]. The stiffness matrix is updated in each time step. The Newmark method is used for updating of displacement at each time interval and then stiffness matrix is solved using direct solver in particular sparse solver, for every time step:

$$[M]\{\ddot{U}\}+[C]\{\dot{U}\}+[K]\{U\}=\{F^a\}, \tag{2}$$

where M is the structural mass matrix, C is the structural damping matrix, K is structural stiffness matrix, F_a is the applied load vector, and $\ddot{U}$, $\dot{U}$, and U represent acceleration, velocity, and displacement vector.

2.1. FSI Algorithm. The FSI analysis is performed using sequentially coupled FSI solver in ANSYS [9, 10]. This solver uses ANSYS FLOTRAN elements for fluid domain and ANSYS structural for solid domain. The FSI algorithm is shown in form of flow chart in Figure 1. In the beginning of the FSI analysis, first fluid domain has to be modeled using appropriate element then required boundary conditions are applied followed by solid domain.

The FSI algorithm solves the fluid and solid domains independently. After fluid domain is solved, the surface load to the solid domain is transferred through fluid-structure interface and solid domain gets solved. After the convergence is achieved, the fluid mesh has to be morphed using Arbitrary Lagrangian-Eulerian (ALE) formulation. The algorithm continues to loop through the solid and fluid domain until convergence is reached for that time step. Later the loop continues to solve for the time period specified.

3. Modeling and Analysis

In the patient, the right side common carotid is normal and left side has 66% eccentric stenosis. The model of stenosed common carotid is generated in ANSYS-11.0 as shown in Figure 2. The model is generated based on the details taken at different sections depending upon the convenience using M-mode scan for velocity and B-mode scan for diastolic diameter [3, 9]. The generated stenosed model has a diastolic diameter of 6.5 mm, total length of 100 mm, throat of stenosis is at a distance of 40 mm from inlet, and artery wall has a thickness of 0.7 mm.

Both the fluid and the structural parts are modelled with 8 nodded brick element and the grid distribution across the cross-section of stenosis is shown in the Figure 3. Initially, a static FSI analysis was carried to perform grid independency test. A cross-section at throat of stenosis was considered to calculate the average pressure and maximum wall shear stress and the observed variation is shown in the Figure 4. Finally, the fluid and structural models with 50000 and 12000 hexahedral elements were considered for further transient FSI simulation.

In this study, the inlet boundary condition is set by a velocity pulse at the inlet for a period of 0.8 sec as shown in Figure 5 and outlet boundary condition is set by a pulse pressure, applied at the outlet as shown in Figure 6. Each pulse cycle is discretized into 100 time steps to simulate the flow behaviour more accurately. The range of pulse pressure is different as the simulation is carried out for both normal pressure and high pressure having 80–125 mm Hg and 100–170 mm Hg, respectively, as shown in Figure 6 [6]. The inlet and outlet of solid model are constrained by specifying zero displacement in all the directions and the rest of the nodes are left free to undergo displacement in any direction [10, 11].

The blood is considered to be Newtonian in the present study as the focus is on large arteries where velocity and shear rate will be high. The density and dynamic viscosity of the blood are considered to be 1050 kg/m^3 and 0.004 N-sec/m^2, respectively. The artery is assumed to be linearly elastic material, having density of 1120 kg/m^3, Poisson's ratio of 0.40, and elastic modulus is 0.9 Mpa [10, 11]. Since the patient is an aged person, the arterial wall is stiffened and elasticity is less when compared with that of healthy young person [12, 13]. A generalized ANSYS Parametric Development Language (APDL) written in ANSYS generates the geometric model solves and performs the postprocessing of the results. The convergence criteria of fluid flow and

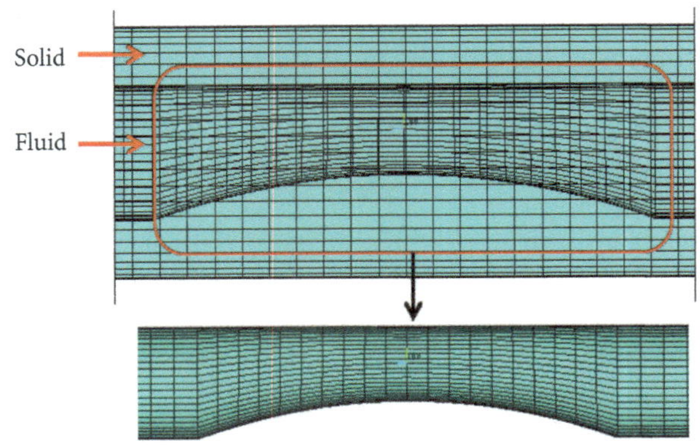

Figure 3: Grid distribution across stenosis.

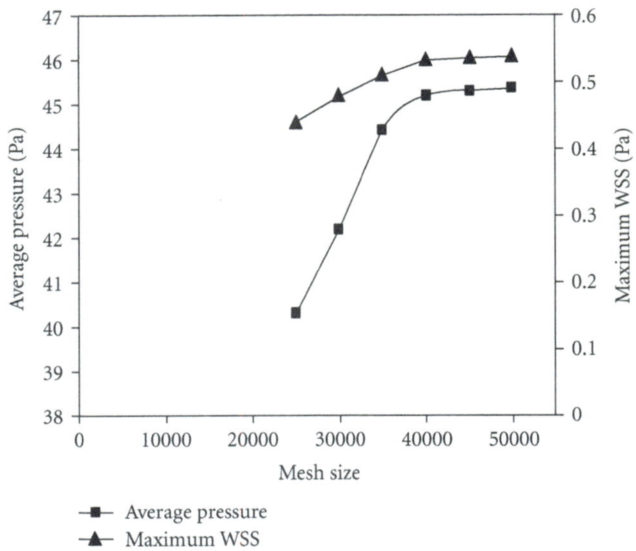

Figure 4: Grid independency results.

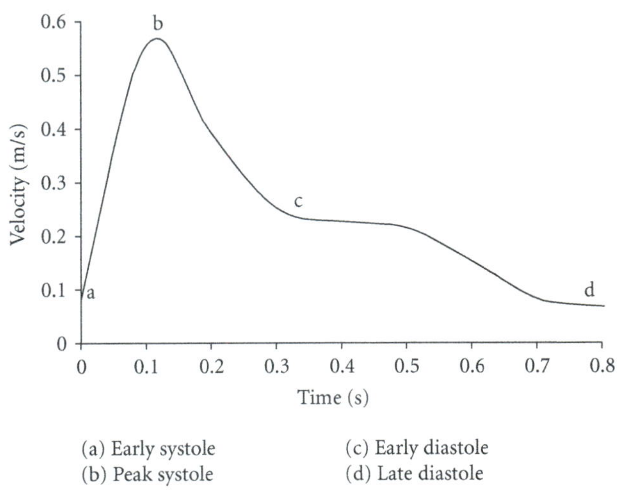

Figure 5: Input velocity BC.

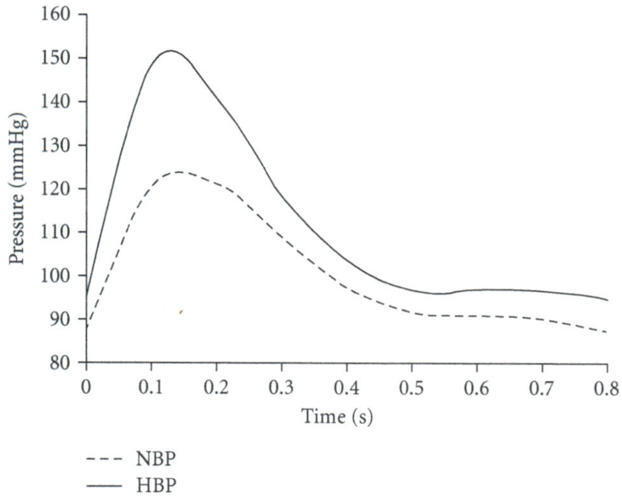

FIGURE 6: Output pressure BC.

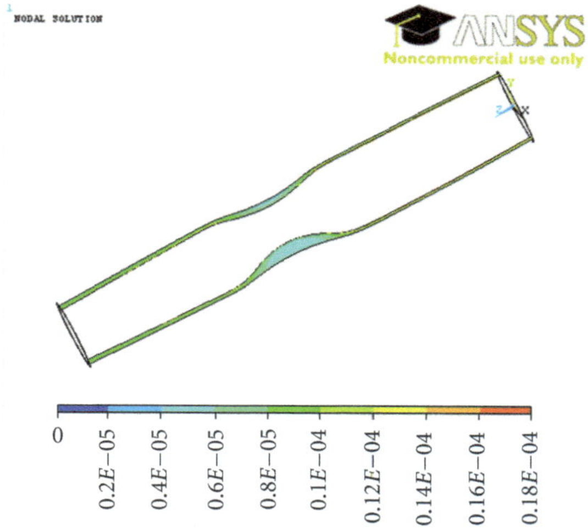

FIGURE 7: Wall deformation at early systole in NBP.

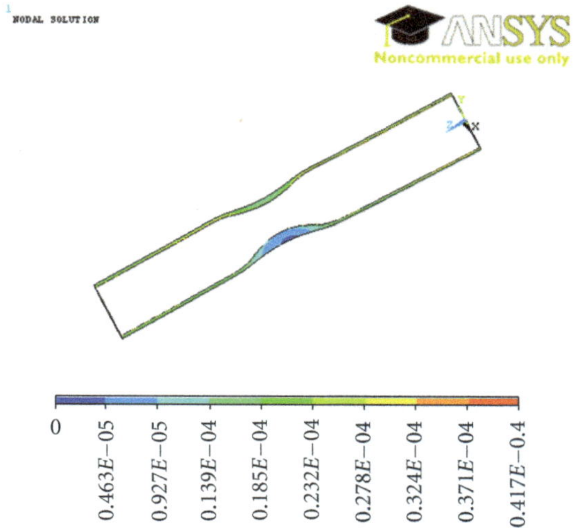

FIGURE 8: Wall deformation at peak systole in HBP.

across the fluid-surface interface are set at 10^{-4} and 10^{-3}, respectively.

4. Results and Discussion

The simulated results obtained for normal blood pressure (NBP) and high blood pressure (HBP) conditions are compared to observe the variations in the flow behaviour. The velocity contours, wall displacement, and wall shear obtained at specified instants of pulse cycle are considered for observation. The computed results like numerical values of deformation, wall shear stress, and velocity patterns are normalized to facilitate the comparison. The velocity, WSS, and wall deformation are found to be maximum during peak systole due to high flow rate.

The wall deformation patterns are shown in Figures 7 and 8 during early systole and peak systole for NBP and HBP conditions, respectively. The maximum wall deformation is within 10% and 16% of diastolic diameter for NBP and HBP conditions, respectively. The wall deformation is uniform in upstream as well as in downstream of the model apart from the throat region during the entire pulse cycle [11]. There is significant difference in wall deformation due to HBP condition, which is maximum during peak systole when compared with NBP case. The nonuniform structural deformation is observed to be severe in poststenotic region than in prestenotic region. Due to the instant increase in the pressure, the wall collapsibility is observed after peak systole and further increases during later part of the pulse cycle because of low pressure gradient.

The velocity contours during peak systole for NBP and HBP conditions are shown in Figures 9 and 10, respectively. The characteristic of stenotic flow shows a sudden increase in velocity and drop in pressure at the throat during the entire pulse cycle [14]. The flow jet covers a longer distance in HBP compared to NBP due to large pressure gradient. The flow is more predominant even during later part of pulse cycle in HBP compared to NBP. However, there is no significant difference in the flow rate for both the NBP and HBP cases. Figure 11 shows the velocity contour during early diastole in HBP and velocity profile during late diastole in NBP is shown in Figure 12. Due to the eccentric stenosis, flow separation in downstream of stenosis is intense in case of HBP than in NBP leading to severe eddy formation.

During the later part of the pulse cycle, the flow in downstream side stabilizes within shorter distance in NBP compared to HBP. Due to flow deceleration, the back flow is observed from early diastole in NBP, but in case of HBP this effect is reduced instantly because of large pressure gradient.

The wall deformation is compared during NBP and HBP as shown in Figure 13. The maximum displacement is found to be in HBP condition. The pattern of the displacement is almost similar to the given input pulse cycle, which shows the pulsatile behaviour. The maximum difference is seen during peak systole and in early part of diastole than in any other part of the cycle. Due the high pressure, the time taken for flow stabilization is high in HBP compared to NBP. Also due to wall collapsible behaviour during

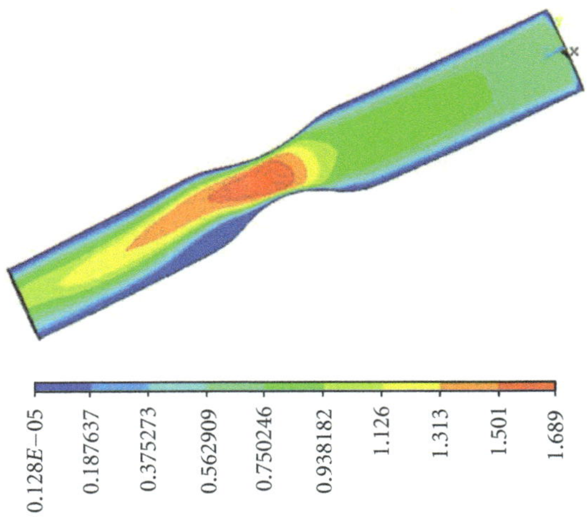

Figure 9: Velocity contour at peak systole in NBP.

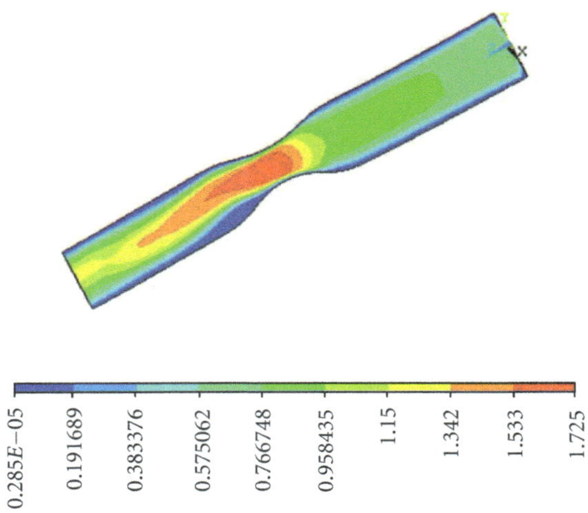

Figure 10: Velocity contour at peak systole in HBP.

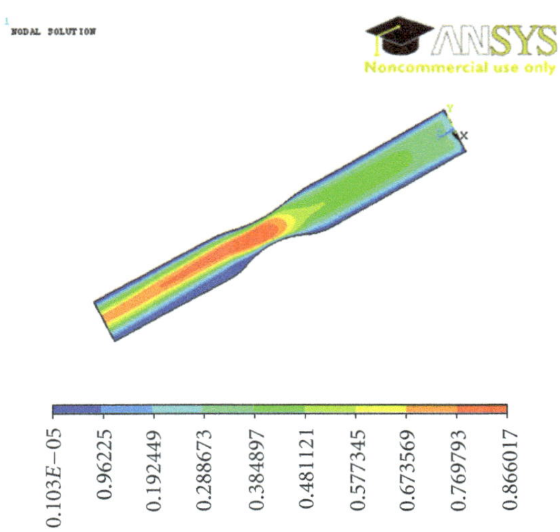

Figure 11: Velocity contour at early diastole in HBP.

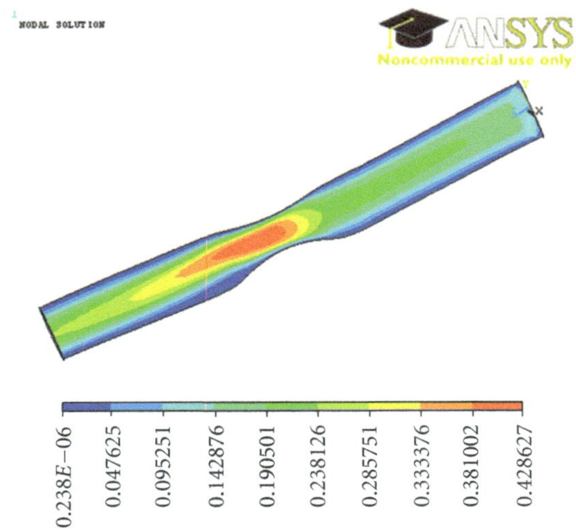

Figure 12: Velocity contour at late diastole in NBP.

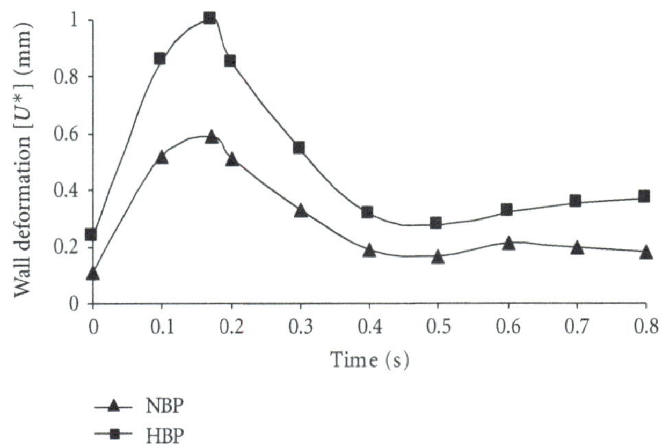

Figure 13: Wall deformation comparison.

late diastole, the difference is less and almost the same between NBP and HBP [6, 14]. Due to this, the effect on flow dynamics in downstream is significant causing high turbulence. Figure 14 shows the comparison of WSS during NBP and HBP conditions and there is a minor difference because of the same flow rate in both cases. The maximum WSS is at the throat region compared to upstream and downstream side of the model. There is a sudden increase in WSS during peak systole in both the HBP and NBP cases. The variation of maximum WSS is similar to pulse cycle, but the gradual reduction lags by few milliseconds as compared to normal input velocity cycle due to increase in velocity in the throat region. The difference in pressure gradient is less in early part of diastole and because of this flow WSS is constant during this phase of the cycle. The results obtained in NBP agree well with the clinically observed results and with available literature. From the obtained results, it is observed that, including the effect of hypertension can simulate the risk or prognosis of arterial disease and it cannot be neglected because it has profound influence on altering the flow dynamics, WSS distribution, stress levels, and wall deformation.

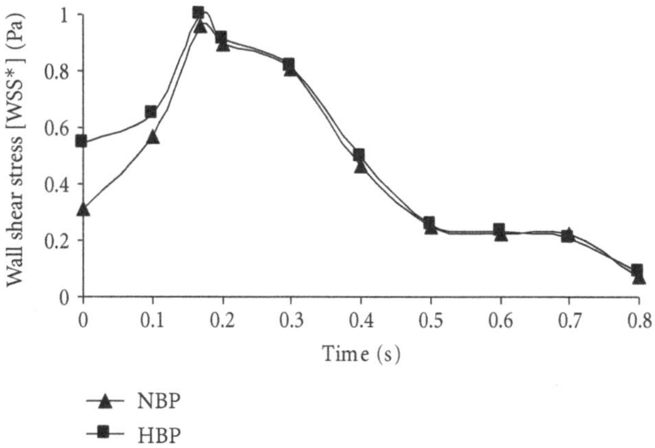

FIGURE 14: Wall shear stress comparison.

5. Conclusions

The effect of hypertension is simulated for an eccentric stenosed model in the present study. The variations in velocity distribution, WSS, wall deformation, pressure are compared for NBP and HBP conditions. The poststenotic deformation and WSS is found to be maximum during peak systole and increased velocity at the throat region. In both the NBP and HBP cases, the wall collapsibility is observed from the early diastole which influences the intense flow separation for later part of pulse cycle. The flow is highly turbulent in downstream especially in the downstream of the throat region. The results demonstrate that HBP causes the significant changes in the wall deformation, stress distribution and flow especially in downstream during early diastole part of the pulse cycle than in NBP condition. The variation of wall deformation in the poststenotic region and the velocity contours obtained under the NBP conditions agree well with the clinically observed results.

Nomenclature

CCA: Common carotid artery
U^*: Normalized wall displacement in mm = $U_t/U_{\max}$
U_t: Radial displacement at time, t in mm
$U_{\max}$: Maximum radial displacement in time period in mm
WSS*: Normalized wall shear stress in Pa = $WSS_t/WSS_{\max}$
WSS_t: WSS at time, t in Pa
$WSS_{\max}$: Maximum WSS in time period in Pa
S: Closed surface
Ω: Control volume.

References

[1] M. Bathe and R. D. Kamm, "A fluid-structure interaction finite element analysis of pulsatile blood flow through a compliant stenotic artery," *Journal of Biomechanical Engineering*, vol. 121, no. 4, pp. 361–369, 1999.

[2] Q. Long, X. Y. Xu, K. V. Ramnarine, and P. Hoskins, "Numerical investigation of physiologically realistic pulsatile flow through arterial stenosis," *Journal of Biomechanics*, vol. 34, no. 10, pp. 1229–1242, 2001.

[3] K. Perktold and G. Rappitsch, "Computer simulation of local blood flow and vessel mechanics in a compliant carotid artery bifurcation model," *Journal of Biomechanics*, vol. 28, no. 7, pp. 845–856, 1995.

[4] D. Tang, C. Yang, S. Kobayashi, J. Zheng, and R. P. Vito, "Effect of stenosis asymmetry on blood flow and artery compression: a three-dimensional fluid-structure interaction model," *Annals of Biomedical Engineering*, vol. 31, no. 10, pp. 1182–1193, 2003.

[5] D. Tang, C. Yang, Y. Huang, and D. N. Ku, "Wall stress and strain analysis using a three-dimensional thick-wall model with fluid-structure interactions for blood flow in carotid arteries with stenoses," *Computers and Structures*, vol. 72, no. 1, pp. 341–356, 1999.

[6] R. Torii, M. Oshima, T. Kobayashi, K. Takagi, and T. E. Tezduyar, "Fluid-structure interaction modeling of aneurysmal conditions with high and normal blood pressures," *Computational Mechanics*, vol. 38, no. 4-5, pp. 482–490, 2006.

[7] J. Ferziger and M. Peric, *Computational Methods for Fluid Dynamics*, Springer, Berlin, Germany, 2002.

[8] Y. Fung, *Biodynamics-Circulation*, Springer, New York, NY, USA, 1984.

[9] *ANSYS Release 11.0 Documentation*, ANSYS Company, Pittsburgh, Pa, USA, 2008.

[10] Z. Li and C. Kleinstreuer, "Fluid-structure interaction effects on sac-blood pressure and wall stress in a stented aneurysm," *Journal of Biomechanical Engineering*, vol. 127, no. 4, pp. 662–671, 2005.

[11] C. A. Figueroa, I. E. Vignon-Clementel, K. E. Jansen, T. J. R. Hughes, and C. A. Taylor, "A coupled momentum method for modeling blood flow in three-dimensional deformable arteries," *Computer Methods in Applied Mechanics and Engineering*, vol. 195, no. 41–43, pp. 5685–5706, 2006.

[12] R. G. Gosling and M. M. Budge, "Terminology for describing the elastic behavior of arteries," *Hypertension*, vol. 41, no. 6, pp. 1180–1182, 2003.

[13] W. A. Riley, R. W. Barnes, G. W. Evans, and G. L. Burke, "Ultrasonic measurement of the elastic modulus of the common carotid artery: the atherosclerosis risk in communities (ARIC) study," *Stroke*, vol. 23, no. 7, pp. 952–956, 1992.

[14] S. Ahmed, I. D. Sutalo, and H. Kavnoudias, "Fluid-structure interaction modelling of a patient specific cerebral aneurysm: effect of hypertension and modulus of elasticity," in *Proceedings of the 16th Australian Fluid Mechanics Conference*, Crown Plaza, Gold Cost, Australia, 2007.

11

Temperature Dependence of Ascending Bubble-Driven Flow Patterns Found in Champagne Glasses as Determined through Numerical Modeling

Fabien Beaumont,[1] Catalin Popa,[1] Gérard Liger-Belair,[2] and Guillaume Polidori[1]

[1] *Laboratoire de Thermomécanique, GRESPI-EA 4301, Université de Reims, 51687 Reims Cedex 2, France*
[2] *Equipe Effervescence, Champagne et Applications, Groupe de Spectrométrie Moléculaire et Atmosphérique UMR CNRS 7331, Université de Reims, 51687 Reims Cedex 2, France*

Correspondence should be addressed to Fabien Beaumont; fabien.beaumont@univ-reims.fr

Academic Editor: Bo Yu

A numerical modeling of bubble-driven flow patterns in a glass of champagne has been carried out for three champagne temperatures, by using the finite-volume method by CFD (computational fluid dynamics). In order to define source terms for flow regime and to reproduce accurately the bubble nucleation process responsible for champagne effervescence, specific subroutines for the gaseous phase have been added to the main numerical model. These subroutines allow the modeling of bubbles behavior based on semiempirical formulas relating to bubble diameter, mass transfer, velocity, and drag force. Both ascending bubble dynamics and bubble-driven flow patterns dynamics were examined, respectively, 60 s, 180 s, and 300 s after pouring champagne into the glass. Details and development of the various steps of modeling are presented in this paper, showing that the bubble-driven flow patterns velocities of the liquid phase significantly vary with the champagne temperature.

1. Introduction

Ascending bubbles are indeed very common in our everyday life. They play a crucial role in many natural as well as industrial processes (in physics, chemical and mechanical engineering, oceanography, food science, and even medicine) [1–6]. Nevertheless, their behavior is often surprising and, in many cases, still not fully understood. In the science of champagne, much interest has been devoted to bubble dynamics in the past decade. A large piece of experimental works has focused on what could be called the bubble trilogy, namely, the bubble nucleation, the bubble rise, and the bubble collapse, as they reach the champagne surface (for a global review see [7] and references therein).

Since the end of the 17th century, champagne has been a worldwide renowned French sparkling wine. From a strictly physicochemical point of view, champagne wines are multicomponent hydroalcoholic systems, with a density close to unity, a surface tension $\gamma \approx 50\,\mathrm{mN\,m^{-1}}$ (indeed highly ethanol dependent), and a viscosity about 50% larger than that of pure water (also mainly due to the presence of 12,5% v/v ethanol) [7]. Champagne wines are also supersaturated with CO_2-dissolved gas molecules formed together with ethanol during the second fermentation process (called *prise de mousse*, and promoted by adding yeasts and a certain amount of sugar inside bottles filled with a base wine and sealed with a cap). It is worth noting that a standard 75-centiliters champagne bottle typically holds about 9 grams of dissolved CO_2, which correspond to a volume close to 5 liters of gaseous CO_2 under standard conditions for temperature and pressure. No wonder champagne and sparkling wine tasting mainly differs from still noneffervescent wine tasting due to their large amount of dissolved carbon dioxide.

When a bottle is uncorked, and champagne is poured into a glass, there are indeed two pathways for progressive CO_2 and volatile organic compounds (VOCs) losses. CO_2 and VOCs escape (i) into the form of heterogeneously nucleated bubbles, the so-called *effervescence* process and (ii) by "invisible" diffusion through the champagne/air interface [8–10]. Glass shape, and especially its open aperture, is therefore also

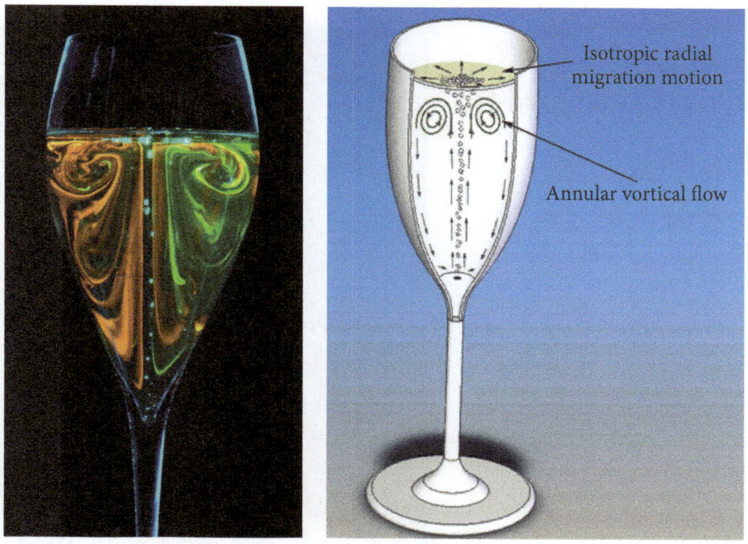

FIGURE 1: Flow visualization evidenced through laser tomography technique (a); and 3D scheme of flow patterns found in a glass (b).

suspected to play an important role as concerns the kinetics of CO_2 and flavor release during champagne tasting. From the taster's point of view, the perception of wine flavors was indeed found to be influenced by glass shape [11, 12]. From the consumer point of view, the role of effervescence is indeed essential in champagne, sparkling wines, beers, and to a great extent, in any other carbonated beverage [13–15]. Without bubbles, champagne would be unrecognizable as such, beers and sodas would be flat. However, the role of effervescence is suspected to go far beyond the solely aesthetical point of view. Effervescence and dissolved CO_2 impact champagne and sparkling wine tasting in terms of (i) visual perception, by the presence of dancing bubbles in the glass, (ii) taste and mouthfeel, by the fizzy and chemosensory excitation of nociceptors in the oral cavity (via the conversion of dissolved CO_2 to carbonic acid) [16–18], and (iii) aromatic perception, as bubbles release their content in gaseous CO_2 and VOCs above the champagne surface. Furthermore, ascending bubbles were strongly suspected to play an indirect role on the release of gaseous CO_2 and VOCs above the champagne surface. Actually, it is worth noting that, at the bubble scale, the lower part of a rising bubble is a low-pressure area which literally attracts the fluid molecules around. A rising bubble is thus able to drain by viscous effect some fluid along its path toward the free surface. The huge number of bubbles released from the several nucleation sites found in a typical champagne glass (of order of several hundreds of bubbles released per second during the first minutes of champagne tasting) was therefore suspected to set the whole liquid bulk in motion. In turn, the release of gaseous CO_2 by diffusion through the free air/champagne interface, as well as the release of the numerous VOCs found in the liquid phase (which both strongly depend on the mixing flow conditions of the liquid medium), should be logically enhanced in case of ascending bubble-driven flow patterns. For all the aforementioned reasons, a very strong coupling was therefore strongly suspected in champagne and sparkling wines tasting, between dissolved CO_2, the presence of rising bubbles, glass shape, CO_2 discharge, and VOCs release [19].

Recently, the simultaneous quantification of CO_2 and ethanol in the headspace of a champagne glass was monitored, in real tasting conditions, all along the first 15 minutes following pouring, depending on whether the glass shows effervescence or not. Both CO_2 and ethanol were found to be enhanced by the presence of ascending bubbles, thus confirming the close link between rising bubbles and the release of gaseous CO_2 and volatile organic compounds [7]. Only quite recently, laser tomography techniques combined with fluorescent continuous dyes or solid discrete tracers were used to evidence ascending bubble-driven flow patterns found in champagne glasses [20–22]. Flow patterns found in champagne glasses were characterized by a vortex ring showing axisymmetry with regard to the ascending bubble column [23–25], as illustrated in Figure 1. Very recently also, numerical modeling was used in order to approach bubble-driven flow patterns found in champagne glasses. The influence of the glass geometry was investigated [26]. From a numerical point of view, the two most common approaches to model bubble columns are the Euler-Euler (called also two-fluid approach) and the Euler-Lagrange (called also discrete phase) ones [1, 2]. In the Euler-Euler approach, the continuous liquid phase and the dispersed gas phase are modeled as two interpenetrating continuous phases. In the Euler-Lagrange approach, the volume-averaged Navier-Stokes equations are used to describe the motion of the liquid phase [2]. In the latter situation, each bubble is tracked on the basis of a balance of forces acting upon bubbles and on a motion equation [26]. Even if the dynamics of individual bubbles are now well understood, it seems that there is a lack of data in the study of bubble-driven flow patterns in beverages supersaturated with dissolved gas. Moreover, temperature is strongly suspected as being a key parameter in champagne

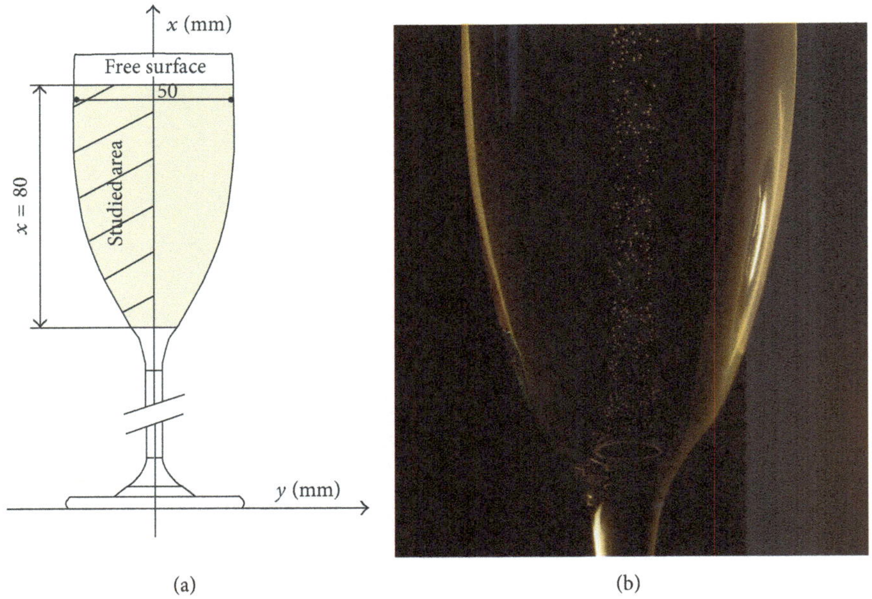

FIGURE 2: Champagne flute model used in our study (a); inspired from a real flute etched by the glassmaker on its bottom to promote bubble formation (b); the small ring etched on its bottom has been done with adjoining laser beam impacts.

tasting. The temperature indeed modifies both the liquid viscosity and the vapor pressure of volatile aromatic compounds.

In this paper, the influence of champagne temperature on ascending bubble-driven flow patterns found in a very typical champagne flute was investigated, through numerical modeling. Both bubble dynamics and bubble-driven flow patterns in the champagne bulk were followed with time, during the first 300 seconds after champagne was poured into the flute, and for three champagne temperatures (namely, 4, 12, and 18°C).

2. Materials and Methods

2.1. Geometry and Mesh Generation. A traditional flute has been considered as a reference glass case. The glass geometry used in this study has been created from the real dimensions measured of the glass used for the experiments (see Figure 2(a)).

To ensure a continuous and perfectly controlled process of effervescence, glassmakers usually consider circulary etched glasses (Figure 2(b)). In such a way, as previously mentioned, the flow structure exhibits a quasi-steady two dimensional behavior [22] (Figure 1).

In this situation, a 2D examination in the axis symmetry plane can be considered as sufficient. For this purpose, only half of the studied area was drawn. The study area has a total height of 80 mm which corresponds to the fill level and its diameter is 50 mm at the liquid surface level (Figure 2(a)).

The ANSYS Workbench Design Modeler software has been used to draw the geometry from the real size (scale 1 : 1) of the numerical study glass used as reference.

The mesh of the domain has been carried out using the ANSYS Workbench meshing software. It consists in a two-dimensional mesh efficient in the case of simulations of axisymmetrical flow features. The body has been meshed with quadrilateral elements (0.2 × 0.2 mm). Along the central part of the flow, the mesh is structured but to follow the wall, the mesh is unstructured (Figure 3).

2.2. Equations and Numerical Scheme. Because champagne is a wine in which gaseous as well as liquid phase is simultaneously present, the flows in a glass of champagne have been simulated numerically with a multiphase model. The flow is supposed laminar [19, 27] and transient, so the liquid-phase hydrodynamics are described with the continuity and momentum conservation equations for laminar flows.

2.2.1. Continuity Equation. The general form of the continuity equation can be written as follows:

$$\frac{\partial \rho}{\partial t} + \nabla \cdot (\rho \vec{v}) = S_m. \tag{1}$$

The source S_m is the mass added to the continuous phase from the dispersed second phase.

For 2D axisymmetric geometries, the continuity equation is given by

$$\frac{\partial \rho}{\partial t} + \frac{\partial}{\partial x}(\rho v_x) + \frac{\partial}{\partial r}(\rho v_r) + \frac{\rho v_r}{r} = S_m, \tag{2}$$

where x is the axial coordinate, r is the radial coordinate, v_x is the axial velocity, and v_r is the radial velocity.

2.2.2. Navier-Stokes Equations. Conservation of momentum is described by

$$\frac{\partial}{\partial t}(\rho \vec{v}) + \nabla \cdot (\rho \vec{v}\vec{v}) = -\nabla p + \rho \vec{g} + \vec{F}, \tag{3}$$

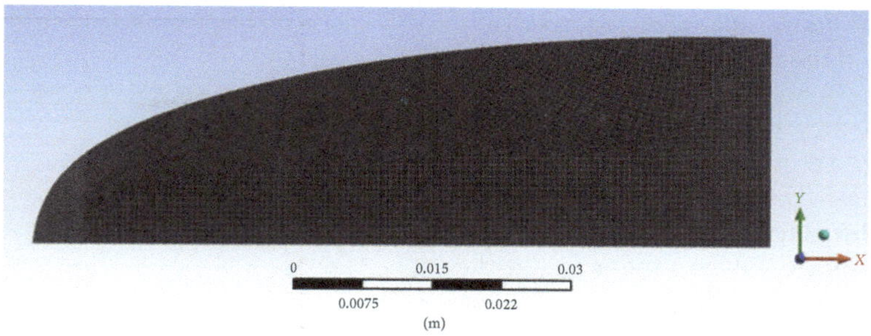

FIGURE 3: Mesh of the half fluid domain.

where ρ is the static pressure and $\rho\vec{g}$ and $\vec{F}$ are the gravitational body force and external body forces (forces that arise from interaction between the liquid phase and the dispersed one).

For 2D axisymmetric geometries, the axial and radial momentum conservation equations are given by

$$\frac{\partial}{\partial t}(\rho v_x) + \frac{1}{r}\frac{\partial}{\partial x}(r\rho v_x v_x) + \frac{1}{r}\frac{\partial}{\partial r}(r\rho v_r v_x) = -\frac{\partial p}{\partial x} + \frac{1}{r}\frac{\partial}{\partial x}\left[r\mu\left(2\frac{\partial v_x}{\partial x} - \frac{2}{3}(\nabla\cdot\vec{v})\right)\right] + \frac{1}{r}\frac{\partial}{\partial r}\left[r\mu\left(\frac{\partial v_x}{\partial r} + \frac{\partial v_r}{\partial x}\right)\right] + F_x, \tag{4}$$

$$\frac{\partial}{\partial t}(\rho v_r) + \frac{1}{r}\frac{\partial}{\partial x}(r\rho v_x v_r) + \frac{1}{r}\frac{\partial}{\partial r}(r\rho v_r v_r) = -\frac{\partial p}{\partial r} + \frac{1}{r}\frac{\partial}{\partial r}\left[r\mu\left(2\frac{\partial v_r}{\partial r} - \frac{2}{3}(\nabla\cdot\vec{v})\right)\right] + \frac{1}{r}\frac{\partial}{\partial x}\left[r\mu\left(\frac{\partial v_r}{\partial x} + \frac{\partial v_x}{\partial r}\right)\right] - 2\mu\frac{v_r}{r^2} + \frac{2}{3}\frac{\mu}{r}(\nabla\cdot\vec{v}) + F_r, \tag{5}$$

where

$$\nabla\cdot\vec{v} = \frac{\partial v_x}{\partial x} + \frac{\partial v_r}{\partial r} + \frac{v_r}{r}. \tag{6}$$

In this work, we have used the Lagrangian-Eulerian approach which analyzes the liquid phase (primary phase) by the Eulerian method and the bubble phase (secondary phase) by Lagrangian assumption allowing the monitoring of bubbles life cycle.

The Euler-Lagrange approach is the basis of the lagrangian discrete phase model. The dispersed phase is solved by tracking the bubbles through the calculated flow domain while the liquid phase is treated continuously by solving the continuity and Navier-Stokes equations. Exchanges of momentum and mass are realized between the dispersed phase and the fluid one.

According to the Lagrangian multiphase model, the volume fraction of the discrete phase (secondary phase) is quite small.

The bubbles trajectories are computed individually at each time step during the liquid phase calculation. This model is perfectly adapted for modeling the flows in a glass of champagne [26]. In order to reproduce as closely as possible the principle of nucleation, subroutines have been used for the gaseous phase. These subroutines have been written based on physical laws that are taken from previous experimental results [19, 27].

2.2.3. Ascending Bubbles Dynamics. The trajectory of a bubble is predicted by integrating the force balance in a Lagrangian reference frame.

During its rise in the liquid, a bubble is subjected to the action of several forces [19, 27]. The driving force of bubble ascent is buoyancy:

$$F_B = \frac{4}{3}\pi R^3 \rho g. \tag{7}$$

The drag force F_D exerted by the fluid around the bubble is expressed as follows:

$$F_D = \frac{1}{2}C_D \rho u^2 \pi R^2, \tag{8}$$

where C_D is the drag coefficient.

The movement of the surrounding fluid leads to an additional force F_{MA} called "added mass" related to the variation in the amount of movement of liquid displaced:

$$F_{MA} = \frac{\rho d}{dt}(Vu). \tag{9}$$

The volume V of liquid entrained in the wake of the bubble is roughly equal to the half of the volume of the bubble. Thus

$$F_{MA} = \frac{2}{3}\rho\pi\frac{d}{dt}\left(R^3 u\right). \tag{10}$$

The equation of motion can be written as follows:

$$\frac{2}{3}\rho\pi R^3\left(\frac{dU}{dt} + \frac{3U}{R}\frac{dR}{dt}\right) = \frac{4}{3}\pi R^3\rho g - \frac{1}{2}C_D\rho U^2\pi R^2. \tag{11}$$

The force of added mass has been compared to the buoyancy along the path of the bubble to the surface. The force of added mass does not exceed 2-3% of the buoyancy, so it can be neglected in the remainder of the study. The equation of motion is finally reduced to a simple equality between the drag force and buoyancy.

During ascent, surface-active materials progressively accumulate at the rear part of the rising bubble, thus increasing the immobile area of the bubble surface.

A rising bubble rigidified by surfactants runs into more resistance than a bubble presenting a more flexible interface free from surface-active materials. The champagne bubbles showed therefore a behavior intermediate between that of a rigid and that of a fluid sphere. To take into account the surfactants accumulation, the two following experimental drag coefficient laws, available in the range of intermediate Reynolds numbers (10^{-1} to 10^{2}) covered by champagne bubbles, have been used. Magnaudet et al. [28] have proposed a semiempirical relationship between the drag coefficient and the Reynolds number:

$$C_D = \frac{16}{\mathrm{Re}}\left(1 + 0.15\sqrt{\mathrm{Re}}\right) \quad (\mathrm{Re} < 50). \tag{12}$$

This experimental determination of the drag coefficient for fluid spheres is available for Reynolds number less than 50.

Since the Reynolds number exceeds the limit of 50 for sufficiently long path, another empirical law has been used available for Re > 50, determined by Maxworthy et al. [29]

$$C_D = 11.1\mathrm{Re}^{-0.74} \quad (1 < \mathrm{Re} < 800). \tag{13}$$

Because bubbles do not exceed a critical diameter of 2 mm, they remain spherical during their ascent. Moreover, the assumption that the bubbles do not coalesce or breakup has been considered.

The Reynolds number is defined by:

$$\mathrm{Re} = \frac{2\rho R u}{\eta}. \tag{14}$$

The density and the viscosity of the champagne wine for the liquid phase and the density and the viscosity of the carbon dioxide for the gaseous phase have been stored in the materials database.

The numerical simulations have been carried out with the ANSYS FLUENT software using finite volume approach. The convergence criteria were based on the residuals resulting from the integration of the conservation equations over finite control volumes. During the iterative calculation process, these residuals were constantly monitored and carefully scrutinized. For all simulations performed in this study, converged solutions were usually achieved with residuals as low as 10^{-5} (or less) for all the governing equations. Calculations have been performed for the three wine's temperatures (4, 12, and 20°C) with the same time steps and the same mesh.

2.3. Boundary Conditions. Models based on classical nucleation theory do not give a satisfactory approach of nucleation in the effervescent wines [30]. The idea has been to create routines to simulate the principle of nucleation and then compare the results with those obtained with experimental data [19, 27]. In order to simulate the bubbles growth, the bubbles velocity, the mass transfer between bubbles, the mass flow rate, and the drag law, we have written user-defined functions (UDF) in C language which defines source terms for

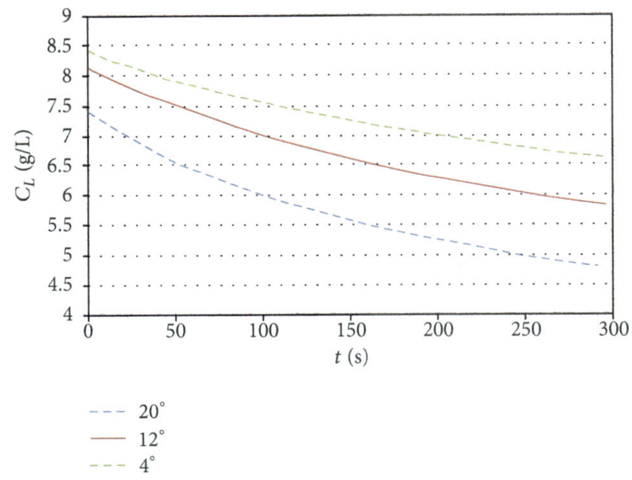

FIGURE 4: Evolution of the rate of dissolved carbon dioxide according to time in a champagne flute for the three service temperatures. The values have been experimentally measured with the same glass and the same filling level, throughout the 30 minutes following the pouring process of the glass.

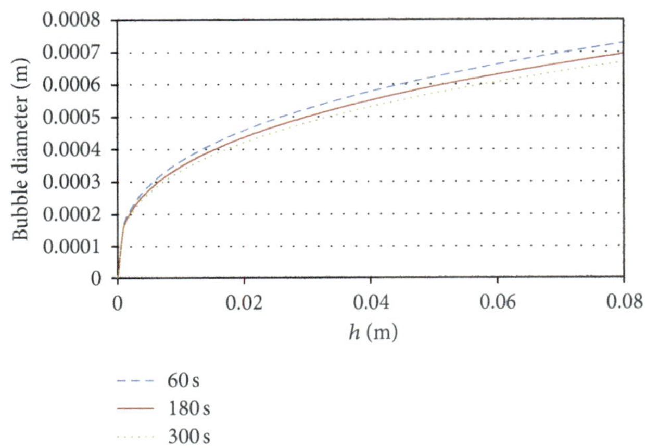

FIGURE 5: Evolution of the diameter according to the height h for a given champagne temperature of 4°C.

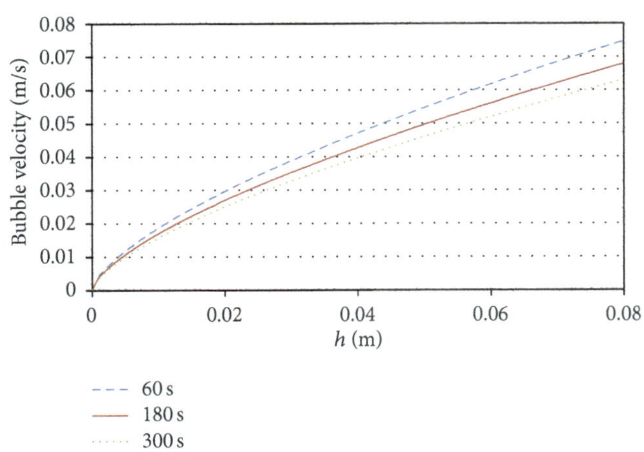

FIGURE 6: Evolution of the bubble velocity according to the height h for a given champagne temperature of 4°C.

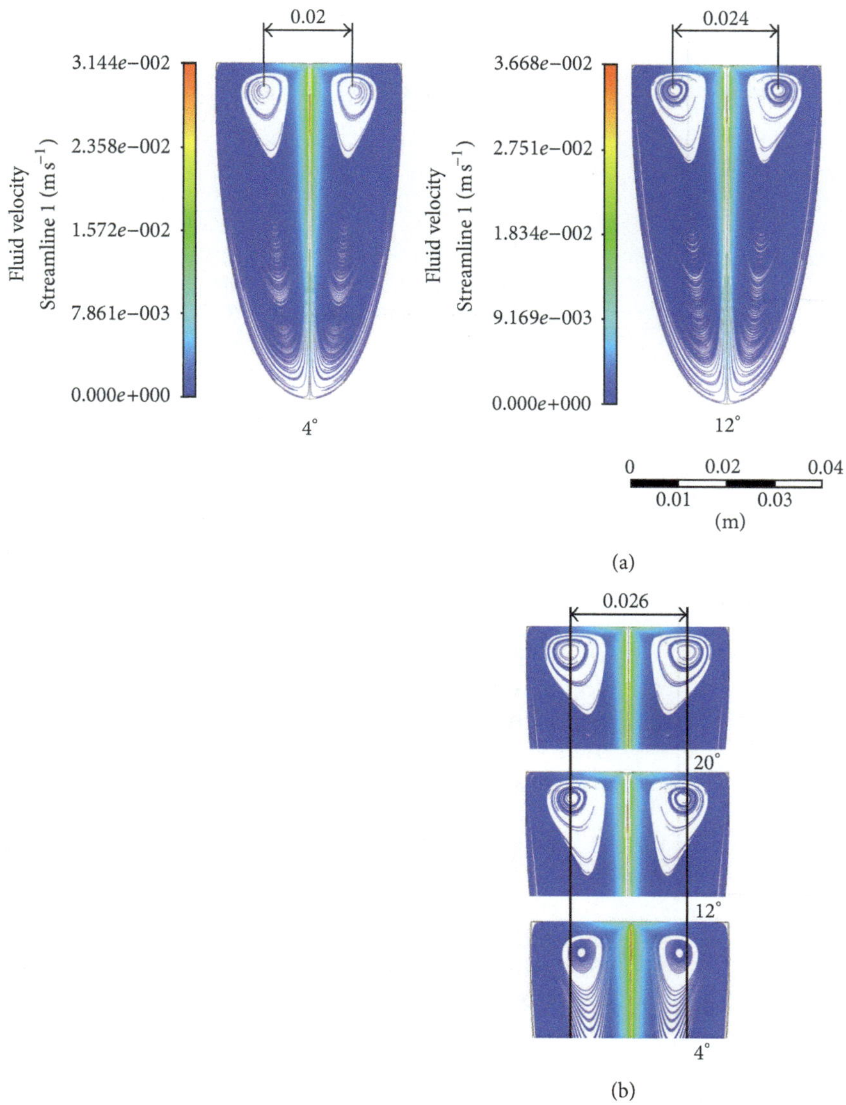

FIGURE 7: Evolution of the location of the vortex cores at 60 s following the pouring process and for three service temperatures (a), detail on the upper part of the glass (b).

the flow regime. The variation of bubbling frequency, which is a function of the CO_2-dissolved concentration, is made possible by changing the time step during the calculation.

The radius R of a bubble increases with time at a constant growth rate $k = dR/dt$, as bubbles rise toward the liquid surface, as experimentally determined by Liger-Belair [19].

Thus, $R(t) = R_0 + kt$, where R_0 is the bubble radius as it detaches from the nucleation site.

The semiempirical growth rate (k, expressed in μm/s) of bubbles rising in champagne was linked with some physicochemical properties of liquids as follows [19]:

$$k = \frac{dR}{dt} \approx 0.63 \frac{\Re\theta}{P_0} D_0^{2/3} \left(\frac{2\alpha\rho g}{9\eta} \right)^{1/3} (C_L - k_H P_0). \quad (15)$$

The bubbles diameter depends on several parameters, including both the distance traveled by the bubble, denoted by h, and the dissolved CO_2 concentration, denoted by C_L which continuously decreases with time (as a result of bubbling). Recently, the rate at which the dissolved CO_2 concentration decreases as time proceeds after pouring champagne into a flute etched on its bottom was experimentally determined for three distinct champagne temperatures (namely, 4, 12, and 20°C) [31]. As illustrated in Figure 4, the rate at which the dissolved CO_2 concentrations escape from champagne is strongly temperature dependent. For more details about the determination of Figure 4, see the article by Liger-Belair et al. [31] and references therein.

The law governing the change in radius R of a bubble is as follows [19]:

$$R \approx 2,7 \times 10^{-3} \times \theta^{5/9} \times \left(\frac{1}{\rho g} \right)^{2/9} \times \left(\frac{C_L - K_H P_0}{P_0} \right)^{1/3} \times h^{1/3}, \quad (16)$$

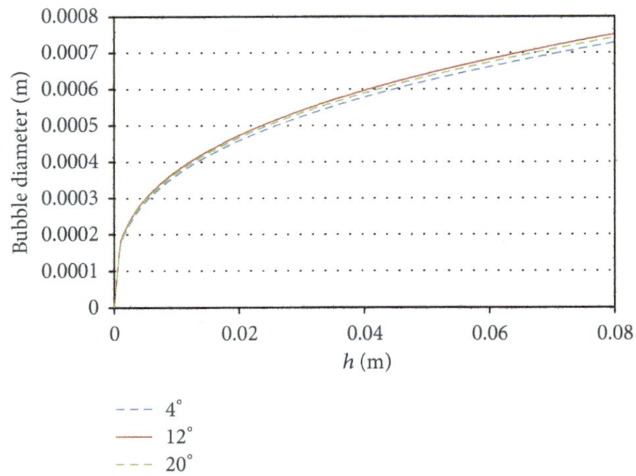

FIGURE 8: Bubbles' diameters in the bubble column, 60 s after the pouring process, and for the three champagne temperatures.

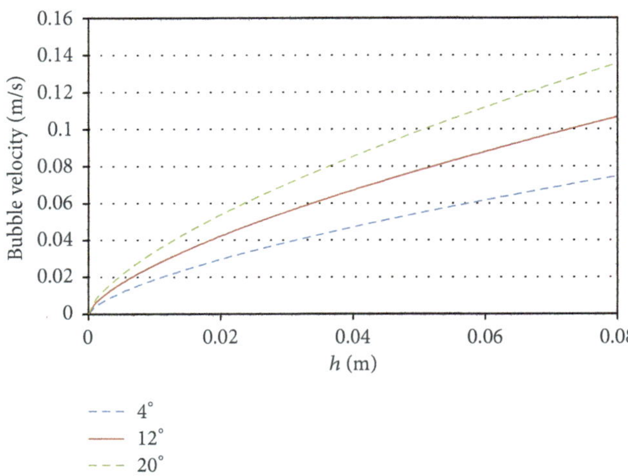

FIGURE 9: Bubbles' ascending velocities, 60 s after the pouring process, and for the three champagne temperatures.

where α is a numerical coefficient that depends on the fluid in question, estimated to be 0.7 in the case of sparkling wines [19].

The bubbles velocity u(m/s) varies according to the following expression [19, 32]:

$$u \approx \frac{2\alpha \rho g}{9\eta} R^2. \tag{17}$$

This factor accounts the rigidity of the bubble and thus the braking effect due to the presence of surfactant molecules on the surface of bubbles. It will be smaller if the bubble is made rigid by a thick shield of surfactant molecules.

The mass flow rate Q_m (kg/s) is defined by

$$Q_m = \frac{N}{t} \times \rho_p \times V_b. \tag{18}$$

The CO_2-dissolved concentration C_L decreases continuously over time once the wine is poured into the glass (Figure 4).

TABLE 1: Physicochemical parameters of champagne wine for the three service temperatures.

t (°C)	θ (°K)	k_H (Kg/m^3/Atm)	η (Kg/m/s)	D (m^2/s)
4	277.15	2.59	0.0027	0.751
12	285.15	1.93	0.0020	1.034
20	293.15	1.44	0.0016	1.385

This parameter also varies with the temperature. In this study, we have used the champagne physicochemical parameters for the three wines temperatures (Table 1).

The fluid viscosity varies in a significant way with the temperature; it tends to decrease when the temperature increases. The dependence of the viscosity of the champagne with the temperature follows an Arrhenius-like equation [7]:

$$\eta(\theta) = 1,08 \times 10^{-7} \exp\left(\frac{2806}{\theta}\right). \tag{19}$$

The temperature appears in (16), directly in the term $\theta^{5/9}$, but it is also included in Henry's coefficient k_H as well as in the viscosity η, which both depend on the temperature. Henry's constant depends strongly on some gas and on some liquid considered, as well as on the temperature. The dependence of Henry's constant on the temperature follows a law described by the equation of Van't Hoff:

$$K_H = K_{298^\circ\mathrm{K}} \exp\left[-\frac{\Delta H_{\mathrm{diss}}}{\Re}\left(\frac{1}{\theta} - \frac{1}{298}\right)\right]. \tag{20}$$

The effects on the bubble growth and on the upward bubbles dynamics are intrinsically connected to the variations of the physico-chemical parameters varying from a temperature to the other one (Table 1). Once (16) solved, we show that the influence of the temperature on the bubbles size is weak. Thanks to the influence of the viscosity and Henry's constant (Table 1), it is worth noting that increasing the temperature of a champagne of about 10°C (from 278° to 288°K) leads to increase of the bubbles size from 5 to 6% (which is totally undetectable in the naked eye). On the contrary, the temperature will have a significant influence on the viscosity of the liquid phase, varying in important way (42% between 4° and 20°). The velocity of the liquid phase, and therefore the mixing mechanisms of the wine, are conditioned by the unsteady evolution of the gaseous phase. Let us call back besides that the evolution of the rate of dissolved carbon dioxide differs for the three champagne temperatures (Figure 4).

3. Results and Discussion

We are going to present the numerical results (diameter and velocities of bubbles and velocity isocontours for the liquid phase) for three champagne temperatures (4°, 12°, and 20°). We will study the influence of these parameters on the liquid phase for 3 moments (60, 180, and 300 s) after the pouring process.

3.1. Ascending Bubbles' Dynamics. During its ascent towards the surface, the bubble grows by accumulating progressively

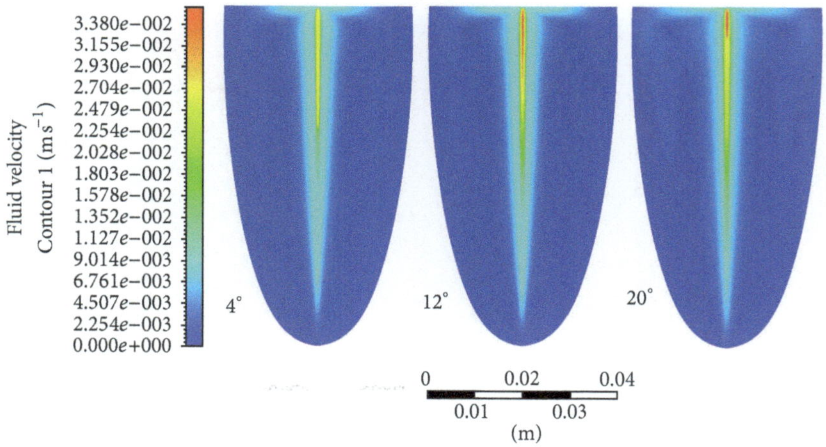

FIGURE 10: Flow patterns velocity magnitude, 60 s after the pouring process, and for the three champagne temperatures.

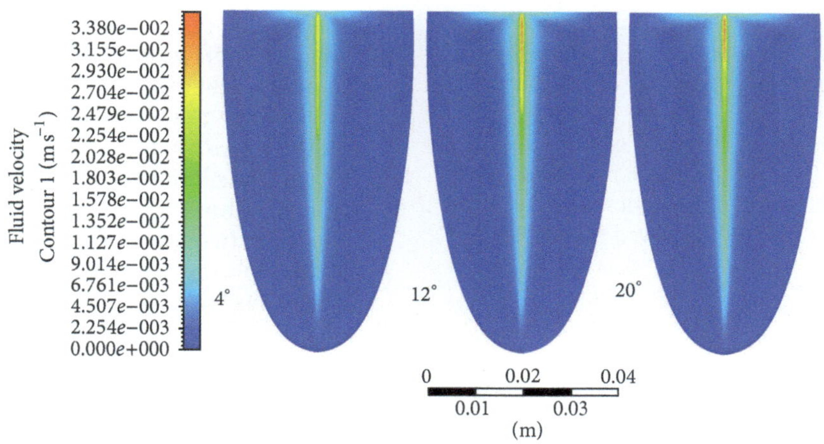

FIGURE 11: Flow patterns velocity magnitude, 180 s after the pouring process, and for the three champagne temperatures.

the CO_2 dissolved in the wine. The Figure 5 shows the unsteady evolution of the bubbles diameter in the case of a 4° service temperature. One may notice clearly a progressive decrease of the bubbles' size according to time. This is due to the progressive decrease with time of dissolved carbon dioxide in the liquid phase (as shown in Figure 4).

Moreover, because bubbles increase in size while rising toward the champagne surface, bubbles accelerate, as illustrated in Figure 6. Besides, it is worth noting in Figure 6 that the bubbles' velocity decreases with time, due to the progressive decrease of carbon dioxide dissolved in the liquid phase. Once again, the temperature will influence the upward dynamics of bubbles via the temperature dependence of the viscosity of the liquid phase.

3.2. Temperature Dependence of Ascending Bubble-Driven Flow Patterns Dynamics. A strong hypothesis in the tasting of the champagne wine is that there is a relation between the convection movements in the wine and the gaseous and aromatic exchanges which are very important in the gustative perception. During this study, we are going to show the impact of the temperature on the dynamics of mixing mechanisms of the wine.

Figure 7 represents the streamlines of the liquid phase for three champagne temperatures, 60 s after the pouring process. One may observe that the vortex core goes away from the axis of the flute when the temperature and the velocity increase. We can already note a dependence of bubble-driven flow patterns velocities and overall dynamics on the champagne temperature.

Both the progressive increases of ascending bubbles' diameters, and bubbles' velocities are presented in Figures 8 and 9, for the three various champagne temperatures, 60 s after pouring champagne into the flute.

By observing Figure 8, it is worth noting that bubbles' diameters are very close to each other, whatever the champagne temperature. More important differences are nevertheless observed concerning the bubbles' velocities, as shown in Figure 9. This is due to the temperature dependence of the liquid phase viscosity (see Table 1) which contributes to the bubbles' velocity, as shown in (17). Generally speaking, the higher the temperature is, the higher the bubble velocity is. Because the liquid phase is driven by ascending bubbles (through viscous effects), the dynamics of the flow patterns found in the flute are therefore expected to be strongly temperature dependent. As expected, the higher the champagne

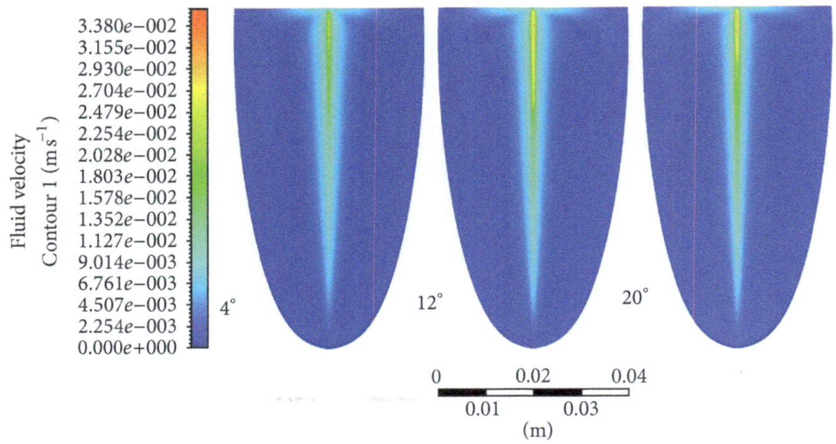

FIGURE 12: Flow patterns velocity magnitude, 300 s after the pouring process, and for the three champagne temperatures.

temperature is, the higher the velocity of bubble-driven flow patterns found in the liquid phase is, as illustrated in Figure 10 (60 s after pouring). The same trend has been observed 180 (Figure 11) and 300 s (Figure 12) after pouring for the three champagne temperatures, but with overall velocities progressively lowering as time proceeds (mainly because both ascending bubbles' diameters and velocities decrease with time due to the progressive loss of dissolved carbon dioxide found in the liquid phase).

4. Conclusion

In this paper, a numerical modeling of bubble-driven flow patterns found in a glass of champagne has been carried out for three champagne temperatures (4, 12, and 20°C) using the finite-volume method by CFD (computational fluid dynamics). The results, for both the gaseous phase (ascending bubbles) and the liquid phase (bubble-driven flow patterns), were systematically compared at several moments following the pouring process. Because ascending bubbles velocities strongly depend on the champagne temperature, the velocities of ascending bubble-driven flow patterns were found to logically vary with the champagne temperature. The champagne viscosity, which strongly depends on its temperature, plays a crucial role in the temperature dependence of ascending bubble-driven flow patterns. As expected, the higher the champagne temperature, the higher the velocities of the liquid phase. In a near future, these numerical results are expected to be confirmed by PIV experiments. Moreover, since a correlation is strongly suspected between ascending bubble-driven flow patterns and the release of aromatic volatile compounds from champagne, further investigations are to be conducted together with sensory analysis experiments.

Nomenclature

C_D: Drag coefficient (dimensionless)
C_l: Lift coefficient (dimensionless)
C_L: CO_2-dissolved concentration in the liquid (g/l)
D_0: Diffusion coefficient of CO_2 molecules (m^2/s)
F_D: Drag force
g: Acceleration due to gravity ($m \cdot s^{-2}$)
h: Liquid height (m)
k: Theoretical growth rate of bubbles (μm/s)
k_H: Henry's constant ($kg/m^3/atm$)
L: Characteristic dimension of a mesh element (mm)
N: Bubbling frequency (Hz)
P_0: Atmospheric pressure (atm)
Q_m: Mass flow rate (kg/s)
R: Bubble radius (m)
Re: Reynolds number (dimensionless)
u: Bubble velocity (m/s)
v: Liquid velocity (m/s)
V_b: Bubble volume (m^3)
$\mathfrak{R}$: Ideal gas constant (8.31 J/mol/K)
ρ: Liquid density (kg/m^3)
ρ_p: Density of CO_2 (kg/m^3)
η: Dynamic viscosity (kg/m/s)
α: Numerical coefficient (dimensionless)
θ: Liquid temperature (°C)
λ: Molecular mean free path (m)
$\emptyset$: Bubble diameter (m).

References

[1] A. Akhtar, V. Pareek, and M. Tadé, "CFD simulations for continuous flow of bubbles through gas-liquid columns: application of VOF method," *Chemical Product and Process Modeling*, vol. 2, no. 1, article 9, 2007.

[2] V. V. Buwa, D. S. Deo, and V. V. Ranade, "Eulerian-Lagrangian simulations of unsteady gas-liquid flows in bubble columns," *International Journal of Multiphase Flow*, vol. 32, no. 7, pp. 864–885, 2006.

[3] E. Delnoij, J. A. M. Kuipers, and W. P. M. Van Swaaij, "A three-dimensional CFD model for gas-liquid bubble columns," *Chemical Engineering Science*, vol. 54, no. 13-14, pp. 2217–2226, 1999.

[4] D. Zhang, N. G. Deen, and J. A. M. Kuipers, "Numerical simulation of the dynamic flow behavior in a bubble column: a study of closures for turbulence and interface forces," *Chemical Engineering Science*, vol. 61, no. 23, pp. 7593–7608, 2006.

[5] F. Özkan, M. Wörner, A. Wenka, and H. S. Soyhan, "Critical evaluation of CFD codes for interfacial simulation of bubble-train flow in a narrow channel," *International Journal for Numerical Methods in Fluids*, vol. 55, no. 6, pp. 537–564, 2007.

[6] G. Liger-Belair, C. Cilindre, R. D. Gougeon et al., "Unraveling different chemical fingerprints between a champagne wine and its aerosols," *Proceedings of the National Academy of Sciences of the United States of America*, vol. 106, no. 39, pp. 16545–16549, 2009.

[7] G. Liger-Belair, "The physics behind the fizz in champagne and sparkling wines," *European Physical Journal Special Topics*, vol. 201, pp. 1–88, 2012.

[8] G. Liger-Belair, "The science of bubbly," *Scientific American*, vol. 288, no. 1, pp. 80–85, 2003.

[9] G. Liger-Belair, "Nucléation, ascension et éclatement d'une bulle de champagne," *Annals of Physics*, vol. 31, pp. 1–133, 2006.

[10] G. Polidori, P. Jeandet, and G. Liger-Belair, "Bubbles and flow patterns in champagne," *American Scientist*, vol. 97, no. 4, pp. 294–301, 2009.

[11] T. Hummel, J. F. Delwiche, C. Schmidt, and K. B. Hüttenbrink, "Effects of the form of glasses on the perception of wine flavors: a study in untrained subjects," *Appetite*, vol. 41, no. 2, pp. 197–202, 2003.

[12] M. Vilanova, P. Vidal, and S. Cortés, "Effect of the glass shape on flavor perception of "toasted wine" from ribeiro (NW Spain)," *Journal of Sensory Studies*, vol. 23, no. 1, pp. 114–124, 2008.

[13] C. G. J. Bisperink and A. Prins, "Bubble growth in carbonated liquids," *Colloids and Surfaces A*, vol. 85, no. 2-3, pp. 237–253, 1994.

[14] Y. Zhang and Z. Xu, "Fizzics of bubble growth in beer and champagne," *Elements*, vol. 4, pp. 47–49, 2008.

[15] N. Shafer and R. Zare, "Through a beer glass darkly," *Physics Today*, vol. 44, pp. 48–52, 1991.

[16] J. Chandrashekar, D. Yarmolinsky, L. Von Buchholtz et al., "The taste of carbonation," *Science*, vol. 326, no. 5951, pp. 443–445, 2009.

[17] A. Dunkel and T. Hofmann, "Carbonic anhydrase IV mediates the fizz of carbonated beverages," *Angewandte Chemie*, vol. 49, no. 17, pp. 2975–2977, 2010.

[18] A. Livermore and T. Hummel, "The influence of training on chemosensory event-related potentials and interactions between the olfactory and trigeminal systems," *Chemical Senses*, vol. 29, no. 1, pp. 41–51, 2004.

[19] G. Liger-Belair, "La physique des bulles de champagne," *Annals of Physics*, vol. 27, pp. 1–106, 2002.

[20] G. Liger-Belair, J. B. Religieux, S. Fohanno, M. A. Vialatte, P. Jeandet, and G. Polidori, "Visualization of mixing flow phenomena in champagne glasses under various glass-shape and engravement conditions," *Journal of Agricultural and Food Chemistry*, vol. 55, no. 3, pp. 882–888, 2007.

[21] G. Liger-Belair, F. Beaumont, P. Jeandet, and G. Polidori, "Flow patterns of bubble nucleation sites (called fliers) freely floating in champagne glasses," *Langmuir*, vol. 23, no. 22, pp. 10976–10983, 2007.

[22] G. Polidori, F. Beaumont, P. Jeandet, and G. Liger-Belair, "Artificial bubble nucleation in engraved champagne glasses," *Journal of Visualization*, vol. 11, no. 4, p. 279, 2008.

[23] G. Liger-Belair, G. Polidori, and P. Jeandet, "Recent advances in the science of champagne bubbles," *Chemical Society Reviews*, vol. 37, no. 11, pp. 2490–2511, 2008.

[24] G. Polidori, F. Beaumont, P. Jeandet, and G. Liger-Belair, "Ring vortex scenario in engraved champagne glasses," *Journal of Visualization*, vol. 12, no. 3, pp. 275–282, 2009.

[25] G. Polidori, F. Beaumont, P. Jeandet, and G. Liger-Belair, "Visualization of swirling flows in champagne glasses," *Journal of Visualization*, vol. 11, no. 3, p. 184, 2008.

[26] F. Beaumont, C. Popa, G. Liger Belair, and G. Polidori, "Numerical modeling of bubble-driven flow patterns in champagne glasses," *International Journal of Numerical Methods For Heat and Fluid Flow*, 2013.

[27] G. Liger-Belair and P. Jeandet, "More on the surface state of expanding champagne bubbles rising at intermediate Reynolds and high peclet," *Langmuir*, vol. 19, no. 3, pp. 801–808, 2003.

[28] J. Magnaudet, M. Rivero, and J. Fabre, "Accelerated flows past a rigid sphere or a spherical bubble—part 1. Steady straining flow," *Journal of Fluid Mechanics*, vol. 284, pp. 97–135, 1995.

[29] T. Maxworthy, C. Gnann, M. Kürten, and F. Durst, "Experiments on the rise of air bubbles in clean viscous liquids," *Journal of Fluid Mech*, vol. 321, pp. 421–441, 1996.

[30] E. Herrmann, H. Lihavainen, A. P. Hyvärinen et al., "Nucleation simulations using the fluid dynamics software FLUENT with the fine particle model FPM," *Journal of Physical Chemistry A*, vol. 110, no. 45, pp. 12448–12455, 2006.

[31] G. Liger Belair, G. Polidori, and V. Zéninari, "Unraveling the evolving nature of gaseous and dissolved carbon dioxide in Champagne wines: a state-of-the-art review, from the bottle to the tasting glass," *Analytica Chimica Acta*, vol. 732, pp. 1–15, 2012.

[32] G. Liger-Belair, R. Marchal, B. Robillard et al., "On the velocity of expanding spherical gas bubbles rising in line in supersaturated hydroalcoholic solutions: application to bubble trains in carbonated beverages," *Langmuir*, vol. 16, no. 4, pp. 1889–1895, 2000.

12

Natural Convection of Nanofluids in a Square Enclosure with a Protruding Heater

J. Guiet,[1] M. Reggio,[1,2] and P. Vasseur[1,3]

[1] *Ecole Polytechnique, Université de Montréal, C.P. 6079, Succ. "Centre Ville", Montréal, QC, Canada H3C 3A7*
[2] *Département de Génie Mécanique, École Polytechnique, Université de Montréal, C.P. 6079, Succ. "Centre Ville", Montréal, QC, Canada H3C 3A7*
[3] *Laboratoire des Technologies Innovantes, Université de Picardie Jules Verne d'Amiens, rue des facultés, le Bailly, 80025 Amiens Cedex, France*

Correspondence should be addressed to M. Reggio, marcelo.reggio@polymtl.ca

Academic Editor: Yogesh Jaluria

This paper reports a numerical study on natural convection from a protruding heater located at the bottom of a square cavity filled with a copper-water nanofluid. The vertical walls of the cavity are cooled isothermally; the horizontal ones are adiabatic, and the heater is attached to the bottom wall. The heat source is assumed either to be isothermal or to have a constant heat flux. The effective viscosity and thermal conductivity of the nanofluid are modeled according to Brinkman and Patel, respectively. Numerical solutions of the full-governing equations, based on the lattice Boltzmann method, are obtained for a wide range of the governing parameters: the Rayleigh number, Ra; the Prandtl number, Pr; the geometrical parameters specifying the heater; the volume fraction of nanoparticles, Φ. For a particular geometry, it has been found that, for a given Ra, heat transfer is enhanced with increasing Φ, independently of the thermal boundary condition applied on the heater.

1. Introduction

The convection of nanofluids, which are a mixture of nanoparticles in a base fluid [1], has recently been an active field of research, because of reports of greatly enhanced thermal properties. Compared with other techniques for enhancing heat transfer in practical applications, nanofluids have the advantage of behaving like pure fluids, because of the small size of nanoparticles. As a result, the possibility of using them as heat transfer fluids for various applications, such as advanced nuclear systems or microchannel and minichannel heat sinks, is currently under consideration.

A review of the literature (see, e.g., Godson et al. [2]) indicates that most studies on this topic are concerned mainly with forced convection applications, but relatively little attention has been devoted to the natural convection of nanofluids. The first study on the natural convection of a nanofluid confined in a differentially heated enclosure seems to be that of Khanafer et al. [3]. The same problem was considered by Jou and Tzeng [4]. The Khanafer et al. model was used to investigate convective heat transfer enhancement in rectangular enclosures filled with an Al_2O_3-water nanofluid. It was reported that increasing the buoyancy parameter and volume fraction causes an increase in the average heat transfer coefficient. Heat transfer in nanofluids by natural convection in a square cavity heated isothermally from the vertical sides has been investigated numerically by Ho et al. [5] and Santra et al. [6]. Two different formulas have been considered by Ho et al. [5], for the effective viscosity, and thermal conductivity of the nanofluids, while the Ostwald-de Waele model for a non-Newtonian shear-thinning fluid has been used by Santra et al. [6] to calculate the shear stress. It was found that the uncertainties associated with the various expressions adopted to model nanofluids have a major influence on the characteristics of heat transfer by natural convection in the enclosure. Heat transfer enhancement in a differentially heated enclosure using the variable thermal conductivity and variable viscosity of Al_2O_3-water and Cu-water nanofluids has been investigated numerically by Abu-Nada et al. [7]. The impact of both variable viscosity and variable thermal conductivity, derived from experimental data, has been studied. It was observed that, at high Rayleigh numbers, the Nusselt number is more sensitive to viscosity models than to thermal conductivity models. Numerical study of natural convection in partially heated rectangular enclosures has been performed by Abu-Nada et al. [8]. Different types of nanoparticles were tested. They found that the heater location affects the flow and temperature fields when nanofluids are used. The same configuration was considered recently by Ghasemi and Aminossadati [9] for the case of an oscillating heat source embedded on the vertical wall of the enclosure. Recently, Alloui et al. [10] investigated natural convection, both analytically and numerically, in a shallow rectangular cavity filled with nanofluids. The critical Rayleigh number for the onset of supercritical convection of nanofluids is predicted explicitly by these authors. Also, results are obtained from the analytical model for finite amplitude convection for which the flow and heat transfer are presented in terms of the governing parameters of the problem.

Natural convection heat transfer in partially divided cavities filled with ordinary fluids has been the subject of many studies in the past, because of its importance in many engineering fields. Applications include thermal control of electric components, nuclear reactors, and so forth. Moukalled and Acharya [11] and Moukalled and Darwish [12] numerically investigated buoyancy-induced heat transfer in partially divided trapezoidal cavities representing attic spaces. The effects of the Rayleigh number, baffle height, and baffle location on heat transfer were discussed for various thermal boundary conditions. An experimental investigation concerning laminar convection in a box, with differentially heated end walls, which is divided by a barrier into two connected chambers, was carried out by Coman et al. [13]. They found that heat transfer diminishes rapidly when the barrier is high and also that it changes little when the lateral position of the barrier varies. Oztop and Bilgen [14] studied heat transfer numerically in a differentially heated and partitioned square cavity containing heat-generating fluid. It was found that the flow field and heat transfer are modified considerably with partial dividers. Natural convection heat transfer from a protruding heater located in a triangular enclosure has been analyzed numerically by Varol et al. [15]. They reported that, in order to obtain better heat removal, the heater must be located at the center of the bottom wall. The same problem was considered by AlAmiri et al. [16] for the case of a square cavity. Their investigation revealed that, as well as location, increasing the height and width of the heater results in the enhancement of heat transfer owing to the increase in the surface area of the heater. All the above investigations are concerned only with partially divided enclosures filled with ordinary fluids. To the author's knowledge, the influence of nanofluids on this flow configuration has not yet been studied.

In our investigation here, we consider natural convection in a square enclosure with a protruding heater subjected to various thermal boundary conditions. A numerical study is carried out to predict the effect of the location and dimension of the heat source on the flow and temperature fields in a cavity filled with a Cu-water nanofluid. The paper is organized as follows. The formulation of the problem and the numerical method are presented in Sections 2 and 3, respectively. In Section 4, the results obtained for the flow and heat transfer are discussed in terms of the geometry of the system. The final last section presents some concluding remarks related to optimum heat transfer enhancement based on the nanoparticle volume fraction, the length and the width of the enclosure, the Rayleigh number, and the location of the protruding heater.

2. Mathematical Formulation

Figure 1(a) schematically represents the flow configuration with a coordinate system. The square enclosure is of height H'. The vertical walls of the enclosure are cooled isothermally, at temperature T'_c, while the horizontal walls are adiabatic. A protruding heater of height l' and width w' is placed at a distance d' from the left wall, as shown in the graph. The cavity is filled with Newtonian, incompressible nanofluid. The thermophysical properties of the mixture considered in this study are assumed constant, except for the density variation, which is determined based on the Boussinesq approximation. Also, it is supposed that the base fluid and the nanoparticles are in thermal equilibrium and that no slip occurs between them.

The governing macroscopic equations that describe the system behavior are those of conservation of mass, momentum, and energy, which are given as

$$\begin{aligned}
&\frac{\partial^2\Psi'}{\partial x'^2} + \frac{\partial^2\Psi'}{\partial y'^2} = -\omega', \\
&\frac{\partial\omega'}{\partial t'} + u'\frac{\partial\omega'}{\partial x'} + v'\frac{\partial\omega'}{\partial y'} \\
&\qquad = \frac{1}{\rho_{\mathrm{nf}}}\left[\mu_{\mathrm{nf}}\left(\frac{\partial^2\omega'}{\partial x'^2} + \frac{\partial^2\omega'}{\partial y'^2}\right) + (\rho\beta')_{\mathrm{nf}}g\frac{\partial T'}{\partial x'}\right], \\
&\frac{\partial T'}{\partial t'} + u'\frac{\partial T'}{\partial x'} + v'\frac{\partial T'}{\partial y'} = \alpha_{\mathrm{nf}}\left(\frac{\partial^2 T'}{\partial x'^2} + \frac{\partial^2 T'}{\partial y'^2}\right),
\end{aligned} \tag{1}$$

in terms of stream function Ψ', vorticity ω', and temperature T'. As usual, in order to satisfy the continuity equation, the stream function Ψ' is defined such that $u' = \partial\Psi'/\partial y'$, $v' = -\partial\Psi'/\partial x'$.

The appropriate boundary conditions applied on the walls of the cavity are

$$x' = 0, H', \quad 0 \le y' \le H',$$

$$\Psi' = \frac{\partial\Psi'}{\partial x'} = 0, \qquad \omega' = -\frac{\partial^2\Psi'}{\partial x'^2}, \qquad T' = T'_c,$$

$$y' = H', \quad 0 \le x' \le H',$$

$$\Psi' = \frac{\partial\Psi'}{\partial y'} = 0, \qquad \omega' = -\frac{\partial^2\Psi'}{\partial y'^2}, \qquad \frac{\partial T'}{\partial y'} = 0,$$

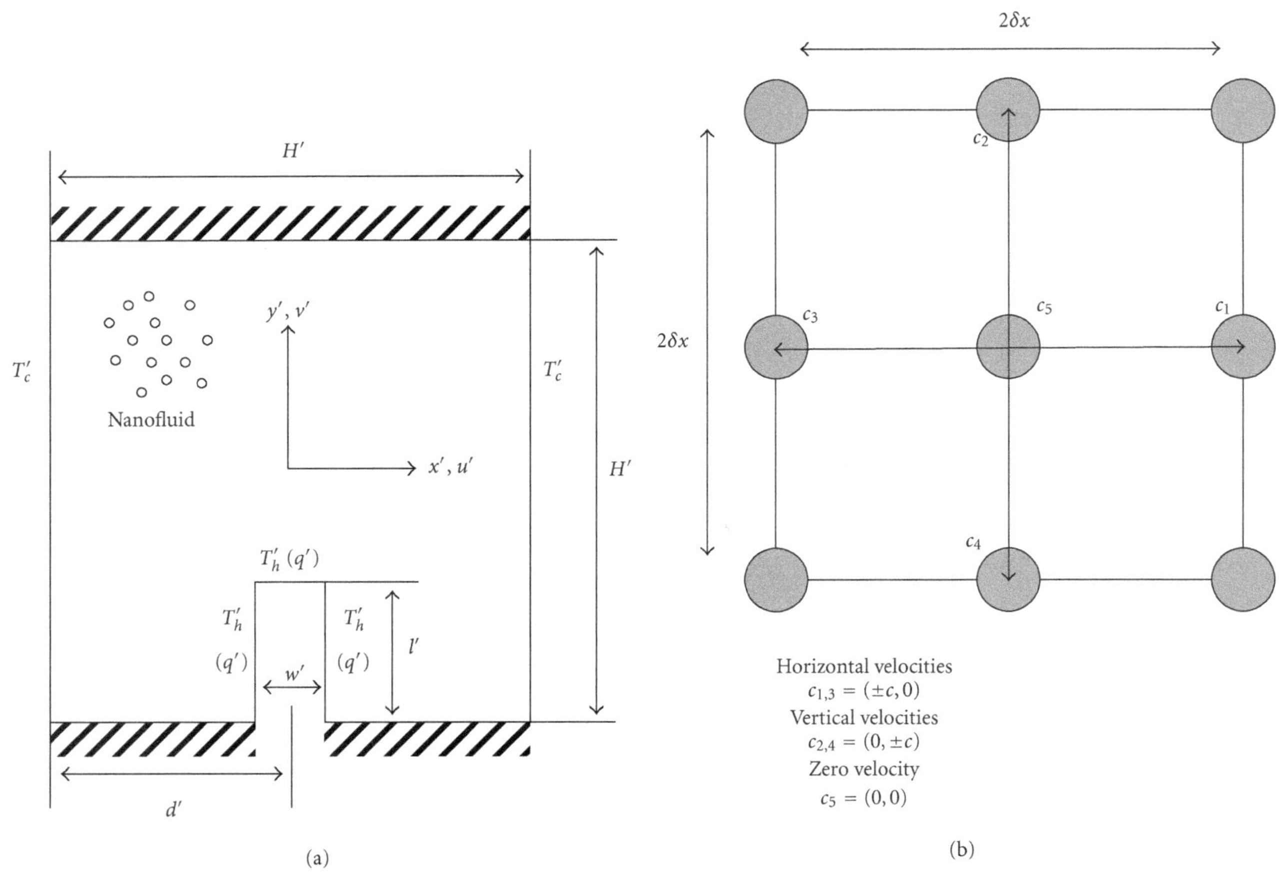

FIGURE 1: (a) Schematic diagram for physical model with coordinate system. (b) Diagram of the discrete D2Q5 phase space.

$$y' = 0 \quad \begin{cases} 0 \le x' \le d' - \dfrac{w'}{2} \\ d' + \dfrac{w'}{2} \le x' \le H', \end{cases}$$

$$\Psi' = \frac{\partial \Psi'}{\partial y'} = 0, \qquad \omega' = -\frac{\partial^2 \Psi'}{\partial y'^2}, \qquad \frac{\partial T'}{\partial y'} = 0. \tag{2}$$

On the heater walls, the hydrodynamic boundary conditions are given by

$$\Psi' = \frac{\partial \Psi'}{\partial n'} = 0, \qquad \omega' = -\frac{\partial^2 \Psi'}{\partial n'^2}, \tag{3}$$

while the two following thermal boundary conditions are considered:

(1) a constant temperature T'_h, such that

$$T' = T'_h, \tag{4}$$

(2) a constant heat flux q', such that

$$\frac{\partial T'}{\partial n'} = -\frac{q'}{k_{\text{nf}}}. \tag{5}$$

In these equations, n' represents the normal to the heater walls, and the constant k_{nf} is the characteristic thermal conductivity of the fluid considered.

The effective density of the nanofluid is given as

$$\rho_{\text{nf}} = (1 - \Phi)\rho_f + \Phi\rho_{\text{np}}, \tag{6}$$

where Φ is the solid volume fraction of nanoparticles.

The thermal diffusivity of the nanofluid is

$$\alpha_{\text{nf}} = \frac{k_{\text{nf}}}{(\rho C_P)_{\text{nf}}}, \tag{7}$$

where the heat capacitance of the nanofluid is given by

$$(\rho C_P)_{\text{nf}} = (1 - \Phi)(\rho C_P)_f + \Phi(\rho C_P)_{\text{np}}. \tag{8}$$

The thermal expansion coefficient of the nanofluid can be determined by:

$$(\rho\beta')_{\text{nf}} = (1 - \Phi)(\rho\beta')_f + \Phi(\rho\beta')_{\text{np}}. \tag{9}$$

The effective dynamic viscosity of the nanofluid is calculated using Brinkman's model [17] as follows:

$$\mu_{\text{nf}} = \frac{\mu_f}{(1 - \Phi)^{2.5}}, \tag{10}$$

which gives an estimation of the viscosity of a nanofluid containing a dilute suspension of small, rigid, spherical particles.

The thermal conductivity of the nanofluid is calculated using a model proposed by Patel et al. [18] as follows:

$$\frac{k_{nf}}{k_f} = 1 + \frac{k_{np}A_{np}}{k_f A_f} + c k_{np} Pe \frac{A_{np}}{k_f A_f}, \quad (11)$$

where $c = 25000$ is a constant determined experimentally, and A_{np}/A_f is defined as

$$\frac{A_{np}}{A_f} = \frac{d_f}{d_{np}} \frac{\Phi}{(1-\Phi)}, \quad (12)$$

with d_{np} being the diameter of the solid nanoparticles in the fluid, assumed here to be equal to $d_{np} = 100\,\text{nm}$, and $d_f = 2\,\text{A}^\circ$ being the size of the liquid molecules, namely, those of water.

The Péclet number in (11) is defined as

$$Pe = \frac{u_{np} d_{np}}{\alpha_f}, \quad (13)$$

with u_{np} expressing the Brownian motion velocity of nanoparticle, defined as

$$u_{np} = \frac{2 k_b T'}{\pi \mu_f d_{np}^2}, \quad (14)$$

where $k_b = 1.38065 \times 10^{-23}$ is the Boltzmann constant.

The governing equations with the above models are made dimensionless by scaling the length by H' and the velocity by $\alpha_f \mathrm{Ra}^{1/2}/H'$, with α_f being the thermal diffusivity of pure fluid, and time by $H'^2/\alpha_f \mathrm{Ra}^{1/2}$. Also, we introduce the reduced temperature $T = (T' - T'_0)/\Delta T'$, where $\Delta T' = (T'_h - T'_c)$ for an isothermal heater and $\Delta T' = q' H'/k_f$ for a heater with a constant heat flux. The wall temperature of the enclosure is set arbitrarily to $T'_0 = 22°\mathrm{C}$, and the characteristic temperature difference is fixed to $\Delta T' = 30°\mathrm{C}$.

The dimensionless equations governing the present problem then read

$$\frac{\partial^2 \Psi}{\partial x^2} + \frac{\partial^2 \Psi}{\partial y^2} = -\omega, \quad (15)$$

$$\frac{\partial \omega}{\partial t} + u\frac{\partial \omega}{\partial x} + v\frac{\partial \omega}{\partial y} = \overline{\nu}\frac{\mathrm{Pr}}{\mathrm{Ra}^{1/2}}\left(\frac{\partial^2 \omega}{\partial x^2} + \frac{\partial^2 \omega}{\partial y^2}\right) + \overline{\beta}\,\mathrm{Pr}\frac{\partial T}{\partial x}, \quad (16)$$

$$\frac{\partial T}{\partial t} + u\frac{\partial T}{\partial x} + v\frac{\partial T}{\partial y} = \frac{\overline{\alpha}}{\mathrm{Ra}^{1/2}}\left(\frac{\partial^2 T}{\partial x^2} + \frac{\partial^2 T}{\partial y^2}\right), \quad (17)$$

where

$$\overline{\nu} = \frac{\nu_{nf}}{\nu_f}, \qquad \overline{\beta} = \frac{(\rho\beta')_{nf}}{(\rho_{nf}\beta'_f)}, \qquad \overline{\alpha} = \frac{\alpha_{nf}}{\alpha_f}. \quad (18)$$

The corresponding boundary conditions are

$$x = 0, 1, \quad 0 \le y \le 1,$$

$$\Psi = \frac{\partial \Psi}{\partial x} = 0, \qquad \omega = -\frac{\partial^2 \Psi}{\partial x^2}, \qquad T = 0,$$

$$y = 1, \quad 0 \le x \le 1,$$

$$\Psi = \frac{\partial \Psi}{\partial y} = 0, \qquad \omega = -\frac{\partial^2 \Psi}{\partial y^2}, \qquad \frac{\partial T}{\partial y} = 0, \quad (19)$$

$$y = 0 \quad \begin{cases} 0 \le x \le D - \dfrac{W}{2} \\ D + \dfrac{W}{2} \le x \le 1, \end{cases}$$

$$\Psi = \frac{\partial \Psi}{\partial y} = 0, \qquad \omega = -\frac{\partial^2 \Psi}{\partial y^2}, \qquad \frac{\partial T}{\partial y} = 0,$$

and the no slip condition on the heater wall yields

$$\Psi = \frac{\partial \Psi}{\partial n} = 0, \qquad \omega = -\frac{\partial^2 \Psi}{\partial n^2}, \quad (20)$$

while the dimensionless thermal boundary conditions are as follows:

(1) the constant temperature applied on the heater

$$T = 1 \quad (21)$$

(2) the constant heat flux on the heater

$$\frac{\partial T}{\partial n} = -\frac{k_f}{k_{nf}}. \quad (22)$$

The heat transfer rates along the vertical walls are expressed in terms of the local and average Nusselt's numbers Nu and $\overline{\mathrm{Nu}}$, respectively. Depending on the thermal boundary conditions applied on the heater, these parameters are computed as follows:

(3) on the left vertical wall of the heater for a constant temperature,

$$\mathrm{Nu} = -\frac{k_{nf}}{k_f}\frac{\partial T}{\partial x}\bigg|_{x=0}, \quad (23)$$

$$\overline{\mathrm{Nu}} = \int_0^1 \mathrm{Nu}\, dy, \quad (24)$$

(4) on the heater for a constant heat flux,

$$Nu = \frac{1}{T}\bigg|_{\text{heater}}, \quad (25)$$

$$\overline{\mathrm{Nu}} = \frac{\int_{\text{Heater}} \mathrm{Nu}\, dl}{\int_{\text{Heater}} dl}, \quad (26)$$

with dl representing an elementary length along the walls of the heater.

From the above equations, we can see that the present problem is governed by six dimensionless parameters, namely, the thermal Rayleigh number $\mathrm{Ra} = g\beta'_f \Delta T' H'^3/\alpha_f \nu_f$, the Prandtl number $\mathrm{Pr} = \nu_f/\alpha_f$, the solid volume fraction of nanoparticles Φ, and the geometrical parameters for the heater, that is, $L = (l'/H')$, $W = (w'/H')$, and $D = (d'/H')$. Naturally, the type of nanoparticles considered must also be specified, which introduces another parameter.

3. Numerical Solution

The solution of the governing equations, (15) to (17), is obtained using the lattice Boltzmann method (LBM). The LBM, developed two decades ago, examines fluids in the molecular state instead of at the classical macroscopic level and makes it possible to simulate flows by solving the Boltzmann transport equation for particle distribution functions on a simplified phase space, called the lattice [19, 20].

In this study, we used the LBM to solve the stream function, vorticity, and temperature equations, instead of the classical Navier-Stokes equations for velocity and temperature. This approach, recently proposed by Chai and Shi [21] and Chen et al. [22, 23], enables us to solve the advection-diffusion equations, (16) and (17), and the Poisson equation, (15), using particle distribution functions and applying the lattice Boltzmann approach on the D2Q5-phase space described in Figure 1(b).

We considered three sets of particle distribution functions. One set describes the temperature T and another the vorticity ω, both of them evolving following the collision and propagation steps described by Chen et al. [22, 23]. The third set of particle distribution functions is used for the stream function Ψ and is computed by iterating the collision and propagation steps, as described by Chai and Shi [21].

The parameters characterizing the nanofluid, that is, to say thermal diffusivity $\overline{\alpha}$, kinematic viscosity $\overline{\nu}$, and thermal expansion coefficient $\overline{\beta}$, are introduced in the model with the definition of characteristic relaxation times. These characteristic parameters, which make it possible to link the macroscopic formulation of the Navier-Stokes equations to the mesoscopic formulation of the LBM, were modified to take into account the influence of the nanoparticles on the fluid in the LBM framework for the various models studied (see e.g., Guiet et al. [24]).

With the present approach, a wide range of simulations were conducted for the simulation of natural convection in partially divided square enclosures. For these simulations, the convergence was considered to be reached when the relative error on variables $V = \Psi$, ω, and T between two successive iterations, t and $t+1$, was smaller than a chosen tolerance:

$$\mathrm{ERR}_V = \frac{\sum_x \sum_y \left| V_{x,y}^{t+1} - V_{x,y}^{t} \right|}{\sum_x \sum_y \left| V_{x,y}^{t+1} \right|} < 10^{-6}. \qquad (27)$$

The calculations were performed on 101×101 regular grids using D2Q5 lattices, this refinement being a good compromise between computational time and accuracy.

To check the ability of this LBM formulation to simulate flows in partially divided cavities, various simulations were conducted for pure fluid and compared with results provided by AlAmiri et al. [16]. The accuracy of the LBM was checked for various Rayleigh's numbers, Ra, in different configurations with various heights L, widths W, and positions D of an isothermal heater. The results are summarized in Table 1. Good agreement is observed, with the maximum deviation being of the order of 1.7% for the 101×101 grid. We can note that, while increasing Ra, the stream function amplitude $\Psi_{\max} - \Psi_{\min}$ reaches a maximum at $\mathrm{Ra} = 10^5$ and then decreases. This can be explained by the fact that, increasing $\mathrm{Ra} < 10^5$, the conduction is not negligible compared to the convection; the presence of the heater affects more and more fluid in the cavity (by conduction) inducing a convection of more and more fluid in the cavity. A maximum is reached where all the fluid in the cavity is convected. Then, increasing Ra, the conduction becomes negligible, and the convection occurs more and more close from the vertical walls (cooled fluid) and above the heater (warmed fluid), leaving larger areas of the cavity where the fluid is not affected by the convection. The stream function amplitude $\Psi_{\max} - \Psi_{\min}$ tends to reduce ($\mathrm{Ra} > 10^5$).

4. Results and Discussion

The results presented in this paper were obtained for pure fluid and a copper (Cu)-water mixture. The thermophysical properties of the base fluid, that is, water, and the nanoparticles, are reported in Table 2. Since this study is limited to water-based solutions, it is assumed that the Prandtl number (Pr) equals 7. As discussed by many authors (see, e.g., Trevisan and Bejan [25]), this type of convective heat transfer is independent of this parameter, provided that this latter is of order one or greater. The results are presented for Rayleigh's numbers varying from 10^3 to 10^6, for various geometries, $0.25 \le L \le 0.8$, $0.1 \le W \le 0.8$, and positions, $0.15 \le D \le 0.5$, of the heater, while the nanoparticle volume fraction covers the $0 \le \Phi \le 0.05$ range.

Typical numerical results are presented in Figure 2 for $\mathrm{Ra} = 10^5$, $D = 0.5$, and $W = 0.2$. On the graphs, streamlines and isotherms are presented from left to right. The calculated maximum stream functions ($|\Psi_{\max}|$) and heat transfer ($\overline{\mathrm{Nu}}$) are also given with each graph for reference. In these figures, the streamlines are equally spaced between $\Psi = 0$ on the boundaries and the maximum (minimum) value $\Psi_{\max}$ ($\Psi_{\min}$). Also, the isotherms are equally spaced between $T_{\max} = 1$ and $T_{\min} = 0$. Figure 2(a) shows the results obtained for a cavity with a protruding isothermal heater of height $L = 0.25$, filled with a pure solution ($\Phi = 0$). The resulting flow and temperature fields are similar to those reported in the literature in the past (see, e.g., AlAmiri et al. [16]). Since the heater is centrally located in the middle of the bottom wall ($D = 0.5$), symmetrical flow and temperature patterns, about a vertical line passing

TABLE 1: Comparison with results provided by Alimiri et al. [16] of the stream function extremes (Ψ_{min}, Ψ_{max}) for various heat source configurations (Pr = 0.71).

Ra	W	L	D	Reference [16]	Present	Difference
10^4	0.2	0.25	0.5	(−0.0333, 0.0333)	(−0.0330, 0.0330)	0.9%
10^5	0.2	0.25	0.5	(−0.0459, 0.0459)	(−0.0458, 0.0458)	0.22%
10^6	0.2	0.25	0.5	(−0.0290, 0.0290)	(−0.0289, 0.0289)	0.34%
10^6	0.1	0.25	0.5	(−0.0292, 0.0292)	(−0.0291, 0.0291)	0.34%
10^6	0.4	0.25	0.5	(−0.0299, 0.0299)	(−0.0299, 0.0299)	0%
10^6	0.2	0.05	0.5	(−0.0331, 0.0331)	(−0.0330, 0.0330)	0.34%
10^6	0.2	0.5	0.5	(−0.0239, 0.0239)	(−0.0235, 0.0235)	1.7%
10^6	0.2	0.25	0.2	(−0.0334, 0.0288)	(−0.0328, 0.0286)	1.3%

TABLE 2: Thermophysical properties of water and Cu-nanoparticles [8].

	ρ (kg m^{-3})	C_p (J kg^{-1} K^{-1})	k (Wm^{-1} K^{-1})	$\beta \times 10^5$ (K^{-1})
H_2O	997.1	4179	0.613	21
Cu	8933	385	401	1.67

through the axis of symmetry of the heated element, are observed in the enclosure. The streamlines are seen to occupy the entire cavity body. The field of isotherms indicates that these are mostly parallel to the heated and cooled boundaries, except above the heated element. As the solid volume fraction is increased to $\Phi = 0.05$, the resulting streamlines and isotherms are depicted as in Figure 2(b). It is found that, as expected, the flow and temperature patterns are influenced by the addition of nanoparticles, which enhances the buoyancy forces, owing to the increase in the effective thermal conductivity of the mixture. As a result, the strength of convective heat transfer is improved with an increase in solid concentration. In fact, the increase in the Nusselt number is about 33.5% for a concentration of nanoparticles as low as $\Phi = 0.05$. Figure 2(c) shows the results obtained when the heater height is increased to $L = 0.50$ for $\Phi = 0.05$. A comparison between Figures 2(b) and 2(c) indicates that, as the height of the heater increases, the intensity of the convective motion diminishes, as indicated by the values of the maximum and minimum stream function patterns. This is because an increase in the length of the heated element results in an increase in resistance to the flow. However, the average Nusselt-number is found to be enhanced from $\overline{\text{Nu}} = 4.356$ to $\overline{\text{Nu}} = 5.839$ as the heater height is increased from $L = 0.25$ to $L = 0.50$. Similar data were obtained for the same conditions, but for the case of a block heated by a constant heat flux. The obtained results (not presented here) indicate that the streamline patterns were qualitatively similar to those reported for the isothermal heater. Naturally, the isotherm patterns are quite different, especially in the vicinity of the heater, where the surface temperature of this latter is not constant.

Figure 3 illustrates the variation in $\overline{\text{Nu}}$ with the solid volume fraction Φ at various Rayleigh's numbers. The specific geometrical configuration of the heating source considered is $D = 0.5$, $L = 0.25$, and $W = 0.2$. The results obtained for the isothermal heater and the constant heat flux heater are depicted in Figures 3(a) and 3(b), respectively. In general, past studies on this topic have demonstrated that the addition of nanoparticles to a base fluid leads to the occurrence of two opposing effects on the convective heat transfer within the cavity. The first effect, resulting from the enhanced thermal conductivity of the mixture owing to the presence of the nanoparticles, improves $\overline{\text{Nu}}$. The second effect,owing to the increase in viscosity caused by the addition of nanoparticles, eliminates the convective motion and thusreduces $\overline{\text{Nu}}$. Which of these two antagonist mechanisms prevails depends on the type of particles used, the convection intensity (Ra), and the models considered to approximate the viscosity and thermal conductivity of the mixture. In this study, the results depicted in Figure 3 show that, for a given Ra value, a monotonic increase in $\overline{\text{Nu}}$ is obtained with the addition of nanoparticles. For the isothermal heater, Figure 3(a) indicates that, at $\Phi = 0.05$, the Nusselt number is enhanced by about 38% for Ra $= 10^3$ and by about 32% for Ra $= 10^6$. The case of a constant heat flux heater, Figure 3(b), is discussed next. For this situation, the temperatures of the surface of the heater, T, are not constant. As a matter of fact, its maximum value, T_{max}, is an important parameter in the thermal design of electronic devices. For a given value of Φ, Figure 3(b) shows that increasing monotonically Ra enhances $\overline{\text{Nu}}$. This follows from the fact that increasing the Rayleigh number enhances convection, that is, heat removal from the heater. As a result, the temperatures on the surface of the heated element are reduced, resulting in an enhancement of the heat transfer. Specifically, it is found that, at $\Phi = 0.05$, the Nusselt number is enhanced by about 36% for Ra $= 10^3$ and by about 22% for Ra $= 10^6$. Note that in that case the computed average Nusselt numbers for Ra $= 10^3$ and 10^4 are almost identical. This is due to the weak impact of natural convection on the heat transfer for these two cases, the heat exchange being due to conduction. The impact of the convection starts appearing for the simulations at Ra $= 10^5$.

This result indicates that the cooling performance of nanofluids is more effective at low Rayleigh's numbers than at

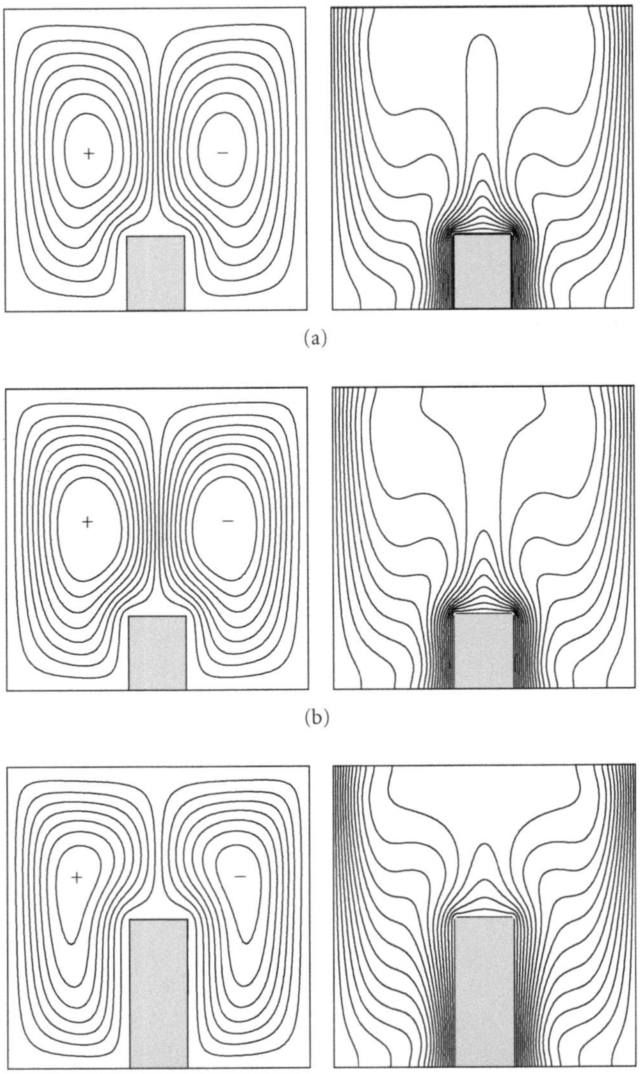

Figure 2: Effect of varying the ratio of nanoparticles, Φ, and the length, L, of an isothermally heated heater on the streamlines (left) and isotherms (right) for $\mathrm{Ra} = 10^5$, $D = 0.5$, $W = 0.2$, and (a) $L = 0.25$ and $\Phi = 0$: $|\Psi_{max}| = 0.0391$, $\overline{\mathrm{Nu}} = 3.264$, (b) $L = 0.25$ and $\Phi = 0.05$: $|\Psi_{max}| = 0.0506$, $\overline{\mathrm{Nu}} = 4.356$, (c) $L = 0.50$ and $\Phi = 0.05$: $|\Psi_{max}| = 0.0392$, $\overline{\mathrm{Nu}} = 5.839$.

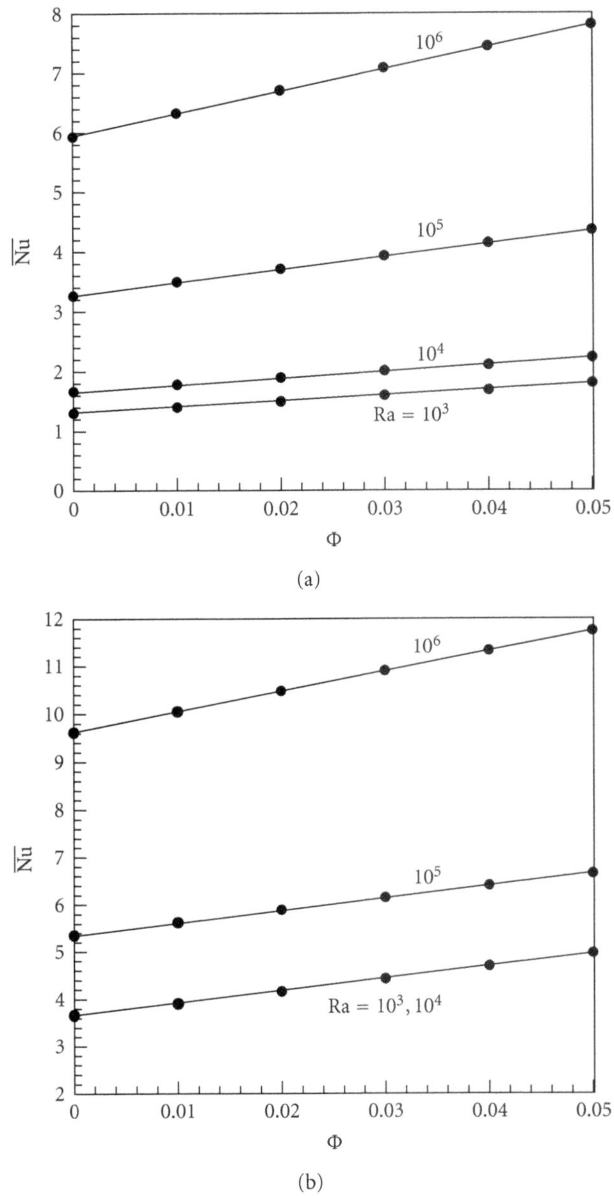

Figure 3: Variation of the average Nusselt number $\overline{\mathrm{Nu}}$ with solid volume fraction Φ at various Rayleigh's number's Ra for $D = 0.5$, $L = 0.25$, $W = 0.2$, and for, (a) isothermal heater; (b) heater with a constant heat flux.

high ones, independently of the heating conditions imposed on the heater. This behavior of $\overline{\mathrm{Nu}}$ with Φ is in good agreement with the results reported in previous studies, especially when the Rayleigh number is small enough that convection is weak ($\mathrm{Ra} \leq 10^4$). For this situation, for which heat transfer by conduction prevails, the increase in $\overline{\mathrm{Nu}}$ with Φ is due to the enhancement in thermal conductivity of the nanofluid. This trend has been found to be independent of the models used to characterize the mixture. It must be mentioned that a review of the literature indicates that, at higher Ra values for which heat transfer is dominated by convection, the effects of the nanoparticle volume fraction on the Nusselt number was found to be either beneficial or detrimental, depending on the formulas adopted to model the effective thermal conductivity and dynamic viscosity of the nanofluid.

Figure 4 illustrates the variation of $\overline{\mathrm{Nu}}$ for $\mathrm{Ra} = 10^5$, in terms of the geometry and position of the heat source and Φ of the mixture, for a heater with a constant heat flux. Figures 4(a) and 4(b) show the variation of $\overline{\mathrm{Nu}}$ with the height L of the heated element. It can be seen that, for a given solid volume fraction, the Nusselt number decreases as the height increases. This follows from the fact that the temperature at the surface of the element increases as the height of the element increases, owing to the higher heat flux generated by the heat source. The influence of the heater width W on $\overline{\mathrm{Nu}}$ is presented in Figure 4(b). For a given Φ, it is found

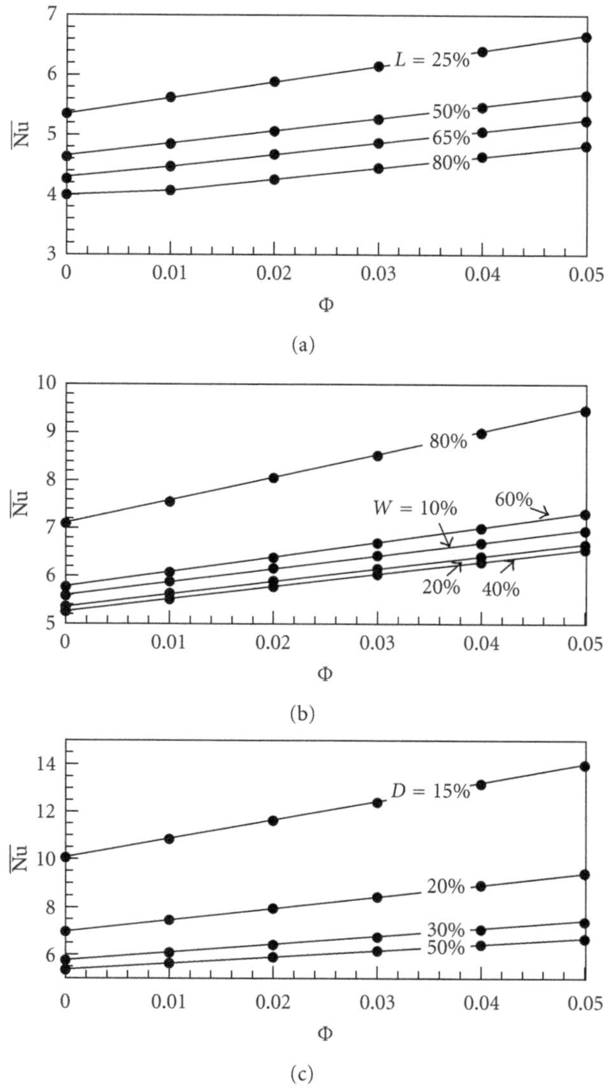

FIGURE 4: Variation of the average Nusselt number $\overline{\text{Nu}}$, for the case of a heater with a constant heat flux, with solid volume fraction Φ for $\text{Ra} = 10^5$: (a) effect of the heater height L for $D = 0.5$ and $W = 0.2$; (b) effect of the heater width W for $D = 0.5$ and $L = 0.25$; (c) effect of the heater position D for $L = 0.25$ and $W = 0.2$.

that, as the width of the heater increases from $W = 0.10$ to $W = 0.80$, the Nusselt number first decreases to reach a minimum at about $W = 0.40$. Above this value, $\overline{\text{Nu}}$ is found to increase sharply with W. Concerning the influence of Φ, Figure 4(a) indicates that, with an increase of 5% in this parameter, the Nusselt number grows by about 24% at $L = 0.25$ and by about 21% at $L = 0.80$. Similarly, it is observed from Figure 4(b) that $\overline{\text{Nu}}$ improves by 24% at $W = 0.10$ and 33% at $W = 0.80$. Figure 4(c) presents the effect of the location of the heat source D on $\overline{\text{Nu}}$. Independently of the concentration Φ of the solution, the average Nusselt number is minimum for $D = 0.50$, that is, when the heater is centrally located at the bottom of the enclosure. Naturally, symmetrical flow and temperature patterns are generated in the enclosure in this case (Figure 5(a)). As the heat source moves towards the left vertical isothermal wall ($D < 0.5$), the flow patterns consist of two nonsymmetrical counterrotating cells of unequal strengths (Figures 5(b) and 5(c)). The isotherm patterns indicate that, as the heater is moved to the left cold wall, the temperature of the heat source, which is not uniform since it is heated by a constant heat flux, decreases. Moving the heat source to the vicinity of the cooled boundary enhances heat removal considerably, which reduces the temperatures on the surface of the heater. As a result, the average Nusselt number increases sharply as it moves away from the central position, $D = 0.50$. Also, Figure 4(c) shows that the average Nusselt number increases monotonically with the solid volume, independently of the position of the heater. With an increase of 5% in solid volume fraction, the Nusselt number grows by about 40% at $D = 0.15$ and by about 24% at $D = 0.5$.

Figure 6 illustrates the results obtained for the same conditions as those of Figure 4, but for the case of an isothermal heater. A bird's eye view of the graphs indicates that, independently of the geometry and position of the heat source, the Nusselt number increases with an increase in the volume fraction Φ, that is, the thermal conductivity of the solution. Thus, as expected, this behavior is independent of the thermal boundary conditions applied on the heat source. The effect of the size of the heater on heat transfer is shown in Figures 6(a) and 6(b), which indicate that $\overline{\text{Nu}}$ improves with an increase in the dimensions of the heating surface. The fact that the average Nusselt number improves with an increase in the height and width of an isothermal heater located at the bottom of a cavity has already been reported by AlAmiri et al. [16] for pure fluid. The influence of the position D of the block heater on the heat transfer is depicted in Figure 6(c). Qualitatively, the results are similar to those reported for the case of the block when heated by a constant heat flux (Figure 4(c)). Quantitatively, Figure 6(c) indicates that, when Φ increases from 0 to 5%, the Nusselt number grows by about 35% at $D = 0.15$ and by about 33% at $D = 0.5$.

5. Conclusions

In this paper, a numerical study of natural convection heat transfer in a square enclosure filled with a suspension of copper-water nanofluid is carried out. The simulations were performed using the lattice Boltzmann approach. The system is heated, by a protruding heater located at the bottom of the system, and cooled isothermally from the two vertical walls. The heater is assumed either to be isothermal or to generate a constant heat flux. The governing parameters of the problem are the thermal Rayleigh number, Ra, the solid volume fraction of nanoparticles, Φ, the size (L and W) of the heater and its position, D. The main conclusions of the present analysis are as follows.

(1) In the range of the governing parameters considered in this study, for a given geometry and position of the heater, the heat transfer is improved by enhancing both the Rayleigh number and the nanoparticle volume fraction.

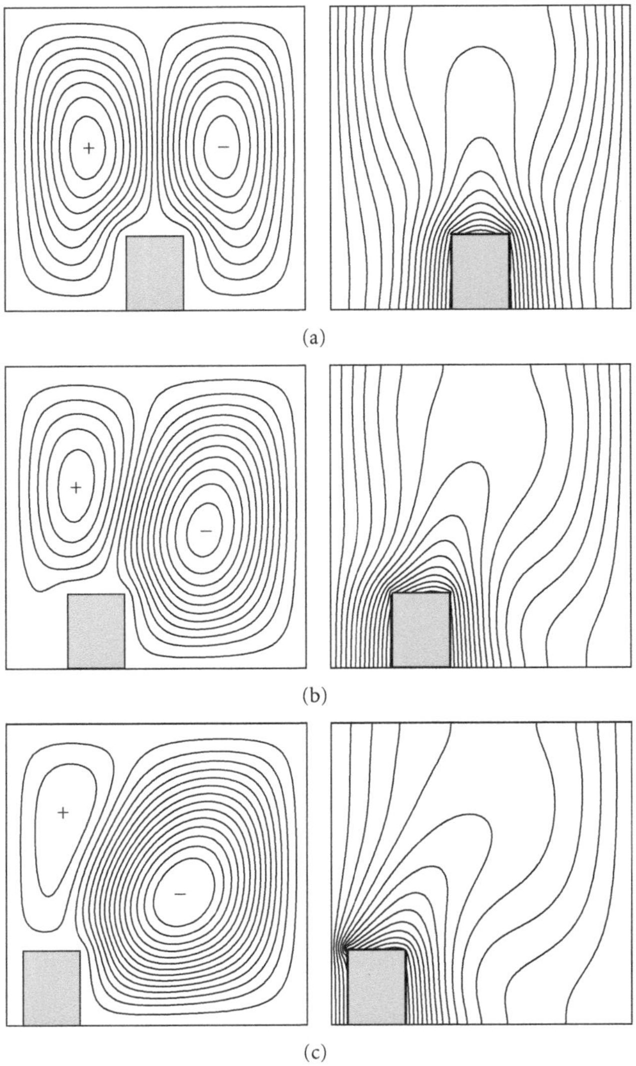

FIGURE 5: Effect of varying the position D of the constant heat flux heater on the streamlines (left) and isotherms (right) for $\mathrm{Ra} = 10^5$, $\Phi = 0.05$, $L = 0.25$, $W = 0.2$ and: (a) $D = 0.50$: $\Psi_{\max} = 0.0153$, $\Psi_{\min} = -0.0153$, $T_{\max} = 0.1677$, $\overline{\mathrm{Nu}} = 6.6556$; (b) $D = 0.30$: $\Psi_{\max} = 0.0092$, $\Psi_{\min} = -0.0190$, $T_{\max} = 0.1526$, $\overline{\mathrm{Nu}} = 7.3719$; (c) $D = 0.15$: $\Psi_{\max} = 0.0036$, $\Psi_{\min} = -0.0199$, $T_{\max} = 0.1476$, $\overline{\mathrm{Nu}} = 13.9672$.

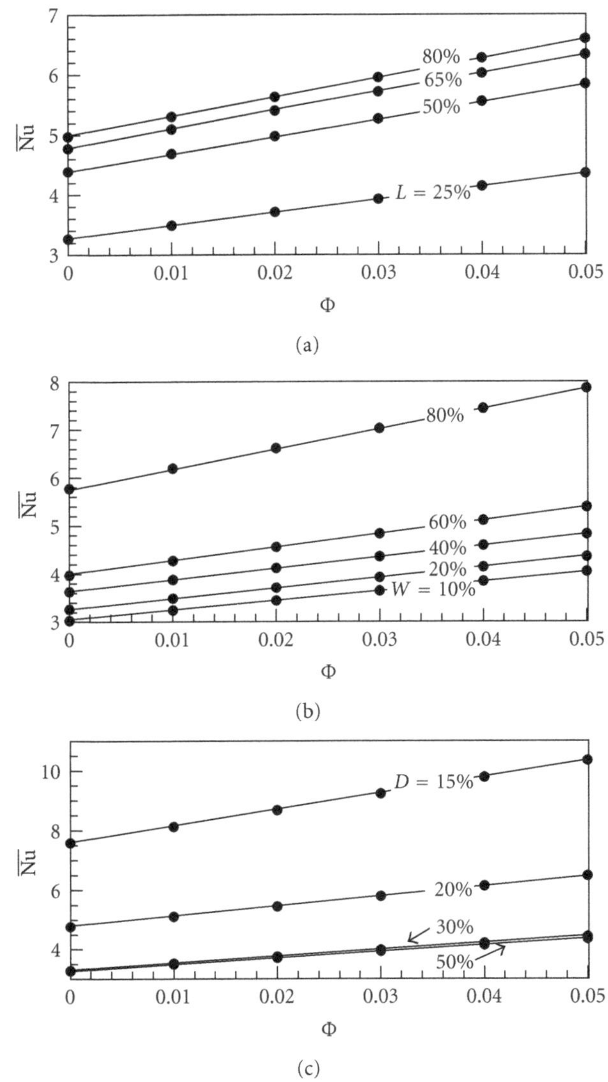

FIGURE 6: Variation of the average Nusselt number $\overline{\mathrm{Nu}}$ (computed on the left vertical wall), for an isothermal heater, with solid volume fraction Φ for $\mathrm{Ra} = 10^5$: (a) effect of the heater height L for $D = 0.5$ and $W = 0.2$; (b) effect of the heater width W for $D = 0.5$ and $L = 0.25$; (c) effect of the heater position D for $L = 0.25$ and $W = 0.2$.

(2) For the case of an isothermal heater, the numerical results indicate that, for a given solid volume fraction of nanoparticles, the heat transfer is improved when the size of the heating source increases. Also, the average Nusselt number is enhanced as the position of the heating block is moved from the center position toward the vertical isothermal boundaries of the enclosure.

(3) For the case of a heater with a constant heat source, for a given Rayleigh number and solid volume fraction of nanoparticles, the increase in the height of the heater results in a rise in the maximum surface temperature of the element. Consequently, the average Nusselt number decreases with the increase in the height of the heat source. The influence of the width, W, on $\overline{\mathrm{Nu}}$ is more complex. When it increases from $W = 0.1$, $\overline{\mathrm{Nu}}$ first decreases to reach a minimum at $W \approx 0.4$, above which the heat transfer increases monotonically with W. As in the case of the isothermal heater, as the position D of the heating block is moved from the center of the bottom wall of the cavity towards the left (right) vertical cooled wall, the maximum temperature of the element decreases, which improves $\overline{\mathrm{Nu}}$.

Nomenclature

C_p: Heat capacitance
D: Dimensionless location of the heater (d'/H')

d': Distance of the heat element from the left wall
d_{np}: Diameter of a solid nanoparticle
d_f: Diameter of a liquid molecule
g: Gravitational acceleration
H': Height of cavity
k: Thermal conductivity
k_b: Boltzmann's constant ($k_b = 1.38065 \times 10^{-23}$)
L: Dimensionless height of the heater (l'/H')
l': Height of the heater
Nu: Nusselt number (23) and(25)
$\overline{\mathrm{Nu}}$: Average Nusselt number (24) and(26)
Pe: Péclet number, ($u_{\mathrm{np}} d_{\mathrm{np}}/\alpha_f$)
Pr: Prandtl number of pure fluid (ν_f/α_f)
q' Constant heat flux per unit area
Ra: Thermal Rayleigh number ($g\beta'_f \Delta T' H'^3/\alpha_f \nu_f$)
t: Dimensionless time ($t'\alpha_f \mathrm{Ra}^{1/2}/H'^2$)
T: Dimensionless temperature ($(T' - T'_0)/\Delta T'$)
T'_0: Reference temperature
u: Dimensionless velocity in x direction ($u'H'/\alpha_f \mathrm{Ra}^{1/2}$)
u_{np}: Brownian motion velocity of nanoparticles
v: Dimensionless velocity in y direction ($v'H'/\alpha_f \mathrm{Ra}^{1/2}$)
W: Dimensionless width of the heater (w'/H')
w': Width of the heater
x: Dimensionless coordinate axis (x'/H')
y: Dimensionless coordinate axis (y'/H').

Greek Symbols

α: Thermal diffusivity
$\overline{\alpha}$: Dimensionless parameter ($\alpha_{\mathrm{nf}}/\alpha_f$)
β': Thermal expansion coefficient
$\overline{\beta}$: Dimensionless parameter ($(\rho\beta')_{\mathrm{nf}}/(\rho\beta')_f$)
ν: Kinematic viscosity of fluid
$\overline{\nu}$: Dimensionless parameter (ν_{nf}/ν_f)
ρ: Density of fluid
Φ: Nanoparticle volume fraction
Ψ: Dimensionless stream function ($\Psi'/\alpha_f \mathrm{Ra}^{1/2}$)
ω: Dimensionless vorticity ($\omega'/\alpha_f \mathrm{Ra}^{1/2}$).

Subscript

0: Reference state
f: Pure fluid
max: Maximum values
min: Minimum values
nf: Nanofluid
np: Nanoparticle.

Superscript

$'$: It refers to a dimensional variable.

References

[1] S. U. S. Choi, "Enhancing thermal conductivity of fluids with nanoparticles," *Flow*, vol. 66, pp. 99–105, 1995.

[2] L. Godson, B. Raja, D. Mohan Lal, and S. Wongwises, "Enhancement of heat transfer using nanofluids—an overview," *Renewable and Sustainable Energy Reviews*, vol. 14, no. 2, pp. 629–641, 2010.

[3] K. Khanafer, K. Vafai, and M. Lightstone, "Buoyancy-driven heat transfer enhancement in a two-dimensional enclosure utilizing nanofluids," *International Journal of Heat and Mass Transfer*, vol. 46, no. 19, pp. 3639–3653, 2003.

[4] R. Y. Jou and S. C. Tzeng, "Numerical research of nature convective heat transfer enhancement filled with nanofluids in rectangular enclosures," *International Communications in Heat and Mass Transfer*, vol. 33, no. 6, pp. 727–736, 2006.

[5] C. J. Ho, M. W. Chen, and Z. W. Li, "Numerical simulation of natural convection of nanofluid in a square enclosure: effects due to uncertainties of viscosity and thermal conductivity," *International Journal of Heat and Mass Transfer*, vol. 51, no. 17-18, pp. 4506–4516, 2008.

[6] A. K. Santra, S. Sen, and N. Chakraborty, "Study of heat transfer augmentation in a differentially heated square cavity using copper-water nanofluid," *International Journal of Thermal Sciences*, vol. 47, no. 9, pp. 1113–1122, 2008.

[7] E. Abu-Nada, Z. Masoud, H. F. Oztop, and A. Campo, "Effect of nanofluid variable properties on natural convection in enclosures," *International Journal of Thermal Sciences*, vol. 49, no. 3, pp. 479–491, 2010.

[8] E. Abu-Nada, Z. Masoud, and A. Hijazi, "Natural convection heat transfer enhancement in horizontal concentric annuli using nanofluids," *International Communications in Heat and Mass Transfer*, vol. 35, no. 5, pp. 657–665, 2008.

[9] B. Ghasemi and S. M. Aminossadati, "Periodic natural convection in a nanofluid-filled enclosure with oscillating heat flux," *International Journal of Thermal Sciences*, vol. 49, no. 1, pp. 1–9, 2010.

[10] Z. Alloui, P. Vasseur, and M. Reggio, "Natural convection of nanofluids in a shallow cavity heated from below," *International Journal of Thermal Sciences*, vol. 50, no. 3, pp. 385–393, 2011.

[11] F. Moukalled and S. Acharya, "Buoyancy-induced heat transfer in partially divided trapezoidal cavities," *Numerical Heat Transfer; Part A*, vol. 32, no. 8, pp. 787–810, 1997.

[12] F. Moukalled and M. Darwish, "Natural convection in a partitioned trapezoidal cavity heated from the side," *Numerical Heat Transfer; Part A*, vol. 43, no. 5, pp. 543–563, 2003.

[13] M. A. Coman, G. O. Hughes, R. C. Kerr, and R. W. Griffiths, "The effect of a barrier on laminar convection in a box with differentially heated endwalls," *International Journal of Heat and Mass Transfer*, vol. 49, no. 17-18, pp. 2903–2911, 2006.

[14] H. Oztop and E. Bilgen, "Natural convection in differentially heated and partially divided square cavities with internal heat generation," *International Journal of Heat and Fluid Flow*, vol. 27, no. 3, pp. 466–475, 2006.

[15] Y. Varol, H. F. Oztop, and T. Yilmaz, "Natural convection in triangular enclosures with protruding isothermal heater," *International Journal of Heat and Mass Transfer*, vol. 50, no. 13-14, pp. 2451–2462, 2007.

[16] A. AlAmiri, K. Khanafer, and I. Pop, "Buoyancy-induced flow and heat transfer in a partially divided square enclosure," *International Journal of Heat and Mass Transfer*, vol. 52, no. 15-16, pp. 3818–3828, 2009.

[17] H. C. Brinkman, "The viscosity of concentrated suspensions and solutions," *The Journal of Chemical Physics*, vol. 20, no. 4, pp. 571–581, 1952.

[18] H. E. Patel, T. Sundararajan, T. Pradeep, A. Dasgupta, N. Dasgupta, and S. K. Das, "A micro-convection model for thermal conductivity of nanofluids," *Pramana—Journal of Physics*, vol. 65, no. 5, pp. 863–869, 2005.

[19] S. Succi, *The Lattice Boltzmann Equation for Fluid Dynamics and Beyond*, Oxford University Press, Oxford, UK, 2001.

[20] D. A. Wolf-Gladrow, *Lattice-Gas Cellular Automata and Lattice Boltzmann Models: An Introduction*, Springer, Berlin, Germany, 2000.

[21] Z. Chai and B. Shi, "A novel lattice Boltzmann model for the Poisson equation," *Applied Mathematical Modelling*, vol. 32, no. 10, pp. 2050–2058, 2008.

[22] S. Chen, J. Tölke, and M. Krafczyk, "Simulation of buoyancy-driven flows in a vertical cylinder using a simple lattice Boltzmann model," *Physical Review E*, vol. 79, no. 1, Article ID 016704, 2009.

[23] S. Chen, J. Tölke, and M. Krafczyk, "Simple lattice Boltzmann subgrid-scale model for convectional flows with high Rayleigh numbers within an enclosed circular annular cavity," *Physical Review E*, vol. 80, no. 2, Article ID 026702, 2009.

[24] J. Guiet, M. Reggio, P. Vasseur, and S. Leclaire, "Application of the lattice Boltzmann method to calculate the natural convection of nanofluids in a shallow cavity," in *Proceedings of the 18th Annual Conference of the CFD Society of Canada*, 2010.

[25] O. V. Trevisan and A. Bejan, "Natural convection with combined heat and mass transfer buoyancy effects in a porous medium," *International Journal of Heat and Mass Transfer*, vol. 28, no. 8, pp. 1597–1611, 1985.

13

Computational Fluid Dynamics Modeling to Improve Natural Flow Rate and Sweet Pepper Productivity in Greenhouse

W. Limtrakarn,[1] P. Boonmongkol,[1] A. Chompupoung,[2] K. Rungprateepthaworn,[3] J. Kruenate,[4] and P. Dechaumphai[5]

[1] *Department of Mechanical Engineering, Thammasat University, Pathumthani 12120, Thailand*
[2] *Royal Project Foundation, Chiang Mai 50200, Thailand*
[3] *National Electronics and Computer Technology Center, Klong Luang, Pathumthani 12120, Thailand*
[4] *National Metal and Materials Technology Center, Klong Luang, Pathumthani 12120, Thailand*
[5] *Department of Mechanical Engineering, Chulalongkorn University, Bangkok 10330, Thailand*

Correspondence should be addressed to W. Limtrakarn, limwiroj@tu.ac.th

Academic Editor: Hakan F. Oztop

Natural flow rate and sweet peppers productivity in tropical greenhouse are improved by CFD simulation is the main objective of this research work. Most of the greenhouse types today are in the arch shape. To develop an improved greenhouse structure for the region, the arch type was built and used as the control model. Mae Sar Mai agriculture research station under the royal project foundation was selected as the field test site. Temperature sensors with data logger were installed to monitor variation of temperature inside the greenhouse. The measured temperature data were used as the boundary conditions for the CFD analysis. A new greenhouse model with two-step roof shape was designed and the air flow behavior was simulated by using CFD. Regarding CFD results the air flow rate of the new model is about 39% higher than that of old model. The maximum temperature of the new model is lower than that of the old one. The sweet paper growths in both greenhouse models were measured and compared. Results show that the new model obtains 4°C lower maximum temperature in day time, 97% in number and 90% in weight higher the first grade pepper productivity than the old one.

1. Introduction

During the past decades, several types of greenhouses with different shapes have been used for crop growing in Thailand. These are the Ridge and Furrow, Saw-tooth, Venlo, Quonset, Gothic arch, and Tunnel shapes. Most of them were originally designed and used in the European countries. These models are introduced and constructed in the northern region of Thailand. In the past, most greenhouses can not yield good production as they do in the cool climate countries. Structure configurations have been modified so that they are suitable for the high-land area. Plant production and growth in tropical countries strongly depend on the greenhouse shapes and their dimensions. They also depend on the temperature, humidity, ventilation, solar heat transmission, covering materials, CO_2, and so forth [1, 2]. Many research works have been performed to improve their performance by studying natural ventilation inside [3–11]. Few of them studied the temperature, pressure and flow patterns by using the CFD analysis [12–14] and effect from the insect-proof screen installation [7, 15].

Plastic greenhouse is mostly used in the northern region of Thailand because of its low cost, ease of installation, and maintenance. There are some research works that study the plant production in plastic greenhouses [2, 16, 17]. Recently, some works have investigated the micro-climate effect for increasing plant productivity in the greenhouses [18, 19].

The main objective of this study is to redesign the roof shape of a greenhouse for reducing overall inside temperature level. The study employs the CFD analysis simulation to help increasing natural flow rate and sweet pepper productivity. Greenhouse structure is made of steel, covered with plastic sheet, and built at Mae Sa Mai village (900 meters above the sea level) of the royal project, Chiang Mai province

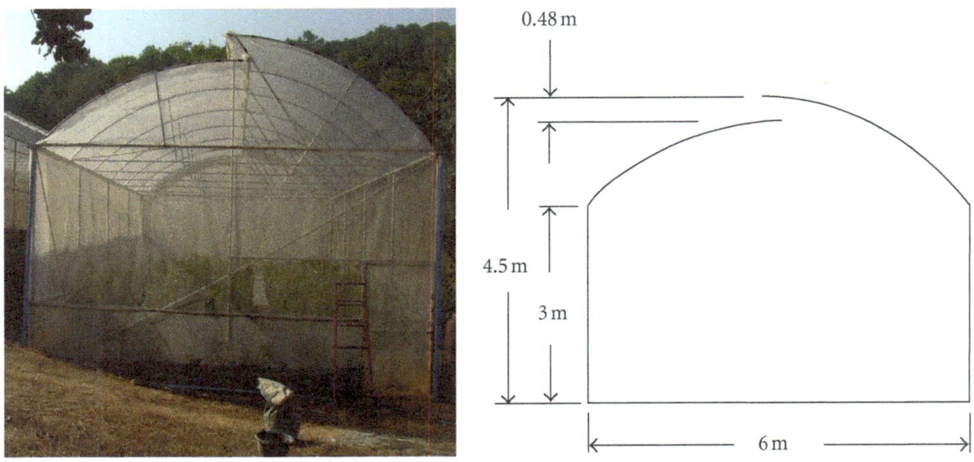

(a) (b)

Figure 1: Arch-type greenhouse configuration.

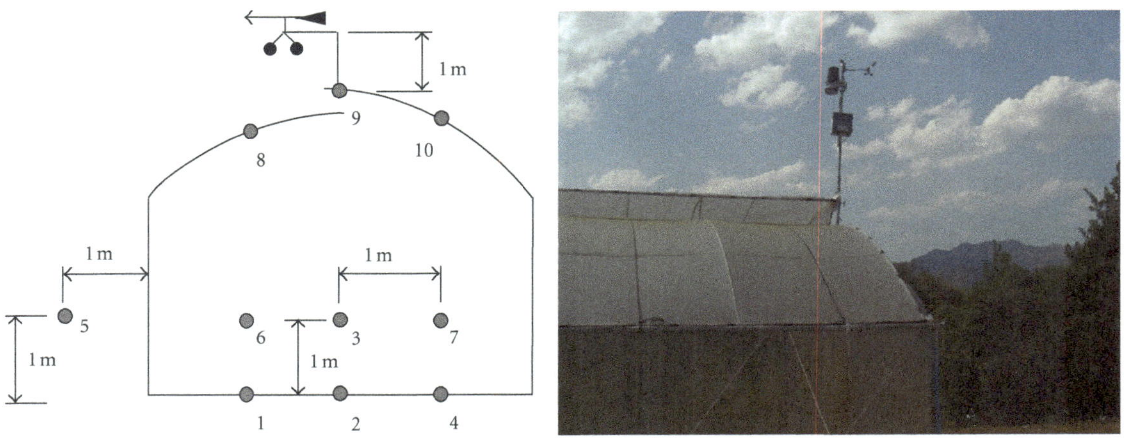

(a) T & H sensors location (b) Weathering station location

Figure 2: T & H sensors and weathering station location.

in Thailand. Sweet pepper crop is a prospect plant, which grows from April to August. In this paper, the arch-type greenhouse is selected as a controlled model. Ten temperature and relative humidity (T&H) sensors are installed to measure the climate change inside the greenhouse. These measured data are used as the boundary conditions for predicting the internal air flow behavior. To arrive at a design shape of the greenhouse that can provide lower interior temperature, the CFD analysis simulation is used. The new model is designed with two-step roof configuration. Prototypes of the old and new models are constructed and with temperature and humidity sensors installed inside. Sweet pepper nurseries are then moved into these two models. Interior temperature and sweet pepper production of two models are examined and recorded.

2. Materials and Methods

2.1. Greenhouse Configuration and Measurements. Survey of existing greenhouses at Mae Sa Mai village in Chiang Mai province shows that there are four types of greenhouses used. About 80% of them are the arch type, 16% are the Saw-tooth type, 2% are the Quonset type, and 2% are the Ridge and Furrow type. Thus, the arc type greenhouse was selected in this research work. Figure 1 shows the arch-type configuration, which was built at the latitude 18° 52′ North, longitude 98° 51′ East, and 900 meters above the sea level. Its dimensions are 6, 5 and 12 m in length, height and width, respectively. The roof is covered by a plastic with UV filter. An insect screen was installed for all sides and the greenhouse door. Air temperature and relative humidity were measured every 5 minutes using the thermistor sensors and data logger DLT-1 10000 M at the roof, the middle, and on the ground. The data collection system sent all data directly to a data server in Bangkok using GSM/General Packet Radio Service (GPRS) gateway connection. Figure 2 shows locations of the nine sensors inside the greenhouse, with one weather station above roof and one sensor outside the greenhouse. The measurement data were collected during January through August.

2.2. CFD Numerical Model. The commercial CFD software, ANSYS/CFX, was used to generate a mesh model for predicting air flow behavior inside. The air flow behavior is governed by the Reynolds-Averaged Navier-Stokes differential equations in the form

$$\frac{\partial \overline{u}}{\partial x}+\frac{\partial \overline{v}}{\partial y}+\frac{\partial \overline{w}}{\partial z}=0, \quad (1)$$

$$\begin{aligned}\rho\left(\frac{\partial \overline{u}}{\partial t}+\overline{u}\frac{\partial \overline{u}}{\partial x}+\overline{v}\frac{\partial \overline{u}}{\partial x}+\overline{w}\frac{\partial \overline{u}}{\partial x}\right)&=-\frac{\partial \overline{p}}{\partial x}+\frac{\partial}{\partial x}\left[\mu\left(\frac{\partial \overline{u}}{\partial x}+\frac{\partial \overline{u}}{\partial x}\right)-\rho u'u'\right]\\&+\frac{\partial}{\partial y}\left[\mu\left(\frac{\partial \overline{u}}{\partial y}+\frac{\partial \overline{v}}{\partial x}\right)-\rho u'v'\right]\\&+\frac{\partial}{\partial z}\left[\mu\left(\frac{\partial \overline{u}}{\partial z}+\frac{\partial \overline{w}}{\partial x}\right)-\rho u'w'\right],\end{aligned} \quad (2)$$

$$\begin{aligned}\rho\left(\frac{\partial \overline{v}}{\partial t}+\overline{u}\frac{\partial \overline{v}}{\partial x}+\overline{v}\frac{\partial \overline{v}}{\partial x}+\overline{w}\frac{\partial \overline{v}}{\partial x}\right)&=-\frac{\partial \overline{p}}{\partial y}+\frac{\partial}{\partial x}\left[\mu\left(\frac{\partial \overline{v}}{\partial x}+\frac{\partial \overline{u}}{\partial y}\right)-\rho v'u'\right]\\&+\frac{\partial}{\partial y}\left[\mu\left(\frac{\partial \overline{v}}{\partial y}+\frac{\partial \overline{v}}{\partial x}\right)-\rho v'v'\right]\\&+\frac{\partial}{\partial z}\left[\mu\left(\frac{\partial \overline{v}}{\partial z}+\frac{\partial \overline{w}}{\partial x}\right)-\rho v'w'\right],\end{aligned} \quad (3)$$

$$\begin{aligned}\rho\left(\frac{\partial \overline{w}}{\partial t}+\overline{u}\frac{\partial \overline{w}}{\partial x}+\overline{v}\frac{\partial \overline{w}}{\partial x}+\overline{w}\frac{\partial \overline{w}}{\partial x}\right)&=-\frac{\partial \overline{p}}{\partial z}+\frac{\partial}{\partial x}\left[\mu\left(\frac{\partial \overline{w}}{\partial x}+\frac{\partial \overline{u}}{\partial z}\right)-\rho w'u'\right]\\&+\frac{\partial}{\partial y}\left[\mu\left(\frac{\partial \overline{w}}{\partial y}+\frac{\partial \overline{v}}{\partial x}\right)-\rho w'v'\right]\\&+\frac{\partial}{\partial z}\left[\mu\left(\frac{\partial \overline{w}}{\partial z}+\frac{\partial \overline{w}}{\partial x}\right)-\rho w'w'\right],\end{aligned} \quad (4)$$

$$\begin{aligned}\frac{\partial}{\partial t}\left[\rho\left(e+\frac{\overline{V}^2}{2}\right)\right]+\vec{\nabla}\cdot\left[\rho\left(e+\frac{\overline{V}^2}{2}\right)\vec{V}\right]&=\rho\overline{Q}+\frac{\partial}{\partial x}\left(k\frac{\partial T}{\partial x}\right)+\frac{\partial}{\partial y}\left(k\frac{\partial T}{\partial y}\right)\\&-\frac{\partial(up)}{\partial x}-\frac{\partial(vp)}{\partial y}+\frac{\partial(\overline{u}\sigma_x)}{\partial x}+\frac{\partial(\overline{u}\tau_{yx})}{\partial y}\\&+\frac{\partial(\overline{v}\tau_{xy})}{\partial x}+\frac{\partial(\overline{v}\sigma_y)}{\partial y}+\mathrm{p}\,\vec{\mathrm{f}}\cdot\vec{V},\end{aligned} \quad (5)$$

where $\overline{u}$, $\overline{v}$, $\overline{w}$ are the mean velocity components in x, y, and z-directions, respectively, while u', v', w' are their fluctuating velocity components. In the above equations, ρ is the fluid density, p is the fluid pressure, e is the total energy, σ and τ are the normal and shear stress components, respectively.

Table 1: Air properties at Mae Sa Mai village.

Air Properties	
Density	1.185 kg/m^3
Specific heat	1004.4 J/kg·k
Thermal conductivity coefficient	0.0261 W/m·k
Viscosity	1.831×10^{-5} kg/m·s

Table 2: Average measured data at different heights on 11 February.

Measured location	Temperature (°C)
Roof	39.6
1 m above ground	26.4
Ground	32
Outside	25

Equation (1) represents the conservation of mass, while (2)–(4) are the conservation of momentums in x, y, and z directions, respectively. The last equation, (5), is the conservation of energy. These equations were solved by the finite volume method. The standard k-ε turbulence model was employed to represent the ventilation phenomena [20, 21]. The instantaneous fluid velocity is sum of the mean and fluctuating components of the turbulent velocity. In the standard κ-ε turbulence model, the turbulent viscosity is computed as a function of the turbulent energy (κ) and dissipation rate of the turbulent kinetic energy (ε).

2.3. Computational Procedure. The air properties, as shown in Table 1, are the density at 900 m above the sea level, the specific heat, thermal conductivity coefficient, and viscosity at outside temperature. Boundary conditions were selected from the measure data of temperature sensors. Table 2 shows the average measure data at 2:00 pm on 11 February at the three heights and outside the model. Roof and ground temperatures were used as Dirichlet boundary conditions, while initial temperature was the outside temperature. The external wind speed and direction were measured from the weather station. It is noted that the wind speed is close to zero about 66% during the day time. Natural ventilation condition was selected in the analysis because the airflow outside the greenhouse is near a stagnation condition.

3. Results and Discussion

3.1. Measurement System and Data. The temperature, relative humidity, and wind speed data were collected at every 5 minutes and sent through the GSM/GPRS system as well as recorded on a computer server in Bangkok. The data bank can be accessed online by the internet system. Figure 3 shows the average greenhouse temperature that varies with time during 24 hours on 11 February. The greenhouse started receiving the solar heating at 7 AM, and the heat was accumulated during the day time. Temperature was above 24°C between 9 AM 5 PM. The maximum temperature of about 31°C occurred between 1 and 4 pm. It is noted that the sweet pepper crop is originated from cold regions where

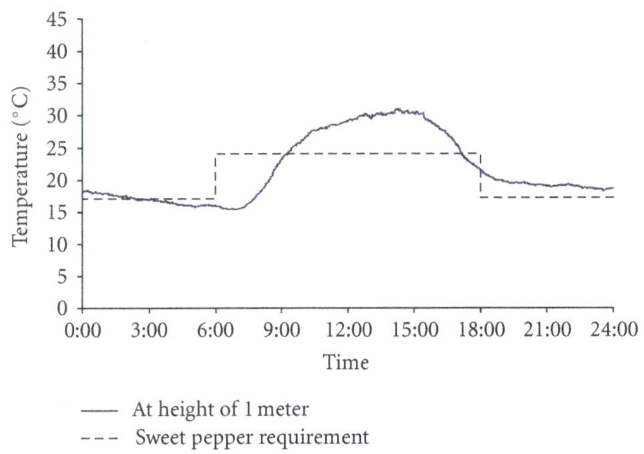

FIGURE 3: Average temperature of arch-type greenhouse on 11 February.

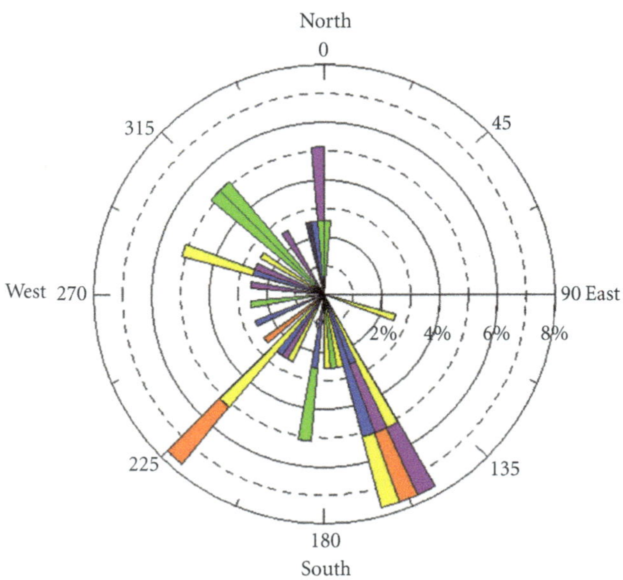

FIGURE 4: Wind speeds and directions on 11 February.

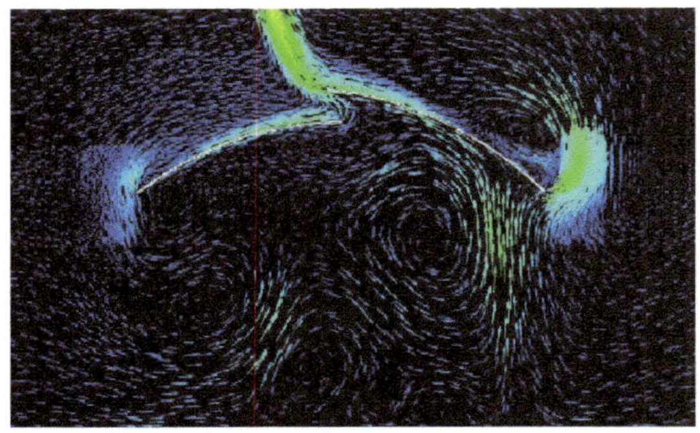

FIGURE 5: Air flow velocity vectors in the arch-type greenhouse.

suitable temperature is about 24°C during the day time and 17°C during the night time. This data implies that the greenhouse needs better ventilation to release accumulated heat during the day time. In this paper, the greenhouse configuration was redesigned to provide lower temperature. Figure 4 shows the wind chart that represents wind speed and its direction. Maximum wind speed is at 3 km/hr and its direction is 347 NW. The wind speed is near zero about 66% during the day time.

3.2. Simulation Results. With the arc-type greenhouse configuration, the CFD analysis for the air flow behavior was performed. CFD mesh is constructed and consists of 11,000 nodes. Figure 5 shows the air flow velocity vectors inside the greenhouse and near the roof opening. Hot air moves upward and flows out from the greenhouse through the roof opening. Because of small opening at roof, there are large circulation under right roof, medium, and small circulation above floor from left to right. Heat is then accumulated all day and increases the temperature level inside the greenhouse. From the CFD result, it is implied that large circulation under right roof is the major factor in heat accumulation inside greenhouse.

Figure 6(a) shows the temperature distribution at 1 meter height above the ground as compared to the measured data. The temperatures at different heights were plotted as shown in Figure 6(b). The temperature results are agreed with the measured data. Regarding high temperature problem in greenhouse, the new design does focus on reduction of accumulated heat using CFD.

Based on this concept the roof opening distance is considered. For the redesign and optimized greenhouse configuration, the two-step roof greenhouse shape was selected as shown in Figure 7. Additional roof or second roof is added above the original roof with higher opening distance. Then redesign model is constructed and generated CFD mesh. CFD analysis is then used to solve flow behavior of redesign GH model. Several redesign models were simulated by using CFD analysis to optimize opening distance. The optimized CFD mesh consists of 14,000 nodes. Grid independence is also considered in all CFD model.

Figure 8 shows the air flow velocity vectors inside and near the roof openings of the new greenhouse shape. Large circulation under right roof in original model is disappeared in redesign model result. At the roof openings, the air flow rate of the redesign model is about 39% higher than that of old model. Temperature at 1 meter height and along the height of the greenhouse were plotted and compared as shown in Figures 6(a) and 6(b), respectively. At 1 meter height, the temperature in the redesign model is lower than that of the old model. The two-step roof model was then constructed and the measurement sensors were installed in the same locations as used in the old model.

Figure 9 shows the comparison of measured temperatures between two models. The redesign model yields a lower temperature than the old model about 4°C during the day time and higher about 1°C during the night time.

3.3. Productivity Results. Sweet pepper crop was plant between May and August. In such period, the plant growth and its productivity were collected 5 times as shown in

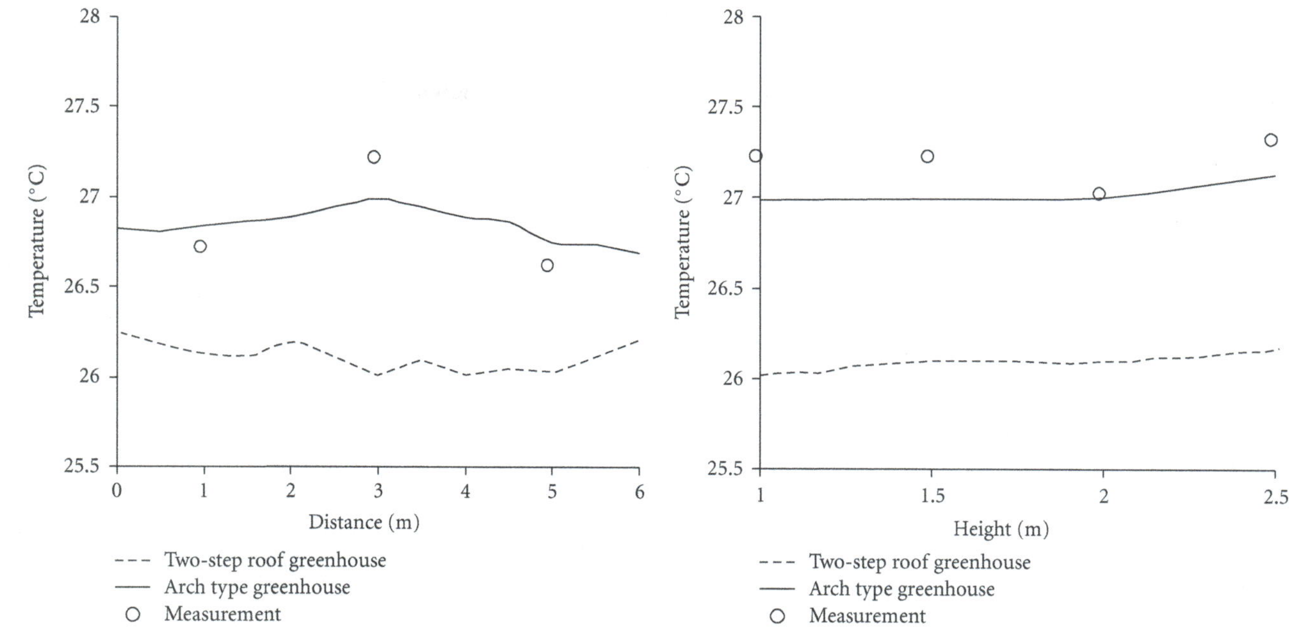

(a) at 1 meter height

(b) Along the height

FIGURE 6: Comparative temperatures between the simulated and measured data.

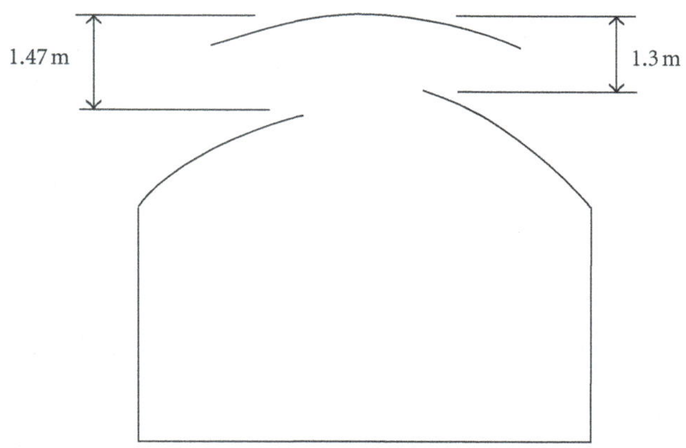

FIGURE 7: Two-step roof greenhouse configuration.

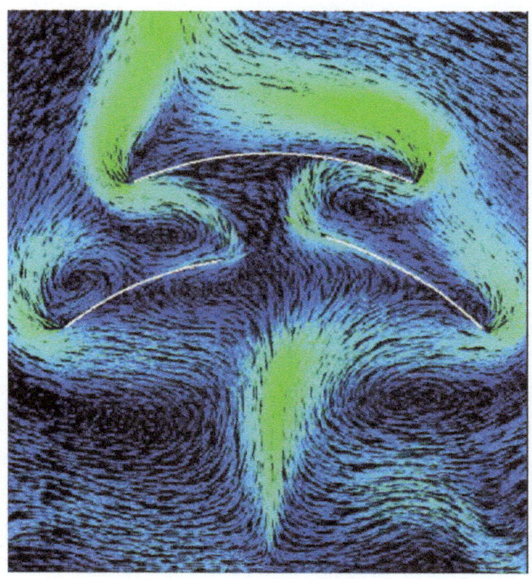

FIGURE 8: Air flow velocity vectors in the two-step roof greenhouse.

Table 3. Sweet pepper crop is classified in 2 grades. The first-grade is for those with high quality, while the second grade is with fair quality. The first grade productivities obtained from the new two-step roof, and the old models are 97% and 90%, respectively. The second-grade productivities obtained from the two models are 64% and 70%, respectively.

4. Concluding Remarks

The CFD simulation is used to enhance natural flow rate and sweet pepper productivity in tropical greenhouses. The CFD technique helped redesigning greenhouse configuration to achieve better ventilation reduce heat storage and temperature inside. The existing and popular greenhouse type, that is, the arch type, at Mae Sa Mai village was selected as the control greenhouse model. The measurement system was installed and the data were transmitted and recorded in a remote server by GSM/GPRS system. A new design configuration with two-step roof was selected. The maximum temperature in new design model is less than that of the old model. At the roof opening section, the air flow rate of new model is 39% higher than that of old model. The two-step roof greenhouse was constructed and a measurement system was installed. Sweet pepper crop was planted inside the old and new models. The maximum temperature measured in the new model is about 4°C less than that of the old model during the day time.

TABLE 3: Plant growth and productivity of sweet pepper crop.

Item	d/m/y	First grade				Second grade			
		Two-step roof GH		Arch-type GH		Two-step roof GH		Arch-type GH	
		Number of fruit	Total Weight	Number of fruit	Total Weight	Number of fruit	Total Weight	Number of fruit	Total Weight
1	26/6/08	76	15.2	40	8	10	2.2	7	1.6
2	3/7/08	54	10.8	37	7.4	63	12.6	42	8.4
3	10/7/08	85	17	38	7.5	65	13	40	8.1
4	17/7/08	18	3	8	2	49	9	20	3
5	27/7/08	35	7	15	3	12	2	13	2
Mean		48	9.45	24	4.975	47	9.15	29	5.375

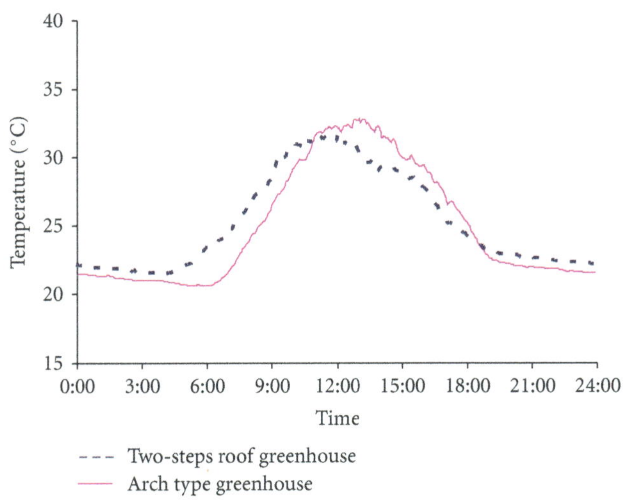

FIGURE 9: Comparative measured temperatures of the two greenhouse models.

The simulated CFD and the measured temperatures are in good agreement. The new design model produced the first-grade sweet pepper productivity about 97% in number and 90% in weight higher than those from the old model. The results have demonstrated the applicability and advantages of the CFD simulation for redesigning the greenhouse configuration to enhance natural flow rate and sweet pepper productivity.

Acknowledgments

The authors are pleased to acknowledge the Thailand Commission on Higher Education of Thailand (the National Research University Project), the National Science and Technology (NRCT), the Thailand Research Fund (TRF), Royal Project Foundation, Thammasat University, and the National Metal and Materials Technology Center (MTEC) for supporting this paper.

References

[1] B. Von Elsner, D. Briassoulis, D. Waaijenberg et al., "Review of structural and functional characteristics of greenhouses in European Union countries, part II: typical designs," *Journal of Agricultural Engineering Research*, vol. 75, no. 2, pp. 111–126, 2000.

[2] B. Cemek, Y. Demir, S. Uzun, and V. Ceyhan, "The effects of different greenhouse covering materials on energy requirement, growth and yield of aubergine," *Energy*, vol. 31, no. 12, pp. 1444–1452, 2006.

[3] T. Boulard, G. Papadakis, C. Kittas, and M. Mermier, "Air flow and associated sensible heat exchanges in a naturally ventilated greenhouse," *Agricultural and Forest Meteorology*, vol. 88, no. 1–4, pp. 111–119, 1997.

[4] A. Mistriotis, C. Arcidiacono, P. Picuno, G. P. A. Bot, and G. Scarascia-Mugnozza, "Computational analysis of ventilation in greenhouses at zero- and low-wind-speeds," *Agricultural and Forest Meteorology*, vol. 88, no. 1–4, pp. 121–135, 1997.

[5] M. Teitel and J. Tanny, "Natural ventilation of greenhouses: experiments and model," *Agricultural and Forest Meteorology*, vol. 96, no. 1–3, pp. 59–70, 1999.

[6] S. Wang, T. Boulard, and R. Haxaire, "Air speed profiles in a naturally ventilated greenhouse with a tomato crop," *Agricultural and Forest Meteorology*, vol. 96, no. 4, pp. 181–188, 1999.

[7] O. F. Nielsen, "Natural ventilation of a greenhouse with top screen," *Biosystems Engineering*, vol. 81, no. 4, pp. 443–451, 2002.

[8] M. Kacira, S. Sase, and L. Okushima, "Effects of side vents and span numbers on wind-induced natural ventilation of a gothic multi-span greenhouse," *Japan Agricultural Research Quarterly*, vol. 38, no. 4, pp. 227–233, 2004.

[9] J. P. Parra, E. Baeza, J. I. Montero, and B. J. Bailey, "Natural ventilation of parral greenhouses," *Biosystems Engineering*, vol. 87, no. 3, pp. 355–366, 2004.

[10] F. J. Jiménez-Hornero, E. Gutiérrez De Ravé, R. Hidalgo, and J. V. Giráldez, "Numerical study of the natural airflow in greenhouses using a two-dimensional lattice model," *Biosystems Engineering*, vol. 91, no. 2, pp. 219–228, 2005.

[11] A. Kumar and G. N. Tiwari, "Thermal modeling of a natural convection greenhouse drying system for jaggery: an experimental validation," *Solar Energy*, vol. 80, no. 9, pp. 1135–1144, 2006.

[12] K. Kim, G. A. Giacomelii, J. Y. Yoon et al., "CFD modeling to improve the design of a fog system for cooling greenhouses," *Japan Agricultural Research Quarterly*, vol. 41, no. 4, pp. 283–290, 2007.

[13] S. A. O. Khaoua, P. E. Bournet, C. Migeon, T. Boulard, and G. Chassériaux, "Analysis of greenhouse ventilation efficiency based on computational fluid dynamics," *Biosystems Engineering*, vol. 95, no. 1, pp. 83–98, 2006.

[14] H. Fatnassi, T. Boulard, C. Poncet, and M. Chave, "Optimisation of greenhouse insect screening with computational fluid

dynamics," *Biosystems Engineering*, vol. 93, no. 3, pp. 301–312, 2006.

[15] T. Bartzanas, T. Boulard, and C. Kittas, "Numerical simulation of the airflow and temperature distribution in a tunnel greenhouse equipped with insect-proof screen in the openings," *Computers and Electronics in Agriculture*, vol. 34, no. 1–3, pp. 207–221, 2002.

[16] T. Boulard and S. Wang, "Experimental and numerical studies on the heterogeneity of crop transpiration in a plastic tunnel," *Computers and Electronics in Agriculture*, vol. 34, no. 1–3, pp. 173–190, 2002.

[17] A. Ghoname and M. R. Shafeek, "Growth and productivity of sweet pepper (*Capsicum annum* L.) grown in plastic house as affected by organic, mineral and bio-N-fertilisers," *Journal of Agronomy*, vol. 4, no. 4, pp. 369–372, 2005.

[18] H. E. S. Fath and K. Abdelrahman, "Micro-climatic environmental conditions inside a greenhouse with a built-in solar distillation system," *Desalination*, vol. 171, no. 3, pp. 267–287, 2005.

[19] C. Kittas and T. Bartzanas, "Greenhouse microclimate and dehumidification effectiveness under different ventilator configurations," *Building and Environment*, vol. 42, no. 10, pp. 3774–3784, 2007.

[20] B. E. Launder and D. B. Spalding, "The numerical computation of turbulent flows," *Computer Methods in Applied Mechanics and Engineering*, vol. 3, no. 2, pp. 269–289, 1974.

[21] M. Kacira, T. H. Short, and R. R. Stowell, "A CFD evaluation of naturally ventilated, multi-span, sawtooth greenhouses," *Transactions of the American Society of Agricultural Engineers*, vol. 41, no. 3, pp. 833–836, 1998.

Modeling and Experimental Investigation of Pressure Field in the Grinding Zone with Nanoparticle Jet of MQL

C. H. Li, Z. L. Han, Q. Zhang, and S. Wang

School of Mechanical Engineering, Qingdao Technological University, Qingdao 266033, China

Correspondence should be addressed to C. H. Li; sy_lichanghe@163.com

Academic Editor: Guan Heng Yeoh

*Solid nano*particles were added in minimum quantity lubrication (MQL) fluid medium to make nanofluids, that is, after the mixing and atomization of nanoparticle, lubricants and high pressure gas, to inject *solid nano*particle in the grinding zone with the form of jet flow. The mathematical model of two-phase flow pressure field of grinding zone with nanoparticle jet flow of MQL was established, and the simulation study was conducted. The results show that pressures in the grinding zone increased with the *acceleration* of grinding wheel, sharply decreased with the increased minimum clearance, and increased with the *acceleration* of jet flow. At three spraying angles of *nozzles*, when the *nozzle* angle was 15°, the pressure of grinding zone along the speed of grinding wheel was larger than the rest two angles. On the experimental platform built by KP-36 precision grinder and nanoparticle jet flow feed way, CY3018 pressure sensor was used to test the regularities of pressure field variations. The impact of the speed of grinding wheel, the gap between workpiece and grinding wheel, jet flow velocity, and spraying angles of *nozzles* on the pressure field of grinding zone was explored. The experimental result was generally consistent with the theoretical simulation, which verified the accuracy of the theoretical analysis.

1. Introduction

Minimum quantity lubrication (MQL) refers to the minimum quantity of lubricants that enters the high temperature grinding zone after being mixed in high pressure gas and atomized with high pressure draft (4.0–6.5 bar). The traditional poured and debridged feed liquid of grinding fluid is 60 L/h for a unit of the width of grinding wheel, while the consumption of MQL grinding fluid is 30–100 mL/h for a unit of the width of the grinding wheel [1–8]. High pressure draft serves as cooling and chip removal. Lubricants are attached to the finished surface of the workpiece, forming a layer of protective film and serving as the lubrication. This technology integrates the advantages of poured and debridged grinding and dry grinding, presenting similar lubrication effects compared with traditional poured and debridged grinding. Lubricants adopt vegetable oil as alkyl ester of base oil, which shows features such as excellent biodegradability, lubricating properties, high viscosity index, low volatility, recycling, short production cycle, and insignificant environmental diffusion. The consumption of lubricants is only parts per thousand or a few hundredths of a percentage point compared with the traditional grinding approaches, which greatly improved the working environment. Thus, high pressure draft is an efficient low-carbon processing technology. However, studies show that cooling effect of high pressure draft is too limited to meet the needs of strengthened high temperature heat transfer of the grinding zone [8–12]. The processing quality of the workpiece and the grinding wheel life is worse than the traditional poured and debridged grinding, indicating that MQL technique requires further improvements.

According to strengthened heat transfer theory, the heat transfer ability of the solid greatly exceeds the liquid and gas. At room temperature, the coefficient of thermal conductivity of solid materials is greater than the fluid material by several orders of magnitude. It can be estimated that the coefficient of thermal conductivity of liquid with suspended metal, nonmetallic, or polymeric solid particles exceeded the pure liquid significantly. If solid particles are added in MQL medium, it is expected to greatly increase the coefficient of thermal conductivity of fluid medium so as to improve the convective heat transfer and offset the defects of insufficient cooling effects of MQL. In addition, nano-particles (referring to ultrafine tiny solid particles with at least one dimension in the three-dimensional space, that is, in the nanoscale range (1–100 nm)) also present tribological features such as special antifriction and high carrying capacity in aspects of lubrication and tribology [13].

In this research, *solid nano* particles were added in MQL fluid medium to make nanofluids, that is, to inject *solid nano* particles after the mixing and atomization of nano-particle, lubricants (oil or oil-water mixture) and high pressure gas in the grinding zone with the form of jet flow. Nanofluids refer to the new heat transfer working medium with nanoscale metal or metallic oxide particles added in fluids at a certain way and ratio. In effect, from the aspect of its compositions, it is the two-phase suspension liquid formed by liquid and nano-particles and can be abbreviated as "nanofluids." Due to the small size effect of nano-particle, its behavior is similar with that of the liquid molecules; the intensive Brownian motion of nano-particle maintains its stable suspension and keeps it from precipitation [5, 6].

There are few differences between Brownian motion of suspended particulates and the thermal motion of molecules in the true solution, and the most significant difference is the form of the thermal motion of molecules. The former involves the integrated effects of the thermal motion impact by many molecules while the latter involves the thermal motion of a single kind of molecules. Hence, in the research of three-phase flow (including compressed air, the grinding liquid, and nano-particles), nano-particles can be made suspending in the grinding liquid to form nanofluids, and transport parameters of nanofluids can be calculated based on related theories, which can thus be abbreviated into a study on two-phase flow.

When the grinding liquid was injected in the wedge-shaped zone between the grinding wheel and the workpiece surface, the fluid dynamic pressure of the grinding liquid was formed. Guo and Malkin [7] applied mass conservation equation and momentum conservation equation in the mathematical modeling, based on the consideration of the impact of the porosity of the grinding wheel. With theoretical calculation, the depth of the grinding liquid in the grinding wheel was obtained, indicating that the grinding wheel with pores has the function to pump the grinding liquid from the grinding zone. Chang [8] and others considered the impact from the porosity of the grinding wheel. According to conservation equation, mathematical model the dynamic pressure of grinding liquid during the gradual feeding was established, and the simulation results show that the larger the supply of grinding liquid is, the larger the dynamic pressure of grinding liquid will be. Ganesan et al. [9] applied Renolds equation and Laplace equation to establish the cooling liquid equation for the infinite narrow grinding wheel and provided the analytical solution for calculating the dynamic pressure of fluid. Furutani et al. [10] considered the impact of degree of permeability of the grinding wheel and established the cooling liquid equation for the infinite wide grinding wheel, so as to solve the lifting force from the fluid pressure and the dynamic pressure. Engineer and others measured the flow of the grinding liquid.

With experimental measurements, the flow in the grinding zone was identified, indicating the significant role of porosity of the grinding wheel and the location of nozzles on the flow. Gviniashvili et al. [11] measured the depth of the grinding liquid in the grinding wheel and identified the relationship between the depth and the maximum temperature of the grinding workpiece. Ebbrell et al. [12], Li et al. [13], and Ganesan et al. [14] considered the impact of inertia force. Based on two-dimensional steady incompressible fluid Navier-Stokes equation, the fluid equation in grinding zone was deducted in planar grinding, and the dynamic pressure of the fluid was simulated and solved, which was verified in experiments. Hryniewicz et al. [15, 16] considered the influence of the roughness and established a two-dimensional mathematical model. The regularities of distribution of the dynamic pressure were simulated and solved, which was verified in experiments. They identified the relationship between the dynamic pressure and the speed of grinding wheel as well as minimum clearance between the grinding wheel and the workpiece. Based on theoretical and experimental analysis, the following conclusions were drawn: with other conditions identical, the larger the roughness, the larger the dynamic pressure of fluid will be. Furthermore, when the grinding wheel left the grinding zone, not all the grinding liquid was thrown from the grinding wheel. Only a part of the liquid left the grinding wheel, and the rest moved with the grinding wheel. In the experiment, Hryniewicz observed the following phenomenon: the grinding liquid left the grinding zone sticking to the grinding wheel surface rather than to the workpiece. Kim [17] and Lee et al. [18] designed experimental apparatus to test the effective flow rate of the grinding liquid in the grinding zone via the feeding with nozzles. They acquired the conclusion that the porosity of the grinding wheel and the position of the nozzles imposed a significant impact on the effective flow rate, while the feeding velocity and grinding depth of the workpiece has little impact over it. Klocke et al. [19] assumed that the radius of the grinding wheel was infinite, and the width of the workpiece and the grinding wheel was finite. Based on the assumption, a two-dimensional mathematical model of pressure distribution in the grinding zone was established according to the Navier-Stokes equation so as to calculate infinite width of the grinding wheel. The normal applied force generated by the dynamic pressure of grinding liquid was defined as Coolant induced force [20]. With the simulating calculation, the larger the velocity of grinding wheel is, the larger the normal force will be. Hryniewicz et al. [15] applied two types of nozzles in the experiment of cooling system. The first was formed by two nozzles, and the main nozzle was used to the feeding of the grinding liquid, and the assistant nozzle was used to break through "airbond" of air flow around the grinding wheel. The second only involves one nozzle for the feeding of the grinding liquid. It can be concluded from comparing the experimental results that the feeding nozzle of two nozzles has lower than that of one nozzle. Klocke [19] studied the percentage experiment of the grinding liquid in the grinding zone that accounts for the entire grinding liquid and summed up that when the flow of the grinding liquid was increased, Coolant induced force was enlarged. Yet when the ratio of

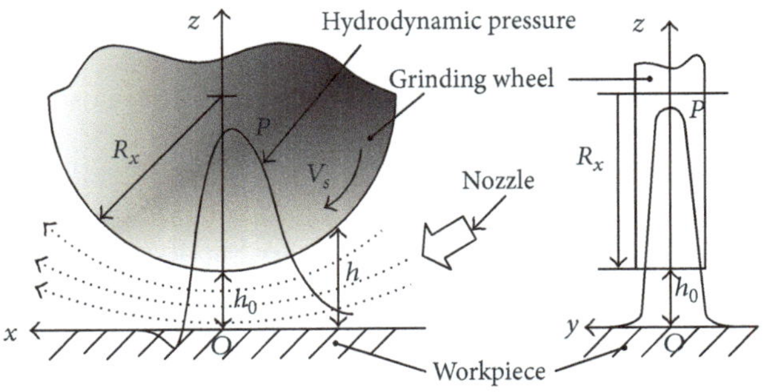

FIGURE 1: The sketch of geometrical model.

the grinding liquid in the grinding zone reached a certain value, it will increase with the increasing flow. Hryniewicz [15] conducted experiments of two grinding liquids (the oiliness grinding liquid and chemical synthetic fluids) and found that in the same experimental conditions, the maximum dynamic pressure generated by the oiliness of the grinding liquid was much larger than that from chemical synthetic fluids. In this research, the modeling and experimental investigation was conducted on nano-particle jet flow of MQL grinding pressure field.

2. Theoretical Modeling of Pressure Field in the Grinding Zone

Figure 1 shows the geometrical model. The three-dimensional system of coordinate was established, and the origin of coordinates was located at the workpiece surface between the grinding wheel and the minimum clearance. The direction of rotation of the workpiece surface along the grinding wheel was taken as X axis, and the direction vertical to the workpiece surface was taken as Y axis, that along the width of the grinding wheel as Z axis.

2.1. Modeling Conditions

(1) *Compressibility of Gas.* The room temperature was taken as $T = 15^\circ$C, the peripheral velocity of grinding wheel as $V_s = 45$ m/s, and it was obtained that sound velocity: $a = 20.05\sqrt{T} = 340$ m/s; Mach number: $M_a = V_s/a = 45/340 = 0.1323 < 0.3$. As $M_a < 0.3$, the compressibility of gas can be excused from consideration in the fluid analysis.

(2) *Identification of Fuid State.* $R_e = (2b_0u_0)/v = (2 \times 45 \times 0.001)/(0.00105 \times 10^{-3}) = 85714.3 > 2300$; where $2b_0$—half width of outlet; u_0—outlet velocity; v—nanofluids dynamic viscosity. $R_e > 2300$, so it is identified as turbulence flow. With the considerations of the following aspects (the fluid evenly flows in the grinding zone, the fluid is incompressible Newtonian flow, and the fluid in intervals is turbulence), for simplified calculation, the model was simplified as [21]:

(i) to neglect the impact of temperature and pressure on fluid viscosity, or that of the inertial effect in the flow;

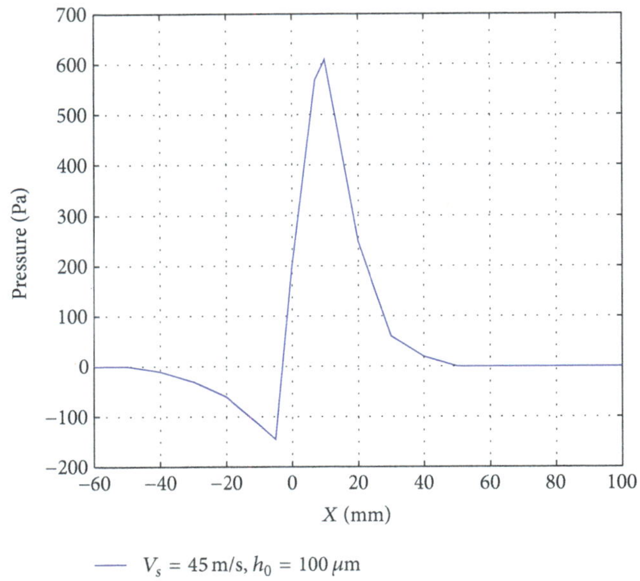

FIGURE 2: Pressure distribution of flow field in the X direction.

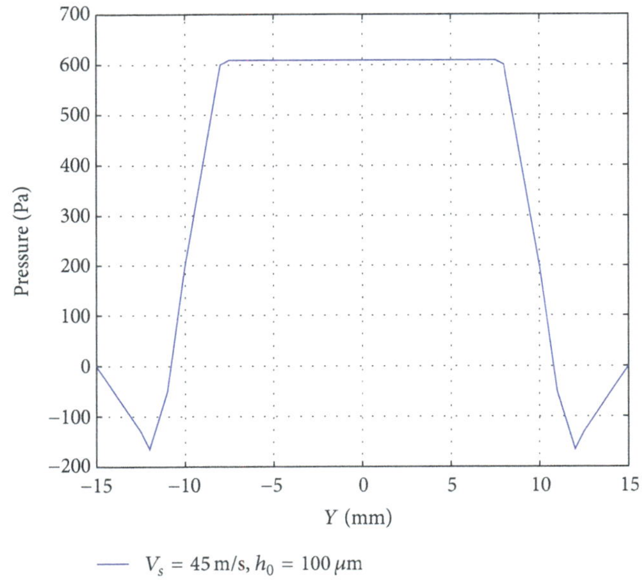

FIGURE 3: Pressure distribution of flow field in the Y direction.

(ii) to not consider the impact of the workpiece surface roughness and elastic deformation;

(iii) to neglect the ratio of feeding velocity of the workpiece and the peripheral velocity of grinding wheel;

(iv) to neglect the ratio of fluid inertia force, body force, and surface force in nanofluid film.

2.2. The Establishment of Mathematical Model. According to the continuity equation, momentum conservation equation and energy conservation equation, of the hydromechanics, the mathematical model of the stress field of the grinding zone was established.

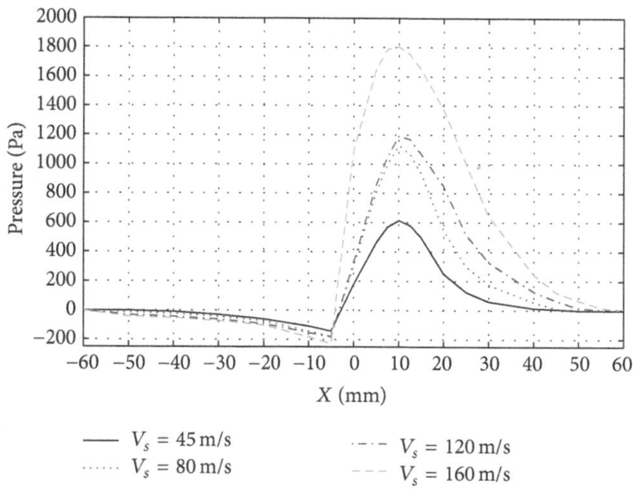

FIGURE 4: Pressure in grinding zone along the velocity change curve of wheel along the velocity direction of grinding wheel.

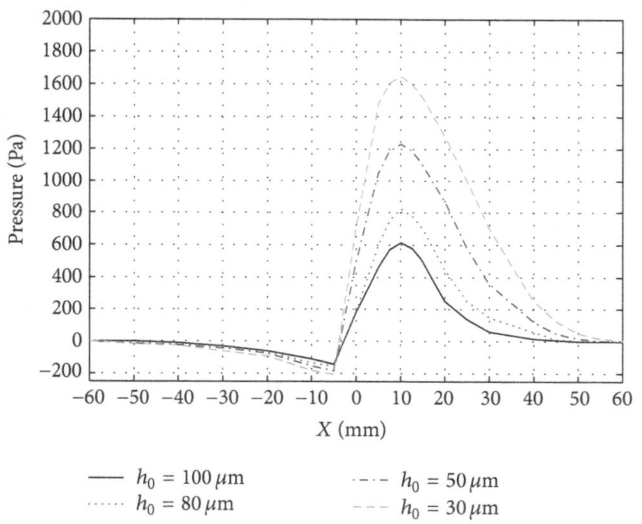

FIGURE 5: Pressure in grinding zone along the velocity change curve of wheel along the changing curve of minimum clearance.

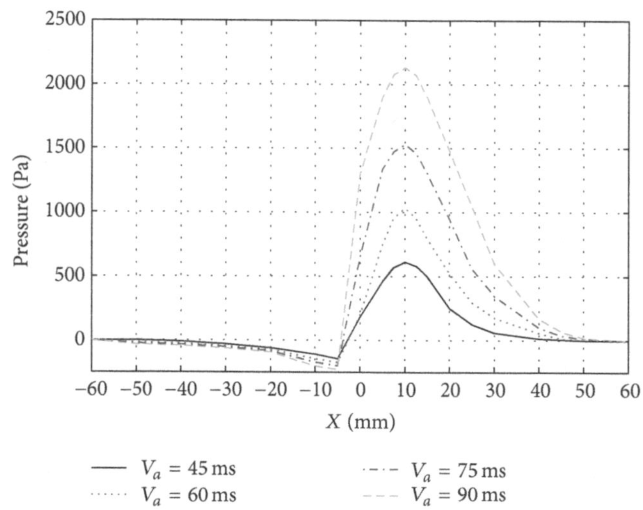

FIGURE 6: Pressure in grinding zone along the velocity change curve of wheel along the changing curve of jet flow velocity.

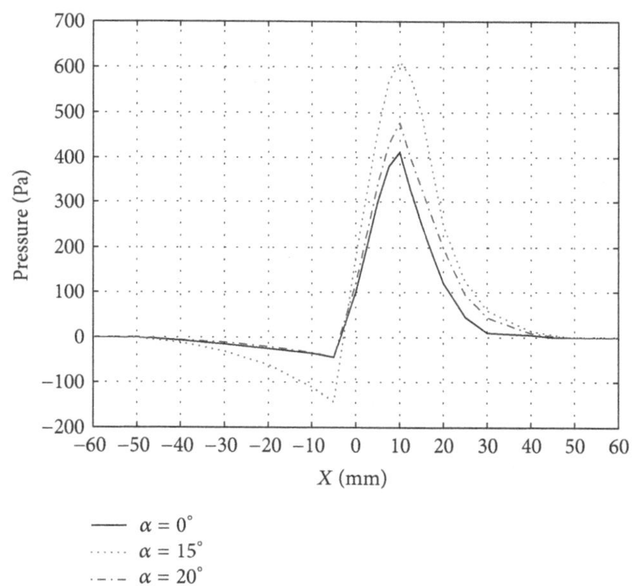

FIGURE 7: Pressure in grinding zone along the velocity change curve of wheel along the changing curve of positions and angles of *nozzle*.

(1) Continuity equation:

$$\frac{\partial \rho}{\partial t} + \frac{\partial u_i}{\partial X_i} = 0. \tag{1}$$

(2) Momentum conservation equation [22]:

$$\frac{\partial u_i}{\partial t} + u_j \frac{\partial u_i}{\partial X_j} = -\frac{1}{\rho}\frac{\partial \rho}{\partial x_i} + \gamma \frac{\partial^2 u_i}{\partial x_j x_j} + \frac{1}{\rho}\frac{\partial \left(\rho \overline{\mu_i' \mu_j'}\right)}{\partial X_j}, \tag{2}$$

where u_i, u_j—average velocity of nanofluids along X_i and X_j direction; ρ—density of nanofluids; $\rho\overline{\mu_i' \mu_j'}$—Reynolds stress.

(3) Energy conservation equation:

$$\left\{\frac{\partial (\rho E)}{\partial t} + \frac{\partial}{\partial x_i}\left[u_i (\rho E + p)\right]\right\} = \frac{\partial}{\partial x_j}\left[\left(k + \frac{c_p \mu_t}{pr_t}\right)\frac{\partial T}{\partial x_j} + u_i \left(\tau_{ij}\right)_{\text{eff}}\right], \tag{3}$$

where E—total energy; k—heat conductivity coefficient; T—temperature;

$$\left(\tau_{ij}\right)_{\text{eff}} = \mu_{\text{eff}}\left(\frac{\partial u_i}{\partial x_j} + \frac{\partial u_j}{\partial x_i}\right). \tag{4}$$

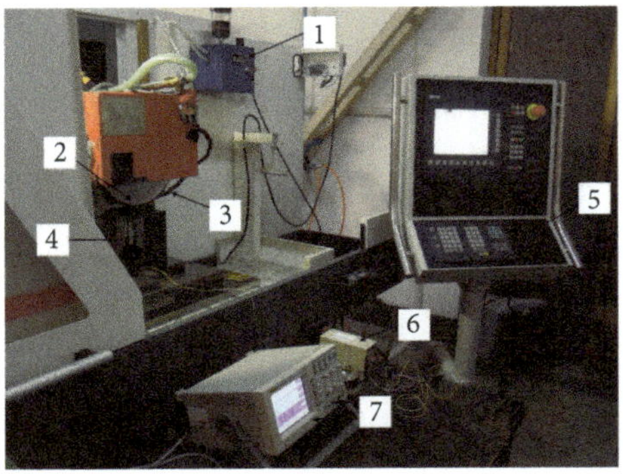

FIGURE 8: Schematic diagram of the experimental device. (1) Nanofluids jet flow device; (2) the CBN grinding wheel; (3) nanofluids jet flow nozzles; (4) operative surface of pressure field test system; (5) numerical control grinder; (6) 24 V DC power supply; (7) pressure data acquisition.

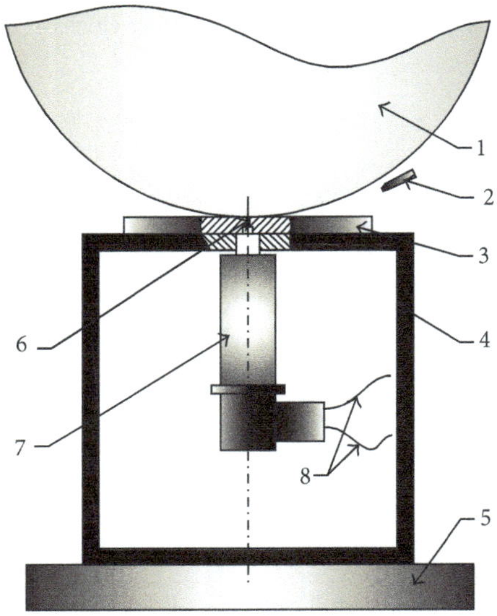

FIGURE 9: Diagram of pressure field test system. (1) The grinding wheel, (2) nozzles, (3) workpiece, (4) support, (5) grinder bench, (6) measuring hole, (7) pressure sensor, (8) cable.

In the previous equation, the amount of unknown numbers exceeded equations. To this end, k-ε turbulence model equation was introduced.

(4) k-ε turbulence model:

(a) K equation [23]:

$$\left[\frac{\partial}{\partial t}(\rho k)+\frac{\partial}{\partial x_i}(\rho k u_i)\right]=\frac{\partial}{\partial x_j}\left[\left(\mu+\frac{\mu_i}{\sigma_k}\right)\frac{\partial k}{\partial x_j}\right] + G_k - \rho\varepsilon - Y_M. \quad (5)$$

(b) ε equation:

$$\left[\frac{\partial}{\partial t}(\rho \varepsilon)+\frac{\partial}{\partial x_i}(\rho \varepsilon u_i)\right]=\frac{\partial}{\partial x_j}\left[\left(\mu+\frac{\mu_i}{\sigma_\varepsilon}\right)\frac{\partial \varepsilon}{\partial x_j}\right] + C_{1\varepsilon}\frac{\varepsilon}{k}G_k - C_{2\varepsilon}\rho\frac{\varepsilon^2}{k}, \quad (6)$$

where G_k shows the turbulence kinetic energy from the velocity gradient of laminar flow, G_b is the turbulence kinetic energy from the buoyancy force, Y_M is the fluctuations from transitional diffusion in the compressible turbulence. C_1, C_2, and C_3 are constants, σ_k and σ_ε are the turbulence Prandtl numbers of k equation and ε equation.

(c) Turbulence velocity model: the turbulence velocity μ_i is determined by the following equation:

$$\mu_i = \rho C_\mu \frac{k^2}{\varepsilon}, \quad (7)$$

where C_μ is a constant; constants of the model include $C_{1\varepsilon} = 1.44$, $C_{2\varepsilon} = 1.92$, $C_\mu = 0.09$, $\sigma_k = 1.0$, $\sigma_\varepsilon = 1.3$.

3. Simulation of Pressure Field in Grinding Zone

Figure 2 shows the pressure distribution of flow field of the grinding zone after the effects of compressed air continuous phase. It can be observed from the figure that the fluid pressure changes were only found at the entrance and exit of the wedge gap. There is a high pressure value, and the rapid pressure gradient occurs near the region where the gap between the grinding wheel and the work surface is minimal. The smaller the gap distance is, the higher the pressure gradient and the peak value pressure become. The pressure distribution is uniform in the direction of the width of the wheel except at the edge of wheel because of the side-leakage and becomes zero at the edge of the wheel, shown in Figure 3.

4. The Simulation Results

4.1. The Impact of the Peripheral Velocity of Grinding Wheel on the Pressure in Grinding Zone. By changing the peripheral velocity of grinding wheel, the pressure changes in the grinding zone can be obtained. The circumference velocity was taken as 45 m/s, 80 m/s, 120 m/s, and 160 m/s, and the corresponding changing curve was shown in Figure 4. In Figure 4, it can be seen that the pressure peak in direction x increased with the peripheral velocity of the grinding wheel. There is a high pressure value, and the rapid pressure gradient which occurred near the region where the gap between the wheel and the work surface was minimal, as shown in Figure 4. Along the X axis (the velocity of grinding wheel), the pressure gradually reached the maximum value from nothing, with the location close to the entrance of wedge gap. After this, it dropped sharply, even showing the negative pressure near minimum clearance. For *different velocities* of grinding wheel, two peak values of the pressure were almost at the same position, that is, close to the minimum clearance between the grinding wheel and the workpiece. Peak values of pressure were observed at the entrance near the wedge-shaped zone. The value was increased with rising velocity of

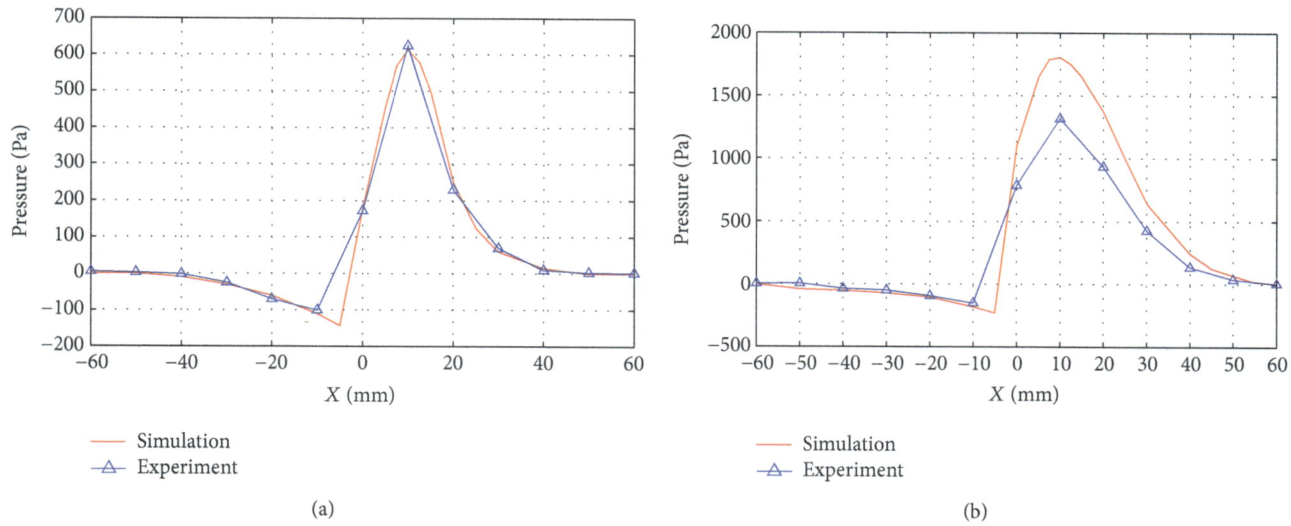

FIGURE 10: Comparison of pressure field theoretical simulation under *different peripheral velocities* of grinding wheel and the experimental results. (a) 45 m/s; (b) 160 m/s.

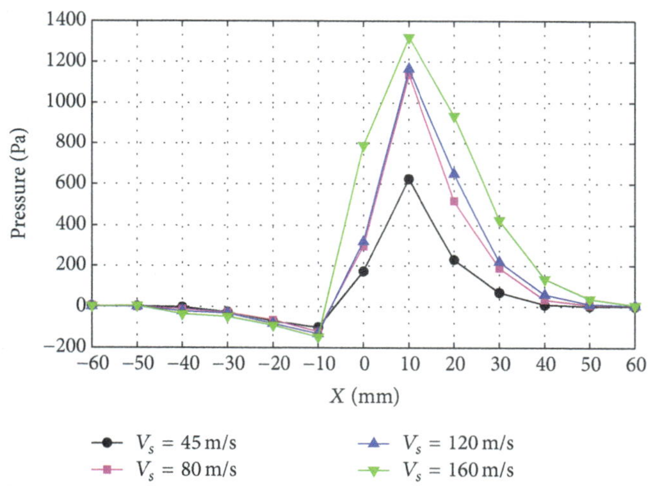

FIGURE 11: Comparison of pressure field experimental result under *different peripheral velocities* of grinding wheel.

grinding wheel, and the peak values of negative pressure were close to the exit of the wedge-shaped zone.

4.2. The Impact of Minimum Clearance between the Grinding Wheel and the Workpiece on Pressure in the Grinding Zone. By changing the minimum clearance between the grinding wheel and the workpiece, pressure changes of the grinding zone were obtained, and minimum clearance was taken as 100 μm, 80 μm, 50 μm, and 30 μm. The corresponding changing curve was shown in Figure 5.

As can be seen from Figure 5, peak values of pressure in grinding zone sharply reduce along the increase of minimum clearance. The smaller the gap distance was the higher the pressure gradient and the peak value pressure attained. It can be inferred that the minimum clearance between the grinding wheel and the workpiece significantly influenced the pressure in the flow field of grinding zone.

4.3. The Impact of Jet Flow Velocity on the Pressure of Grinding Zone. By changing the jet flow velocity, pressure changes of the grinding zone were obtained. The jet flow velocity V_a was taken as 45 m/s, 60 m/s, 70 m/s, and 90 m/s. The corresponding changing curve was shown in Figure 6.

As can be seen from Figure 6, peak values of the pressure in grinding zone increase with increasing jet flow velocity, indicating that jet flow velocity influences the pressure in the flow field of grinding zone.

4.4. The Impact of Positions and Angles of Nozzle on the Pressure in Grinding Zone. By changing the positions and angles of *nozzle*, pressure changes of the grinding zone were obtained. The positions and angles of *nozzle* were taken as the intersection between the workpiece surface as 0°, 15°, and 20°, and corresponding changing curve was shown in Figure 7.

It can be seen from Figure 7 that at 15° the pressure in grinding zone was larger than the rest two angles because at 15°, the sprayed nanofluids were over the boundary of the return flux, when nanofluids can easily enter the wedge gap for cooling and lubricating. Hence, in the grinding with nano-particle jet flow of MQL, the position of nozzles should be located beyond the boundary of the return flux.

5. The Experimental Study of Pressure Field in Grinding Zone

5.1. Experimental Device. Numerical control precision surface grinding machine and minimum quantity nano-particle jet flow feedway were used to establish minimum quantity

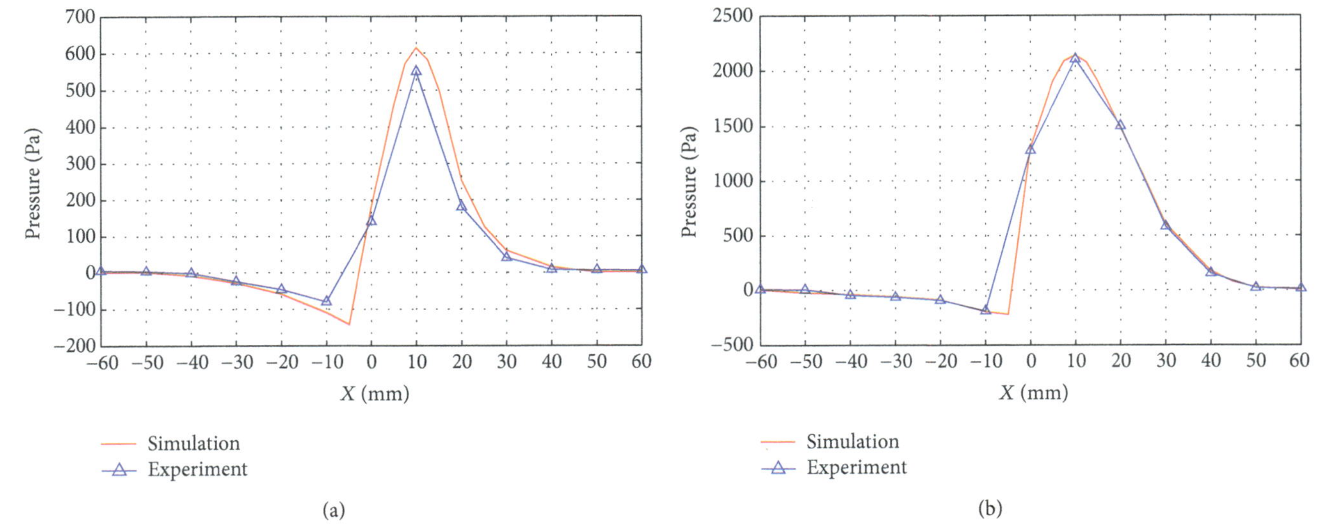

FIGURE 12: Comparison between pressure field theoretical simulation and experimental results at different jet flow velocities; (a) 45 m/s; (b) 90 m/s.

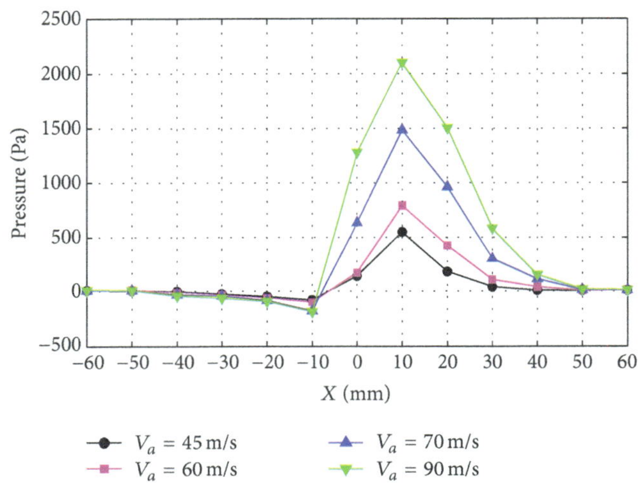

FIGURE 13: Comparison of pressure field experimental result under different at different jet flow velocities.

grinding bench. The jet flow pressure field applied workpiece-embedded CY3018 pressure sensor to measure and collect data. The schematic diagram of the experimental device was shown in Figure 8, and the diagram of pressure field test system was shown in Figure 9. The machining system of the experiment is composed of a rotating grinding wheel and a work surface, and the gap between them is MQL filled in with grinding coolant by a nozzle. The varied parameters included the grinding wheel velocity, nozzle jet velocity, minimum clearance between the grinding wheel and the work surface, and different angles of *nozzle*. A reciprocating table grinder with variable speed was used for the experiments. The wheels used were all of diameter $D = 300$ mm and width $b = 25$ mm. Each work specimen was 300 mm in length along the grinding direction, 100 mm width, and 50 mm thickness. Parameters of nano-particle jet flow of MQL are carbon nanotube particles with the average diameter of 10–20 nm; oil base (vegetable oil); volume fraction: 1%; the MQL flow rates: 615 mL/h.

In the measurement of parameters, the sensor and transmitter transferred the measured values to the tester. The data was uploaded to the host computer via the serial port. After the conversion, the data was displayed in the form of tables and curves. Each experiment was repeated three times for averaging.

In the experiment, the numerical control ultrasonic oscillator was used to prepare stable oil-based nanofluids grinding liquid. The procedures are as follows: with the example of the preparation of 300 mL of oil-based nanofluids, the volume fraction was 1%. Assume the volume of nano-particle as V_n and the volume of base fluid as V_f. The volume fraction of nano-particles of nanofluids is

$$\phi = \frac{V_n}{V_n + V_f}. \tag{8}$$

According to the above equation, it can be calculated that the mass of nano-particles should be 0.74 g, with the volume of vegetable oil as 279 mL, the proportion of the dispersing agent as 7% (28 g). With the combination of nano-particle, base oil, and the dispersant, the ultrasonic vibration was used for the combination, with the vibration frequency as 70 Hz and the time as 15 min. After the mixture stabilized, the viscosity was measured.

6. Analysis of Experimental Results

Four *different peripheral velocities* of grinding wheel were adopted for the experiment. Among them, the experimental results of two grinding wheel velocities were compared with the theoretical simulation as shown in Figure 10. The contrast

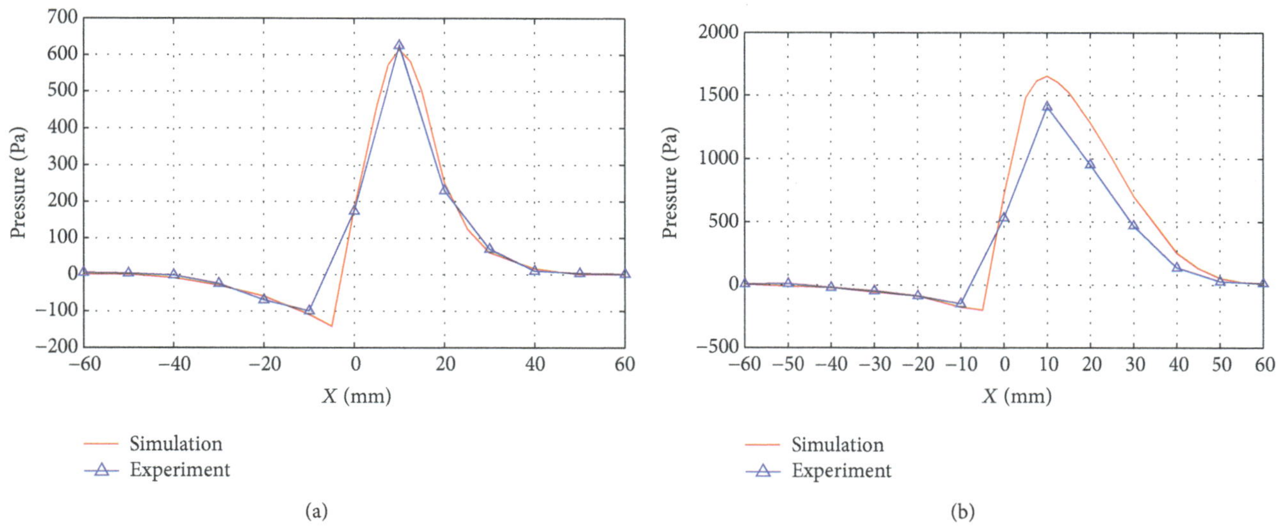

FIGURE 14: Comparison between pressure field theoretical simulation and experimental results with different minimum clearance; (a) 100 μm; (b) 30 μm.

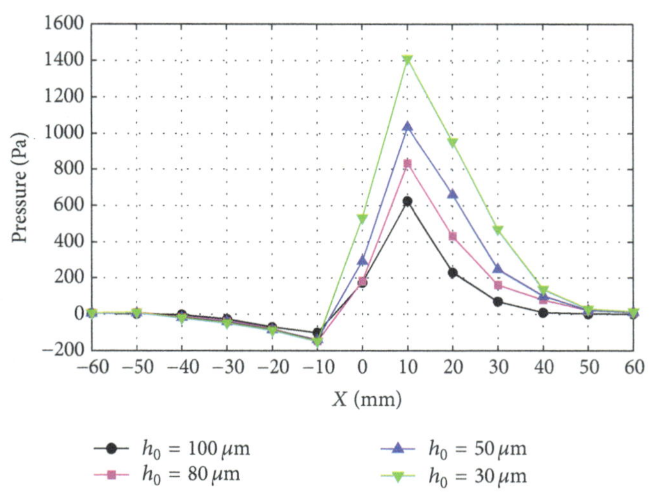

FIGURE 15: Comparison of pressure field experimental result with different minimum clearance.

of experimental result of the pressure in the grinding zone at two *different peripheral velocities* of grinding wheel was shown in Figure 11.

As shown in Figure 10, with lower the peripheral velocity of grinding wheel, the experimental result was relatively consistent with the theoretical simulation, when the velocity was 45 m/s; when the velocity was increased to 160 m/s, the experimental result at the entrance section of the wedge-shaped zone was lower than the theoretical simulation, which may be attributed to the increased difficulties of nanofluids breaking through the airbond with enlarged thickness. Figure 11 provides the contrast of experimental results of four *different peripheral velocities* of grinding wheel. The results show that the pressure in the grinding zone increased with the increases of the peripheral velocity of grinding wheel. With the velocity of 120 m/s, as the airbond layer prevented nanofluids from entering the grinding zone, the pressure change was close to the curve when the grinding wheel velocity was 80 m/s.

Four different jet flow velocities were adopted for the experiment. Among them, the experimental results of two jet flow velocities were compared with the theoretical simulation as shown in Figure 12. The contrast of experimental result of the pressure in the grinding zone from different jet flow velocities from the nozzles was shown in Figure 13.

As shown in Figure 12, with higher jet flow velocity, the experimental result was relatively consistent with the theoretical simulation, when the velocity was 90 m/s; when the velocity was reduced to 45 m/s, the experimental result of the pressure in the grinding zone was lower than the theoretical simulation, which may be attributed to the influence from the airbond around the grinding wheel. With lower jet flow velocity, the sprayed nanofluids were insufficient to break through the airbond and enter the grinding zone. Figure 13 provides the contrast of experimental results of four different jet flow velocities, indicating that the pressure in the grinding zone increases with the jet flow velocity, and its regularities are basically consistent with those in the simulation analyze.

Four different minimum clearances were adopted for the experiment. Among them, the experimental results of two minimum clearances were compared with the theoretical simulation as shown in Figure 14. The contrast of experimental result of the pressure in the grinding zone with different minimum clearances was shown in Figure 15.

It can be observed from Figure 14 that with larger minimum clearance, the experimental result was relatively consistent with the theoretical simulation, when the minimum clearance was 100 μm; yet when the minimum clearance was gradually reduced to 30 μm, the effective flow of nanofluids that enters the wedge gap was reduced, leading to lower experimental result of the pressure in the grinding zone than the theoretical simulation. Figure 15 shows the contrast of experimental curves with different minimum clearances. It

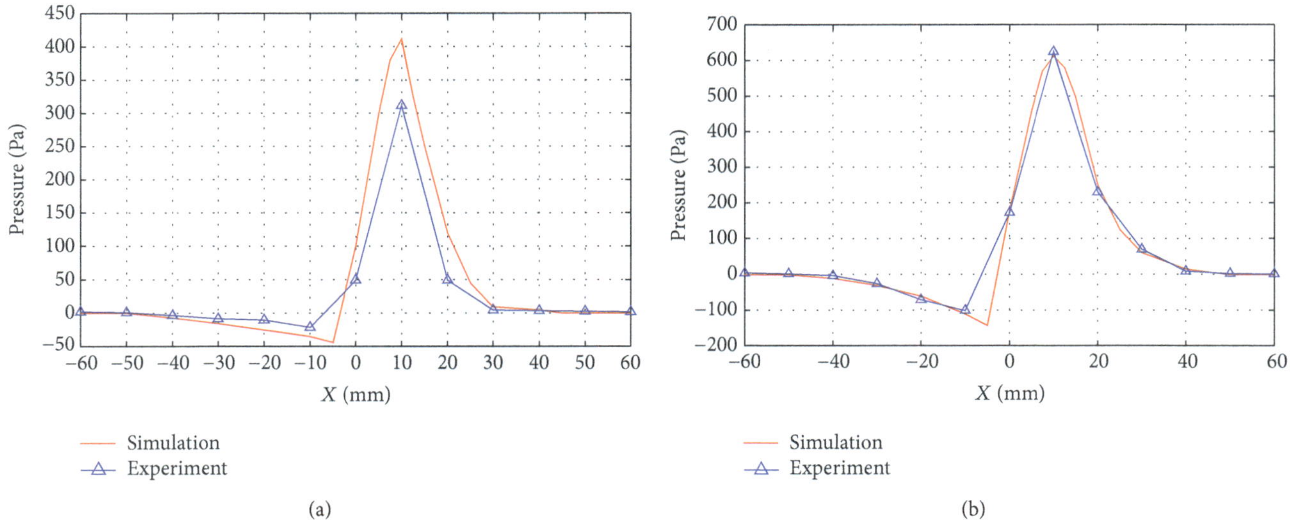

FIGURE 16: Comparison between pressure field theoretical simulation and experimental results with different positions and angles of *nozzle*; (a) 0°; (b) 15°.

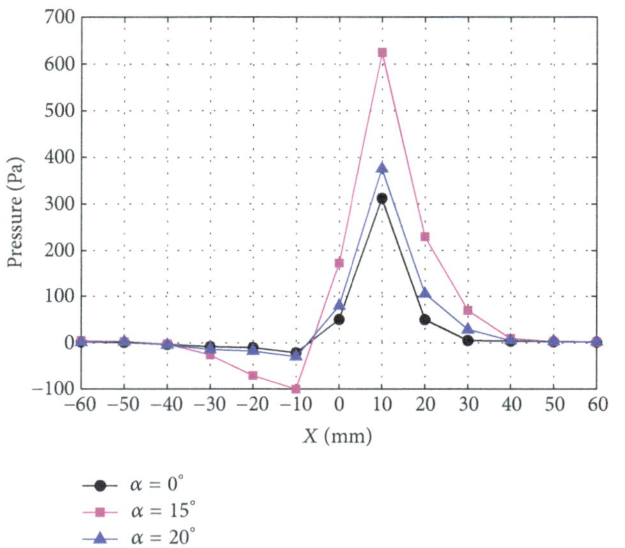

FIGURE 17: Comparison of pressure field experimental result with different positions of nozzles.

can be found that the regularities are basically consistent with the finding in our simulation. In other words, the pressure increases as wedge gap narrows. Along the velocity direction of grinding wheel, the pressure gradually increases, almost reaching the entrance of the wedge gap and then showing the tendency of declining. The negative pressure with a certain value was observed close to the wedge gap.

Three different spraying angles of *nozzles* were adopted for the experiment. The intersection between *nozzle* and the workpiece surface was taken as 0°, 15°, and 20°, respectively. Among them, the experimental results of two spraying angles of *nozzle* were compared with the theoretical simulation as shown in Figure 16. The contrast of experimental result of the pressure in the grinding zone at different positions of nozzles was shown in Figure 17.

It can be observed from Figure 16 that when the spraying angle of *nozzles* was 15°, the experimental result was relatively consistent with the theoretical simulation; when the angle was 0°, the pressure in the grinding zone was lower than the theoretical simulation. It may be attributed to the lower spraying track of nanofluids at 0°. Figure 17 shows the contrast of experimental result of pressure at different spraying angles of *nozzles*. It can be observed that when the spraying angle of *nozzles* was 15°, the pressure was obviously higher than that at two different positions of nozzles, as the nanofluids were higher than the boundary of the return flux. At this angle, the nanofluids can easily enter the wedge-shaped zone.

7. Conclusions

In this paper, the theoretical model of nano-particle jet flow of MQL grinding zone pressure field was established for simulation and experiments. The results show the following.

(1) With the velocity increase of grinding wheel, the pressure in the grinding zone gradually increases with a substantial range. For *different velocities* of grinding wheel, the peak values of the pressure were almost at the same position, that is, close to the minimum clearance between the grinding wheel and the workpiece. Peak values of pressure were observed at the entrance near the wedge-shaped zone. The value increases along rising velocity of grinding wheel. When the velocity was increased to 160 m/s, the experimental result at the entrance section of the wedge-shaped zone was lower than the theoretical simulation, which may be attributed to the increased difficulties of nanofluids breaking through the air-bond with enlarged thickness.

(2) Peak values of pressure in grinding zone sharply reduce along the increase of minimum clearance with a substantial range, indicating that the minimum clearance between the grinding wheel and the workpiece significantly influences the pressure at the wedge gap of the flow field in grinding zone. When the minimum clearance was gradually reduced to 30 μm, the effective flow of nanofluids that enters the wedge gap was reduced, leading to lower experimental result of the pressure in the grinding zone than the theoretical simulation.

(3) Peak values of the pressure in grinding zone increase with increasing jet flow velocity with a substantial range, indicating that jet flow velocity influences the pressure in the flow field of grinding zone to a certain degree. For *different velocities* of jet flow, the peak values of the pressure were almost at the same position, that is, close to the minimum clearance between the grinding wheel and the workpiece. When the velocity was reduced to 45 m/s, the experimental result of the pressure in the grinding zone was lower than the theoretical simulation, which may be attributed to the influence from the airbond around the grinding wheel. With lower jet flow velocity, the sprayed nanofluids were insufficient to break through the airbond and enter the grinding zone.

(4) When the spraying angle of *nozzles* was 15°, the pressure at the wedge gap was obviously higher than that at two different positions of nozzles. When the spraying angles of *nozzles* were 0° and 20°, the spraying track of nanofluids was lower than the boundary of the return flux. In this way, the nanofluids would encounter great obstacles to enter the wedge-shaped zone. Thus, the pressure became lower than the simulation results.

Acknowledgments

This research was financially supported by the National Natural Science Foundation of China (50875138; 51175276) and the Shandong Provincial Natural Science Foundation of China (Z2008F11; ZR2009FZ007).

References

[1] J. Kopac and P. Krajnik, “High-performance grinding—a review,” *Journal of Materials Processing Technology*, vol. 175, no. 1-3, pp. 278–284, 2006.

[2] E. Brinksmeier and E. Minke, “High-performance surface grinding—the influence of coolant on the coolant process,” *CIRP Annals*, vol. 42, no. 1, pp. 367–370, 1993.

[3] E. Brinksmeier, Y. Mutlugünes, F. Klocke, J. C. Aurich, P. Shore, and H. Ohmori, “Ultra-precision grinding,” *CIRP Annals*, vol. 59, no. 2, pp. 652–671, 2010.

[4] J. F. G. Oliveira, E. J. Silva, C. Guo, and F. Hashimoto, “Industrial challenges in grinding,” *CIRP Annals*, vol. 58, no. 2, pp. 663–680, 2009.

[5] J. H. Liu, Z. J. Pei, and G. R. Fisher, “Grinding wheels for manufacturing of silicon wafers: a literature review,” *International Journal of Machine Tools and Manufacture*, vol. 47, no. 1, pp. 1–13, 2007.

[6] C. H. Li, G. Q. Cai, and S. C. Xiu, “Auto-correlation study on the surface profile finished by abrasive jet with grinding wheel as restraint,” *International Journal of Computer Applications in Technology*, vol. 29, no. 2–4, pp. 262–266, 2007.

[7] C. Guo and S. Malkin, “Analysis of fluid flow through the grinding zone,” *Journal of engineering for industry*, vol. 114, no. 4, pp. 427–434, 1992.

[8] C. C. Chang, “An application of lubrication theory to predict useful flow-rate of coolants on grinding porous media,” *Tribology International*, vol. 30, no. 8, pp. 575–581, 1997.

[9] M. Ganesan, C. Guo, R. A. Rafael, and S. Malkin, “Analysis of hydrodynamic forces in grinding,” *Transactions of the North American Manufacturing Research Institute of SME*, vol. 24, pp. 105–110, 1996.

[10] K. Furutani, N. Ohguro, N. T. Hieu, and T. Nakamura, “In-process measurement of topography change of grinding wheel by using hydrodynamic pressure,” *International Journal of Machine Tools and Manufacture*, vol. 42, no. 13, pp. 1447–1453, 2002.

[11] V. K. Gviniashvili, N. H. Woolley, and W. B. Rowe, “Useful coolant flowrate in grinding,” *International Journal of Machine Tools and Manufacture*, vol. 44, no. 6, pp. 629–636, 2004.

[12] S. Ebbrell, N. H. Woolley, Y. D. Tridimas, D. R. Allanson, and W. B. Rowe, “Effects of cutting fluid application methods on the grinding process,” *International Journal of Machine Tools and Manufacture*, vol. 40, no. 2, pp. 209–223, 2000.

[13] C. H. Li, J. Y. Li, S. Wang, and Q. Zhang, “Modeling and numerical simulation of the grinding temperature field with nanoparticle jet of MQL,” *Advances in Mechanical Engineering*, vol. 2013, Article ID 986984, 9 pages, 2013.

[14] M. Ganesan, C. Guo, and S. Malkin, “Measurement of hydrodynamic forces in grinding,” *Transactions of the North American Manufacturing Research Institute of SME*, vol. 23, pp. 103–107, 1995.

[15] P. Hryniewicz, A. Z. Szeri, and S. Jahanmir, “Application of lubrication theory to fluid flow in grinding: part I—flow between smooth surfaces,” *Journal of Tribology*, vol. 123, no. 1, pp. 94–100, 2001.

[16] P. Hryniewicz, A. Z. Szeri, and S. Jahanmir, “Application of lubrication theory to fluid flow in grinding: part II—influence of wheel and workpiece roughness,” *Journal of Tribology*, vol. 123, no. 1, pp. 101–107, 2001.

[17] J.-D. Kim, “Motion analysis of powder particles in EEM using cylindrical polyurethane wheel,” *International Journal of Machine Tools and Manufacture*, vol. 42, no. 1, pp. 21–28, 2002.

[18] S.-K. Lee, Y. Miyamoto, T. Kuriyagawa, and K. Syoji, “Effects of minimizing hydrodynamic pressure in ultra-precision mirror grinding,” *International Journal of Machine Tools and Manufacture*, vol. 44, no. 10, pp. 1031–1036, 2004.

[19] F. Klocke, E. Brinksmeier, and K. Weinert, “Capability profile of hard cutting and grinding processes,” *CIRP Annals*, vol. 54, no. 2, pp. 557–580, 2005.

[20] S. C. Xiu, C. X. Chao, and S. Y. Pei, “Experimental research on surface integrity with less or non fluid grinding process,” *Key Engineering Materials*, vol. 487, pp. 89–93, 2011.

[21] C. Li, Y. Hou, Y. Ding, and G. Cai, “Feasibility investigations on compound process: a novel fabrication method for finishing with grinding wheel as restraint,” *International Journal of Computational Materials Science and Surface Engineering*, vol. 4, no. 1, pp. 55–68, 2011.

[22] Y. Hou, C. Li, Z. Han, J. Li, and H. Zhao, "Examination of the material removal mechanisms during the abrasive jet finishing of 45 steel," *Advanced Science Letters*, vol. 4, no. 4-5, pp. 1478–1484, 2011.

[23] Y. L. Hou, C. H. Li, and Q. Zhang, "Investigation of structural parameters of high speed grinder spindle system on dynamic performance," *International Journal Materials and Product Technology*, vol. 44, no. 1-2, pp. 92–114, 2012.

15

Effect of Viscoelasticity on the Natural Frequencies of Axially Moving Continua

B. Burak Özhan[1,2] and Mehmet Pakdemirli[1,2]

[1] *Department of Mechanical Engineering, Celal Bayar University, Muradiye, 45140 Manisa, Turkey*
[2] *Applied Mathematics and Computation Center, Celal Bayar University, Muradiye, 45140 Manisa, Turkey*

Correspondence should be addressed to B. Burak Özhan; burakozhan@gmail.com

Academic Editor: Jianqiao Ye

Linear models of axially moving viscoelastic beams and viscoelastic pipes conveying fluid are considered. The natural frequencies of the models are calculated. For both models, viscoelasticity terms are assumed to be of order one. Natural frequencies corresponding to various beam and pipe parameters are presented. Effects of viscoelasticity on the natural frequencies are discussed.

1. Introduction

High-speed magnetic tapes, fiber windings, band saw blades, thread lines, power transmission chain and belts, aerial cable tramways, paper sheets, fluid conveying pipes, and so forth are technological examples of axially moving continua. Understanding the dynamics of axially moving continua is very important to design and to prevent such systems from hazards and accidents. Both linear and nonlinear models exist for vibrations of axially moving systems in the literature. A comprehensive review of the literature was presented by Ulsoy et al. [1] and Wickert and Mote Jr. [2]. Transverse vibrations of travelling beam and string models were investigated by Wickert and Mote Jr. [3]. In that study, equations of motion for string and beam were represented in a canonical form. To analyse the models, modal analysis and Green's function method were used. Wickert and Mote Jr. [4] further investigated complex model analysis for axially moving materials. Ulsoy and Mote Jr. [5] studied bandsaw vibrations and their stability. The study emphasized that the natural frequencies of the bands decrease by increasing axial velocity and increase by increasing axial tension or strain. To analyse axially moving continua vibrations, the transport speed can be modeled as constant or variable. The equation of motion of axially moving continua with variable speed was first derived by Miranker [6]. Miranker analysed transverse vibrations of a tape moving between a pair of pulley by using a variational method. The problem of an axially accelerating string with harmonic excitation at one end was investigated by Mote Jr. [7] who determined the stability by using Laplace transformation method. Pakdemirli et al. [8] rederived the equations by using Hamilton's principle. They analysed the stability and response of the system using Floquet theory. Transport velocity variation was assumed to be sinusoidal about a mean velocity of zero. Mockensturm et al. [9] presented an axially moving string vibration model with constant transport velocity and variable tension. Stabilities for different resonance conditions and limit cycle were analysed. Pakdemirli and Batan [10] considered periodic constant acceleration and deceleration type of velocity variations for axially accelerating strips. Öz and Pakdemirli [11] modelled an axially moving Euler-Bernoulli type beam vibration with harmonically varying velocities. The work covered principal parametric resonances, sum and difference type combination resonances. Pakdemirli and Öz [12] gave a detailed resonance analysis for which up to four modes were involved and obtained stability regions for axially moving beams.

Vibrations of pipes conveying fluid possess a similar mathematical model and hence were considered within the context of axially moving continua. Those models were analyzed by many researchers. Païdoussis and Li [13] analyzed in detail pipes conveying fluids. They discussed both linear and nonlinear dynamics of such systems. Öz and Boyacı [14] and Öz [15] examined transverse vibrations of tensioned pipes conveying fluid with fluctuating fluid velocities. In the mentioned papers [14, 15] linear and nonlinear equations of motion were presented.

Axially moving viscoelastic continua vibrations were studied by many researchers. Chen et al. [16] investigated bifurcation and chaos in transverse vibrations of axially moving Kelvin type viscoelastic strings. They analysed the bifurcation and chaos in 1-term and 2-term truncated systems using Galerkin's method. Regular and chaotic vibrations of axially moving strings were investigated for 1st, 2nd, 3rd, and 4th order truncated systems by the same authors [17]. Asymptotic analysis of axially accelerating strings was done by Chen et al. [18]. The authors analysed Mote's and Kirschoff's nonlinear models with variable velocity using asymptotic perturbation method. Natural frequencies of the systems were found and principal parametric resonances were analyzed. Effects of stiffness, viscosity, initial stress, and axial speed fluctuation on the string vibrations were treated and compared for two models. Lee and Oh [19] considered dynamics of axially moving beams subjected to axial tension. They developed spectral element method and investigated the effects of viscoelasticity and moving speed. First, they found the lowest five eigenvalues of simply supported, stationary, one and two span beams using spectral element method, finite element method, and exact theory. Then they obtained the effects of moving speed and viscoelasticity on the variation of the lowest two eigenvalues. The Method of Multiple Scales (a perturbation method) was used by Chen and Yang [20]. They conducted a stability analysis for parametric resonances of axially moving viscoelastic beams. Stability boundaries were investigated for different boundary conditions. The method of averaging was applied to the 2-term Galerkin truncation of the equation of motion of axially accelerating viscoelastic beam by Chen et al. [21]. They analyzed effects of dynamic viscosity, the mean axial speed, and the tension in parametric vibrations of beams. The Method of Multiple Scales was also applied to the nonlinear vibration model of axially moving viscoelastic string supported by a partial viscoelastic guide by Ghayesh [22]. Ghayesh obtained stability and bifurcation point variations on frequency response curves for different string and guide parameters (i.e., viscosity, guide length, speed, nonlinearity coefficient, damping coefficient, and guide stiffness). Panda and Kar [23] studied nonlinear dynamics of viscoelastic pipes conveying fluids. They analyzed combination, principal, parametric, and internal resonances of pipes.

Özhan and Pakdemirli [24, 25] proposed a general solution algorithm to solve an arbitrary cubic nonlinear vibration model which covers gyroscopic systems. In the mentioned studies, axially moving beam and axially moving viscoelastic beam models were taken as application problems. Forced vibrations of axially moving continua were analyzed. Primary resonances and three-to-one internal resonances of forced vibrations were obtained. Özhan and Pakdemirli [26] obtained primary parametric resonances of a generalized cubic nonlinear vibration model. Pipe conveying fluid model was taken as an application problem.

The objective of this study is to show the effect of viscoelasticity on the natural frequencies of axially moving beams and pipes conveying fluids. For both models, the linear equations of motion are considered. Numerical results are discussed and the natural frequencies are obtained. The effects of various beam and pipe parameters on the natural frequencies of the system are investigated in detail in the figures. A detailed analysis of the natural frequencies in the case of viscoelasticity is presented in this study for the first time.

2. Equations of Motion

In this section, equations of motion of axially moving viscoelastic beams and viscoelastic pipes conveying fluid will be presented.

2.1. Axially Moving Viscoelastic Beam Model. Equation of motion of a uniform axially moving viscoelastic beam and boundary conditions in linear nondimensional form are [11, 27]

$$\ddot{w} + \left(v_0^2 - 1\right) w'' + 2v_0\dot{w}' + v_f^2 w^{IV} + \varepsilon\alpha\dot{w}^{IV} = 0, \quad (1)$$

$$w(0,t) = w(1,t) = w''(0,t) = w''(1,t) = 0. \quad (2)$$

In (1) $w(x,t)$ is transverse displacement of the beam. v_0 is the constant axial transport velocity, dot denotes differentiation with respect to time, and prime denotes differentiation with respect to the spatial variable. ε is a small perturbation parameter. The dimensionless parameters are

$$x = \frac{\overline{x}}{L}, \quad w = \frac{\overline{w}}{L}, \quad t = \overline{t}\sqrt{\frac{P}{\rho A L^2}}, \quad v_0 = \frac{\overline{v}_0}{v_c},$$
$$v_c = \sqrt{\frac{P}{\rho A}}, \quad v_f = \sqrt{\frac{EI}{PL^2}}, \quad \alpha = \frac{I\eta}{L^3\sqrt{\rho A P}}, \quad (3)$$

where x and t are nondimensional spatial and time variables, $\overline{x}$ and $\overline{t}$ are dimensional forms of spatial and time variables. $\overline{w}(\overline{x},\overline{t})$ is dimensional form of displacement. η represents viscosity. Viscoelastic material obeys the Kelvin model. α is the dimensionless parameter related to the viscosity. L is the beam length, A is the cross sectional area, ρ is the density of the beam, ρA is the mass per unit length of the beam, and $\overline{v}_0$ and v_c are the dimensional and critical axial velocities of the beam, respectively. P is the axial tension force and I is the second moment of area of the beam cross section.

2.2. Viscoelastic Pipe Conveying Fluid Model. The nondimensional linear equation of motion and boundary conditions of viscoelastic pipes conveying fluid are [15, 23]

$$\ddot{w} + w^{IV} + \left(v_0^2 - P_0\right) w'' + 2\sqrt{\beta}v_0\dot{w}' + \varepsilon\alpha\dot{w}^{IV} = 0, \quad (4)$$

$$w(0,t) = w(1,t) = w''(0,t) = w''(1,t) = 0, \quad (5)$$

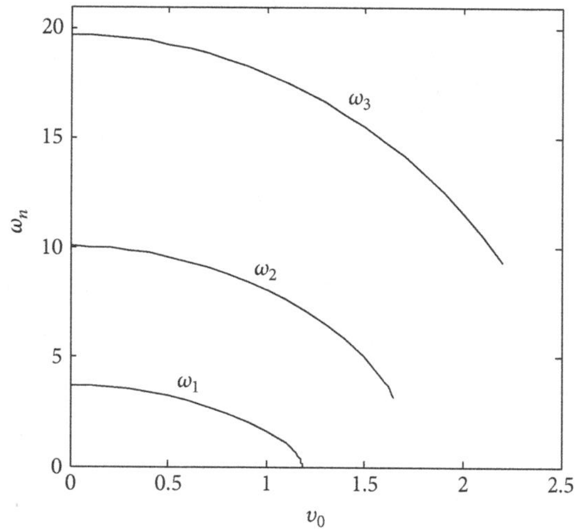

FIGURE 1: First three fundamental frequencies versus axial transport velocities of axially moving viscoelastic beam ($v_f = 0.2, \alpha = 0.001$).

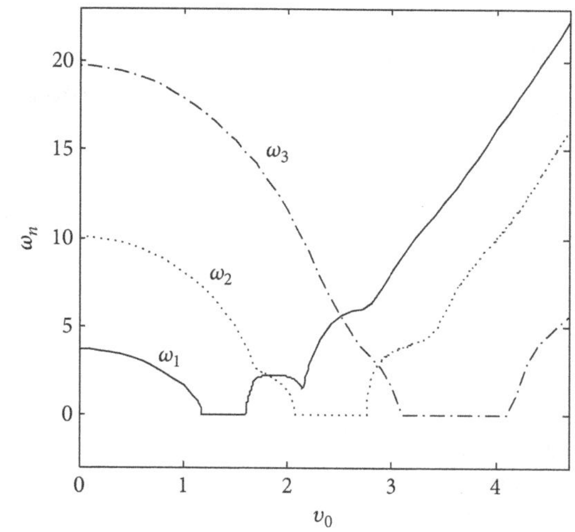

FIGURE 2: Real parts of the first three fundamental frequencies versus axial transport velocities of axially moving viscoelastic beam ($v_f = 0.2, \alpha = 0.001$).

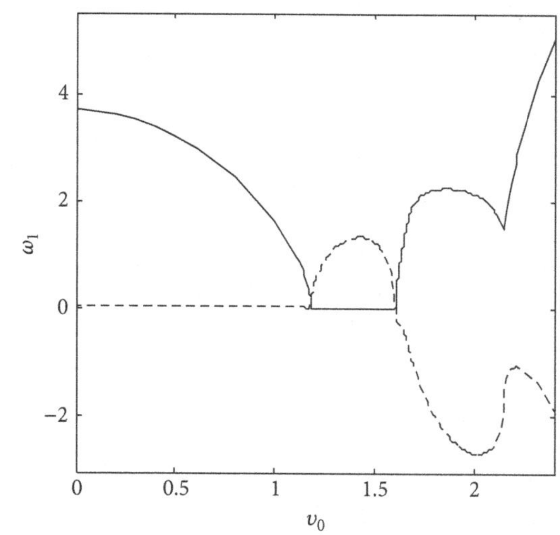

FIGURE 3: Real (solid line) and imaginary parts (dashed line) of the first fundamental frequencies versus axial transport velocities of the viscoelastic beams ($v_f = 0.2, \alpha = 0.001$).

where

$$x = \frac{\overline{x}}{L}, \qquad w = \frac{\overline{w}}{L}, \qquad t = \omega_0 \overline{t},$$

$$\omega_0 = \frac{1}{L^2}\sqrt{\frac{EI}{m_{\text{fluid}} + m_{\text{pipe}}}}, \qquad v_0 = \overline{v}_0 L \sqrt{\frac{m_{\text{fluid}}}{EI}},$$

$$\alpha = \frac{E^*}{L^2}\sqrt{\frac{I}{\left(m_{\text{fluid}} + m_{\text{pipe}}\right)E}}, \qquad \beta = \frac{m_{\text{fluid}}}{m_{\text{fluid}} + m_{\text{pipe}}}. \tag{6}$$

$w(x,t)$ is the transverse displacement of pipe, x and t are the nondimensional spatial and time variables, respectively. $\overline{x}$ and $\overline{t}$ are the dimensional forms of spatial and time variables. $\overline{w}(\overline{x},\overline{t})$ is the dimensional form of the displacement. P_0 is the tensile force, L is the length, EI is the flexural rigidity, A is the area, and ρ is the density of the pipe. E^* is the coefficient of internal dissipation, m_{fluid} and m_{pipe} are the masses of fluid and pipe, and α is the dimensionless parameter related to the viscosity.

3. Natural Frequencies of Beam and Pipe Models

In (1) and (4) $\varepsilon\alpha\dot{w}^{IV}$ represent internal damping for beams and pipes. Although, in the usual modeling, this term is assumed to be small in the literature (i.e., $\varepsilon \ll 1$); in this study, viscoelasticity term is assumed to be of order one. Therefore, the influence of the viscoelasticity on the natural frequencies can be directly analyzed for a wide range of parameters.

To find the natural frequencies of beams and pipes, the linear models with $O(1)$ viscoelasticity are considered

$$\begin{aligned} &\ddot{w} + \left(v_0^2 - 1\right)w'' + 2v_0\dot{w}' + v_f^2 w^{IV} + \alpha\dot{w}^{IV} = 0, \\ &w(0,t) = w(1,t) = w''(0,t) = w''(1,t) = 0, \\ &\ddot{w} + \left(v_0^2 - P_0\right)w'' + 2v_0\sqrt{\beta}\dot{w}' + w^{IV} + \alpha\dot{w}^{IV} = 0, \\ &w(0,t) = w(1,t) = w''(0,t) = w''(1,t) = 0, \end{aligned} \tag{7}$$

for which a solution of the below form can be given [11],

$$w(x,t) = A_n(t)e^{i\omega_n t}Y_n(x) + \overline{A}_n(t)e^{-i\omega_n t}\overline{Y}_n(x), \tag{8}$$

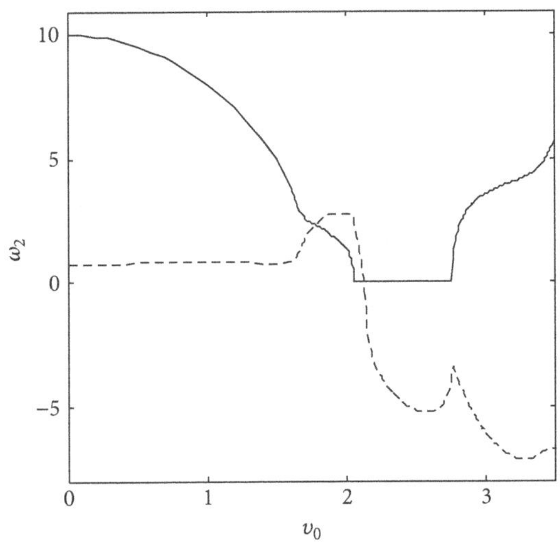

FIGURE 4: Real (solid line) and imaginary parts (dashed line) of the second fundamental frequencies versus axial transport velocities of the viscoelastic beams ($v_f = 0.2$, $\alpha = 0.001$).

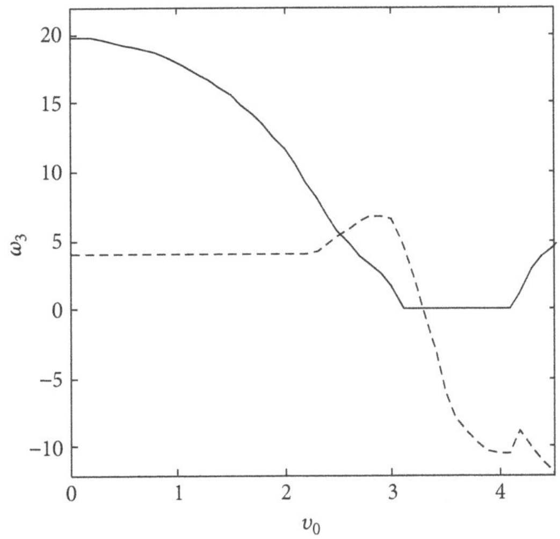

FIGURE 5: Real (solid line) and imaginary parts (dashed line) of third fundamental frequencies versus axial transport velocities of the viscoelastic beams ($v_f = 0.2$, $\alpha = 0.001$).

where ω_n and $Y_n(x)$ represent natural frequencies and mode shapes, respectively. The mode shapes are [11]

$$\begin{aligned} Y_n(x) = c_1 \Bigg\{ & e^{i\beta_{1n}x} - \frac{\left(\beta_{4n}^2 - \beta_{1n}^2\right)\left(e^{i\beta_{3n}} - e^{i\beta_{1n}}\right)}{\left(\beta_{4n}^2 - \beta_{2n}^2\right)\left(e^{i\beta_{3n}} - e^{i\beta_{2n}}\right)} e^{i\beta_{2n}x} \\ & - \frac{\left(\beta_{4n}^2 - \beta_{1n}^2\right)\left(e^{i\beta_{2n}} - e^{i\beta_{1n}}\right)}{\left(\beta_{4n}^2 - \beta_{3n}^2\right)\left(e^{i\beta_{2n}} - e^{i\beta_{3n}}\right)} e^{i\beta_{3n}x} \\ & + \left[-1 + \frac{\left(\beta_{4n}^2 - \beta_{1n}^2\right)\left(e^{i\beta_{3n}} - e^{i\beta_{1n}}\right)}{\left(\beta_{4n}^2 - \beta_{2n}^2\right)\left(e^{i\beta_{3n}} - e^{i\beta_{2n}}\right)} \right. \\ & \left. + \frac{\left(\beta_{4n}^2 - \beta_{1n}^2\right)\left(e^{i\beta_{2n}} - e^{i\beta_{1n}}\right)}{\left(\beta_{4n}^2 - \beta_{3n}^2\right)\left(e^{i\beta_{2n}} - e^{i\beta_{3n}}\right)} \right] e^{i\beta_{4n}(x)} \Bigg\}. \end{aligned} \tag{9}$$

In (9) β_{in} are the eigenvalues of the vibrated systems which were obtained from frequency equations and support conditions as later.

3.1. Axially Moving Viscoelastic Beam Model. Consider the following:

$$\begin{gathered} \left(v_f^2 + i\omega_n\alpha\right)\beta_{in}^4 + \left(1 - v_0^2\right)\beta_{in}^2 - 2v_0\omega_n\beta_{in} - \omega_n^2 = 0 \\ i = 1,2,3,4, \quad n = 1,2, \\ \left(e^{i(\beta_{1n}+\beta_{2n})} + e^{i(\beta_{3n}+\beta_{4n})}\right)\left(\beta_{1n}^2 - \beta_{2n}^2\right)\left(\beta_{3n}^2 - \beta_{4n}^2\right) \\ + \left(e^{i(\beta_{1n}+\beta_{3n})} + e^{i(\beta_{2n}+\beta_{4n})}\right) \\ \times \left(\beta_{2n}^2 - \beta_{4n}^2\right)\left(\beta_{3n}^2 - \beta_{1n}^2\right) \\ + \left(e^{i(\beta_{2n}+\beta_{3n})} + e^{i(\beta_{1n}+\beta_{4n})}\right) \\ \times \left(\beta_{1n}^2 - \beta_{4n}^2\right)\left(\beta_{2n}^2 - \beta_{3n}^2\right) = 0. \end{gathered} \tag{10}$$

3.2. Viscoelastic Pipe Conveying Fluid Model. Consider the following:

$$\begin{gathered} \left(1 + i\omega_n\alpha\right)\beta_{in}^4 + \left(P_0 - v_0^2\right)\beta_{in}^2 + 2v_0\sqrt{\beta}\,\omega_n\beta_{in} - \omega_n^2 = 0 \\ i = 1,2,3,4, \quad n = 1,2, \\ \left(e^{i(\beta_{1n}+\beta_{2n})} + e^{i(\beta_{3n}+\beta_{4n})}\right) \\ \times \left(\beta_{1n}^2 - \beta_{2n}^2\right)\left(\beta_{3n}^2 - \beta_{4n}^2\right) \\ + \left(e^{i(\beta_{1n}+\beta_{3n})} + e^{i(\beta_{2n}+\beta_{4n})}\right) \\ \times \left(\beta_{2n}^2 - \beta_{4n}^2\right)\left(\beta_{3n}^2 - \beta_{1n}^2\right) \\ + \left(e^{i(\beta_{2n}+\beta_{3n})} + e^{i(\beta_{1n}+\beta_{4n})}\right) \\ \times \left(\beta_{1n}^2 - \beta_{4n}^2\right)\left(\beta_{2n}^2 - \beta_{3n}^2\right) = 0, \end{gathered} \tag{11}$$

where ω_n represents the natural frequencies. Numerical values of ω_n as well as β_{in} can be calculated by using (10) for viscoelastic beams and (11) for viscoelastic pipes.

4. Numerical Solutions

Numerical solutions of the natural frequencies appear in the complex form:

$$\omega_n = \omega_{nR} + i\omega_{nI}. \tag{12}$$

If the internal damping is assumed to be of order ε, imaginary parts of (12) will be negligibly small. Thus, the natural frequencies of the system can be obtained from the real part ($\omega_n = \omega_{nR}$). However, if the internal damping is assumed to be of order one, imaginary part of (12) will not be negligible.

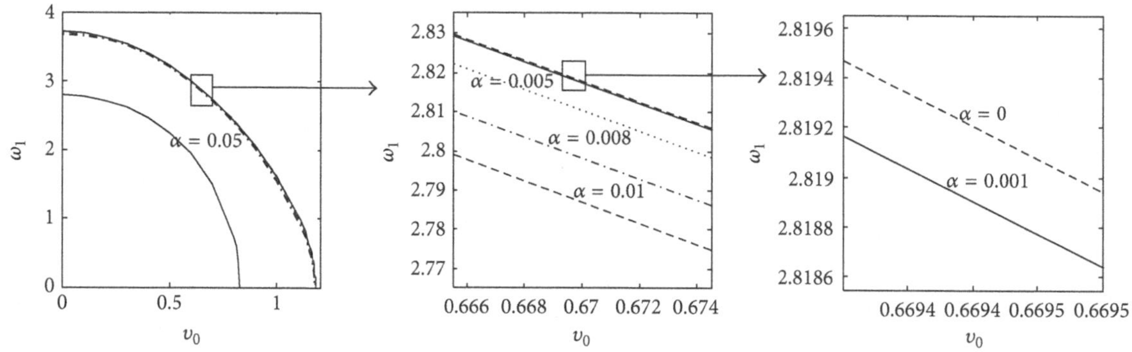

FIGURE 6: First fundamental frequencies versus axial transport velocities of the viscoelastic beams for different α values (small areas are zoomed) ($v_f = 0.2$).

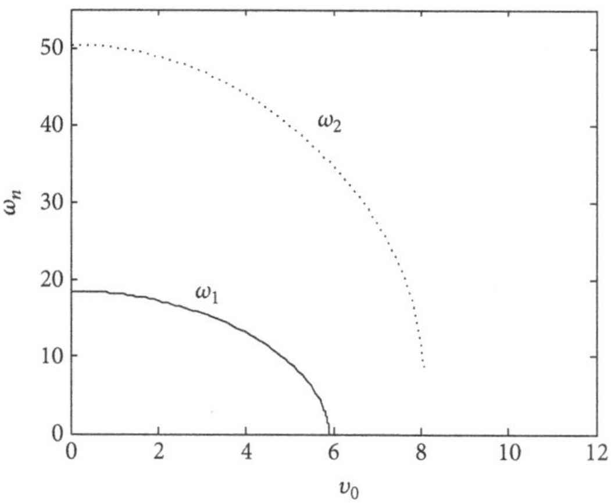

FIGURE 7: First two fundamental frequencies versus fluid velocities of the pipes ($\alpha = 0.001$, $\beta = 0.3$, and $P_0 = 25$).

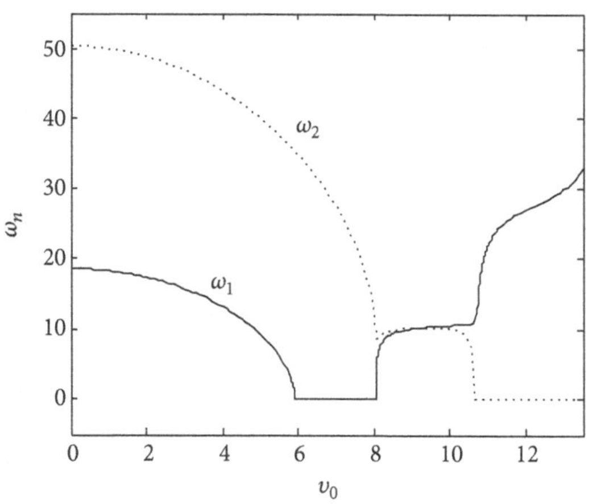

FIGURE 8: Real parts of first two fundamental frequencies versus fluid velocities of the pipes ($\alpha = 0.001$, $\beta = 0.3$, and $P_0 = 25$).

In that case, real parts can be considered as natural frequencies and imaginary parts can be considered as damping effects.

In Figure 1, the first three natural frequencies versus transport velocities are shown. Natural frequencies of the

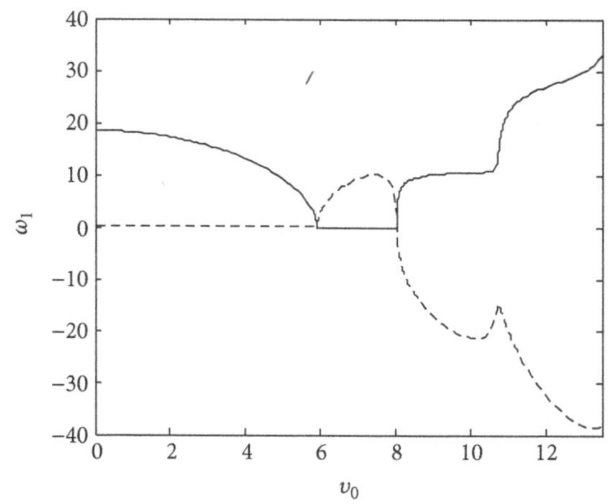

FIGURE 9: Real (solid line) and imaginary parts (dashed line) of the first fundamental frequencies versus fluid velocities of the pipes ($\alpha = 0.001$, $\beta = 0.3$, and $P_0 = 25$).

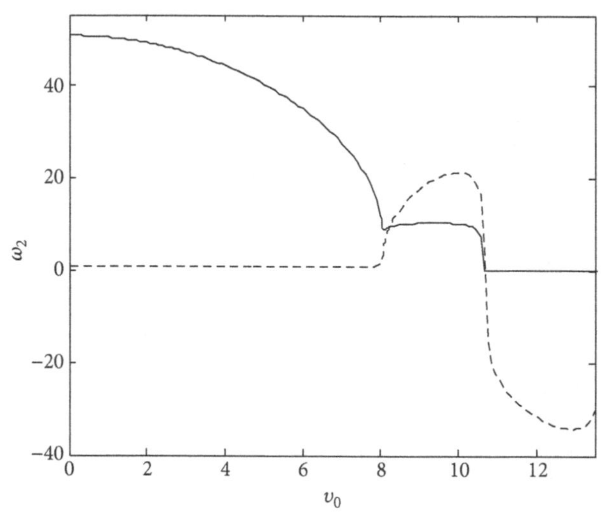

FIGURE 10: Real (solid line) and imaginary parts (dashed line) of the second fundamental frequencies versus fluid velocities of the pipes ($\alpha = 0.001$, $\beta = 0.3$, and $P_0 = 25$).

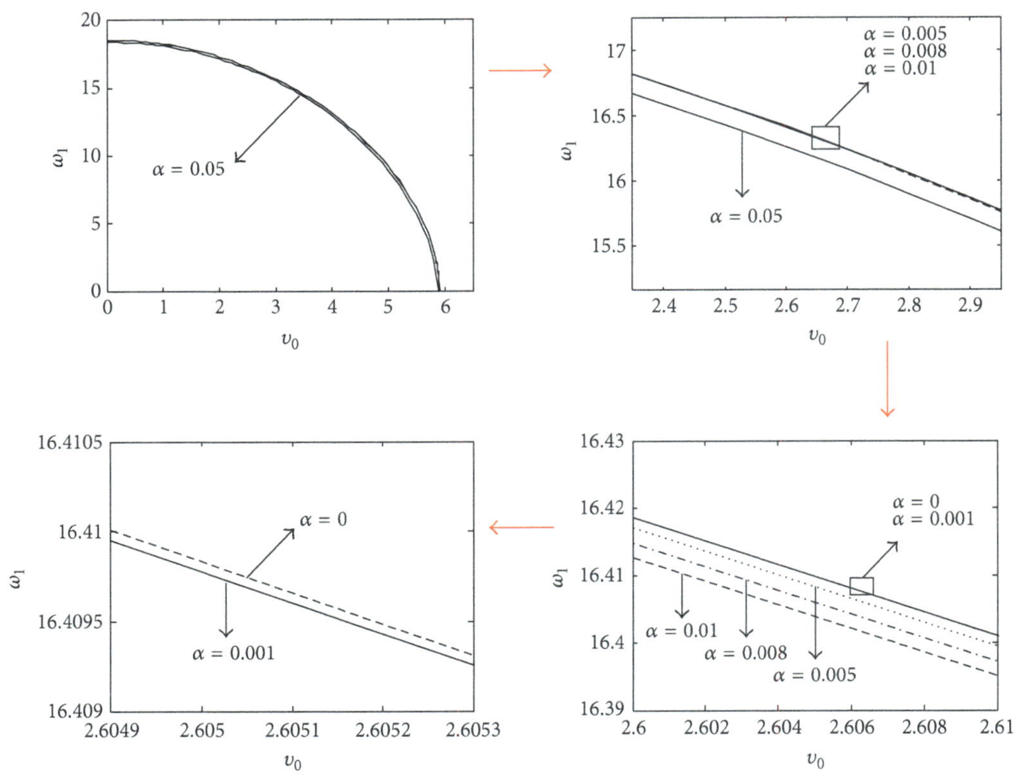

FIGURE 11: First fundamental frequencies versus fluid velocities of the pipes for different α values (small areas are zoomed) ($\beta = 0.3$, $P_0 = 25$).

beam decrease by increasing transport velocity. As seen in the figure, second and third natural frequency (high frequency modes) curves do not cross the horizontal axis. For specific values of velocities (i.e., critical velocity) divergence instability occurs. Figure 2 shows the variation of the real parts of the first three fundamental frequencies with respect to transport velocities. In Figures 3, 4, and 5, real and imaginary parts of the first three natural frequencies are shown. In Figures 4 and 5, the real parts of the fundamental frequencies reach to zero by iterations.

In Figure 6, the effect of the viscoelasticity coefficient on the natural frequencies is shown. The coefficient related to viscoelasticity α is chosen as 0, 0.001, 0.005, 0.008, 0.01, and 0.05. To distinguish the curves, a small fraction of the figure is zoomed. By increasing the viscoelasticity coefficients, natural frequencies decrease for the same transport velocity. Note that $\alpha = 0$ refers to an elastic Euler-Bernoulli beam.

Results of pipes conveying fluids are given in Figures 7–11. Figures 7 and 8 show the variation of the natural frequencies with fluid velocities for the first two modes. Natural frequencies of the pipes decrease by increasing fluid velocity. In Figures 9 and 10, first and second natural frequencies versus fluid velocity including real and imaginary parts that are shown. Figure 11 depicts the effect of the viscoelasticity in detail. In order to distinguish the curves, a small area is zoomed and shown. Similar to the axially moving beams, natural frequencies decrease for increasing α values.

The detailed variations of the natural frequencies in the case of viscoelasticity are given therefore in this study.

5. Concluding Remarks

Vibrations of axially moving beams and pipes conveying fluid are analyzed. Linear problem is considered to calculate the natural frequencies of the vibrated systems. Contrary to some previous work which considered viscoelasticity to be small, viscoelasticity terms are assumed to be of order one in this study. Natural frequencies for different beam and pipe parameters are shown. Effects of viscoelasticity on the natural frequencies are discussed. It is found that the viscoelasticity reduces the natural frequencies. Effects of dissipation mechanism on the frequency and critical velocities are discussed also by others [28–31]. The qualitative behaviours of our results and theirs agree with each other. For a wide range of parameters, the natural frequencies are calculated and can be used in further studies.

References

[1] A. G. Ulsoy, C. D. Mote Jr., and R. Szymni, "Principal developments in band saw vibration and stability research," *Holz als Roh- und Werkstoff*, vol. 36, no. 7, pp. 273–280, 1978.

[2] J. A. Wickert and C. D. Mote Jr., "Current research on the vibration and stability of axially moving materials," *Shock and Vibration Digest*, vol. 20, no. 5, pp. 3–13, 1988.

[3] J. A. Wickert and C. D. Mote Jr., "Classical vibration analysis of axially moving continua," *Journal of Applied Mechanics*, vol. 57, no. 3, pp. 738–744, 1990.

[4] J. A. Wickert and C. D. Mote Jr., "Response and discretization methods for axially moving materials," *Applied Mechanics Reviews*, vol. 44, no. 11, pp. 279–284, 1991.

[5] A. G. Ulsoy and C. D. Mote Jr., "Analysis of bandsaw vibration," *Wood Science*, vol. 13, no. 1, pp. 1–10, 1980.

[6] W. L. Miranker, "The wave equation in a medium motion," *IBM Journal of Research and Development*, vol. 4, pp. 36–42, 1960.

[7] C. D. Mote Jr., "Stability of systems transporting accelerating axially moving materials," *Journal of Dynamic Systems, Measurement and Control*, vol. 97, no. 1, pp. 96–98, 1975.

[8] M. Pakdemirli, A. G. Ulsoy, and A. Ceranoğlu, "Transverse vibration of an axially accelerating string," *Journal of Sound and Vibration*, vol. 169, no. 2, pp. 179–196, 1994.

[9] E. M. Mockensturm, N. C. Perkins, and A. G. Ulsoy, "Stability and limit cycles of parametrically excited, axially moving strings," *Journal of Vibration and Acoustics*, vol. 118, no. 3, pp. 346–351, 1996.

[10] M. Pakdemirli and H. Batan, "Dynamic stability of a constantly accelerating strip," *Journal of Sound and Vibration*, vol. 168, no. 2, pp. 371–378, 1993.

[11] H. R. Öz and M. Pakdemirli, "Vibrations of an axially moving beam with time-dependent velocity," *Journal of Sound and Vibration*, vol. 227, no. 2, pp. 239–257, 1999.

[12] M. Pakdemirli and H. R. Öz, "Infinite mode analysis and truncation to resonant modes of axially accelerated beam vibrations," *Journal of Sound and Vibration*, vol. 311, no. 3–5, pp. 1052–1074, 2008.

[13] M. P. Païdoussis and G. X. Li, "Pipes conveying fluid: a model dynamical problem," *Journal of Fluids and Structures*, vol. 7, no. 2, pp. 137–204, 1993.

[14] H. R. Öz and H. Boyacı, "Transverse vibrations of tensioned pipes conveying fluid with time-dependent velocity," *Journal of Sound and Vibration*, vol. 236, no. 2, pp. 259–276, 2000.

[15] H. R. öz, "Non-linear vibrations and stability analysis of tensioned pipes conveying fluid with variable velocity," *International Journal of Non-Linear Mechanics*, vol. 36, no. 7, pp. 1031–1039, 2001.

[16] L. Q. Chen, N. H. Zhang, and J. W. Zu, "Bifurcation and chaos of an axially moving viscoelastic string," *Mechanics Research Communications*, vol. 29, no. 2-3, pp. 81–90, 2002.

[17] L. Q. Chen, N. H. Zhang, and J. W. Zu, "The regular and chaotic vibrations of an axially moving viscoelastic string based on fourth order Galerkin truncation," *Journal of Sound and Vibration*, vol. 261, no. 4, pp. 764–773, 2003.

[18] L. Q. Chen, H. Chen, and C. W. Lim, "Asymptotic analysis of axially accelerating viscoelastic strings," *International Journal of Engineering Science*, vol. 46, no. 10, pp. 976–985, 2008.

[19] U. Lee and H. Oh, "Dynamics of an axially moving viscoelastic beam subject to axial tension," *International Journal of Solids and Structures*, vol. 42, no. 8, pp. 2381–2398, 2005.

[20] L. Q. Chen and X. D. Yang, "Stability in parametric resonance of axially moving viscoelastic beams with time-dependent speed," *Journal of Sound and Vibration*, vol. 284, no. 3–5, pp. 879–891, 2005.

[21] L. Q. Chen, X. D. Yang, and C. J. Cheng, "Dynamic stability of an axially accelerating viscoelastic beam," *European Journal of Mechanics A*, vol. 23, no. 4, pp. 659–666, 2004.

[22] M. H. Ghayesh, "Nonlinear transversal vibration and stability of an axially moving viscoelastic string supported by a partial viscoelastic guide," *Journal of Sound and Vibration*, vol. 314, no. 3–5, pp. 757–774, 2008.

[23] L. N. Panda and R. C. Kar, "Nonlinear dynamics of a pipe conveying pulsating fluid with combination, principal parametric and internal resonances," *Journal of Sound and Vibration*, vol. 309, no. 3–5, pp. 375–406, 2008.

[24] B. B. Özhan and M. Pakdemirli, "A general solution procedure for the forced vibrations of a continuous system with cubic nonlinearities: primary resonance case," *Journal of Sound and Vibration*, vol. 325, no. 4-5, pp. 894–906, 2009.

[25] B. B. Özhan and M. Pakdemirli, "A general solution procedure for the forced vibrations of a system with cubic nonlinearities: three-to-one internal resonances with external excitation," *Journal of Sound and Vibration*, vol. 329, no. 13, pp. 2603–2615, 2010.

[26] B. B. Özhan and M. Pakdemirli, "Principal parametric resonances of a general continuous system with cubic nonlinearities," *Applied Mathematics and Computation*, vol. 219, pp. 2412–2423, 2012.

[27] L. Q. Chen and X. D. Yang, "Transverse nonlinear dynamics of axially accelerating viscoelastic beams based on 4-term Galerkin truncation," *Chaos, Solitons & Fractals*, vol. 27, no. 3, pp. 748–757, 2006.

[28] L. Q. Chen and X. D. Yang, "Vibration and stability of an axially moving viscoelastic beam with hybrid supports," *European Journal of Mechanics A*, vol. 25, no. 6, pp. 996–1008, 2006.

[29] L. Q. Chen and B. Wang, "Stability of axially accelerating viscoelastic beams: asymptotic perturbation analysis and differential quadrature validation," *European Journal of Mechanics A*, vol. 28, no. 4, pp. 786–791, 2009.

[30] M. P. Païdoussis, *Fluid-Structure Interactions, Slender Structures and Axial Flow*, vol. 1, Academic Press, New York, NY, USA, 1998.

[31] O. Doaré, "Dissipation effect on local and global stability of fluid-conveying pipes," *Journal of Sound and Vibration*, vol. 329, no. 1, pp. 72–83, 2010.

Choosing Actuation Scheme for Optimal Performance of 3-DOF Planar Parallel Manipulators

S. Ramana Babu,[1] V. Ramachandra Raju,[2] and K. Ramji[3]

[1] *Department of Mechanical Engineering, G.V.P. College of Engineering, Visakhapatnam 530048, India*
[2] *Department of Mechanical Engineering, JNTU College of Engineering, Vizianagaram 535003, India*
[3] *Department of Mechanical Engineering, AU College of Engineering, Visakhapatnam 530003, India*

Correspondence should be addressed to S. Ramana Babu, ramanababu_sabbavarapa@yahoo.co.in

Academic Editor: Adib Becker

This paper describes how the actuation scheme affects the size and shape of the workspace of 3RPR and 3PRR, which are the two popular planar parallel manipulators. This study helps in achieving the required singularity-free zones for a specific task by adopting a suitable actuation scheme without resorting to changing the architecture of the manipulator. In addition, the performance index GCI for each actuation scheme has been evaluated and compared with each other. Finally, the GCI for each actuation scheme has been optimized subjected to the geometric constraints for the 3PRR planar parallel manipulator.

1. Introduction

Parallel manipulators are robots that consist of separate serial chains that connect the fixed link to the moving end-effectors. Even though parallel manipulators possess several advantages like, high stiffness, low inertia, and large payload capacity, they are suffered from limited workspace, complicated singularities, and design difficulties. Many researchers have analyzed kinematics, dynamics, workspace, and control of parallel manipulators. Mohammadi Daniali et al. [1] presented an in-depth study of velocity relationships and singular conditions for general planar parallel robots. Gosselin [2] presents computational algorithms for kinematics and dynamics of planar and spatial parallel robots. Merelt et al. [3] solved the forward pose kinematics for a broad class of planar parallel manipulators. Y. Jin et al. [4] described a 3-limb selectively actuated parallel manipulator (SA-PM) in which the end-effecter exhibits either a 3-DOF spherical, 3-DOF translational, or a complete 6-DOF spatial motion depending on the types of the actuation chosen for the actuators. They also presented singularity analysis of the SA-PM based on geometry for all actuation schemes to facilitate the kinematic design of the manipulator. Arakelian et al. [5] proposed the legs of variable structure for the increase of reachable workspace of a spatial manipulator. Su et al. [6] studied trajectory planning and actuation schemes, which are crucial for their control. Ebrahimi et al. [7] presented a model of an actuation scheme and its effects on the singularities of parallel manipulators for a given path in the workspace. Actuation schemes and actuator relocation for serial planar manipulator have also been studied in detail by Matone and Roth [8] and Ramji et al. [9] respectively. Li and Xu [10] designed a new XYZ novel micromanipulator, which was called a 3-PRC manipulator in order to generate a cuboids' shape workspace by using the (PZT) piezoelectric actuator to drive the prismatic active joint due to its major advantages of high stiffness and compact size; however, stroke of the adopted PZT cannot meet the application requirements so in order to enlarge the workspace the stroke of the PZT is amplified by using a lever amplification mechanism and making the leg as an equivalent PRPR linkage. Xu and Li [11] proposed a new PRC translational parallel manipulator with orthogonally arranged fixed actuators for producing a singularity-free workspace through an extensive reduction in cost in building and controlling such a manipulator. Li and Xu [12] analyzed the mobility of the 3-PRC translational parallel manipulator via screw theory and also illustrated the effect of architectural parameters on the shape and size of the workspace; further more dexterity characteristics are investigated in the local and global sense for real machine design.

Actuation scheme depends on the Jacobian matrix that transforms the active input joint rates into end-effecter task space velocities. Earlier researchers on PKMs studied the kinematic analysis and performance of the manipulator for only one set of active joint rates (consists of only one joint from one leg) preferably the base joints. Where the kinematic and workspace analysis is not only confined to one set of active joints and also tried to investigate the changes in reachable workspace and performance of manipulators for the other set of joints, which are chosen as active joints. We have to enumerate the optimal geometric parameters like link lengths, joint variables, and orientation of the prismatic joints, which greatly influence the workspace and performance of the manipulator for each possible actuation scheme. Though actuation schemes have been proposed by various researchers, these have not been applied to any specific case except by Rakotomanga et al. [13]. This paper is an attempt to apply the concept of actuation schemes to two popular planar manipulators, namely, 3RPR and 3PRR.

The organization of the paper is as follows: Method explains the basic theory of architecture, kinematic modeling, and actuation schemes of 3RPR manipulator first, followed by a similar treatment for the 3PRR manipulator. Results describes through graphs and tables, comparison of the performance of a manipulator for all possible actuation schemes. Also included is the optimal geometric design of 3PRR planar parallel manipulator. These results are followed by Discussion.

2. Method

2.1. Mechanism Architecture of 3RPR Planar Manipulator. The manipulator considered here is symmetric and consists of three identical legs connecting the fixed base to the end-effecter as shown in Figure 1. The three revolute joints are located on the fixed base triangle at A_i and another three revolute joints are located on the moving triangle at C_i where $i = 1, 2, 3$. The prismatic joint variables L_i correspond to the leg lengths A_iC_i. The moving frame {H} is located at the centroid of the end-effecter, which is an equilateral triangle as shown in Figure 2. The fixed frame {B} is chosen at the origin of one of the legs on the fixed base, which is also an equivalent triangle.

For the 3RPR manipulator, the workspace limitations are due to the limitations of prismatic actuators, and the maximum and minimum lengths of the prismatic actuators of the ith chain are denoted by $L_{\max i}$, $L_{\min i}$, respectively, as the prismatic joint variables. The maximum possible geometric workspace of 3RPR manipulator is the intersection of three annular regions A_iP centered in A_i whose radii are $(L_{\min i} + \|PC_i\|)$ and $(L_{\max i} - \|PC_i\|)$ as shown in Figure 3. This annular region will exist if and only if $(L_{\max i} - L_{\min i}) \geq 2\|PC_i\|$. The determination of the maximal workspace has been addressed by Kumar [14] and Pennock and Kassner [15] who pointed out that the boundary of this workspace is composed of circular arcs and portions of sextic curves.

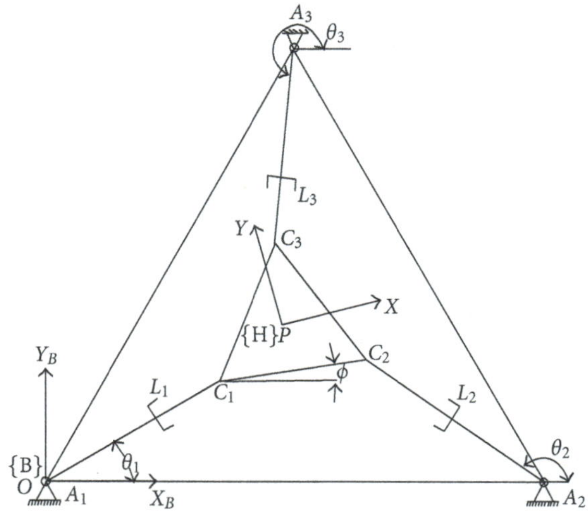

FIGURE 1: 3RPR planar manipulator.

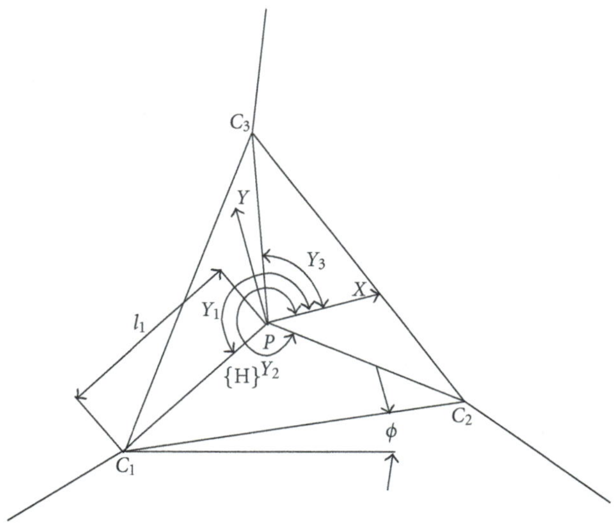

FIGURE 2: End-effecter.

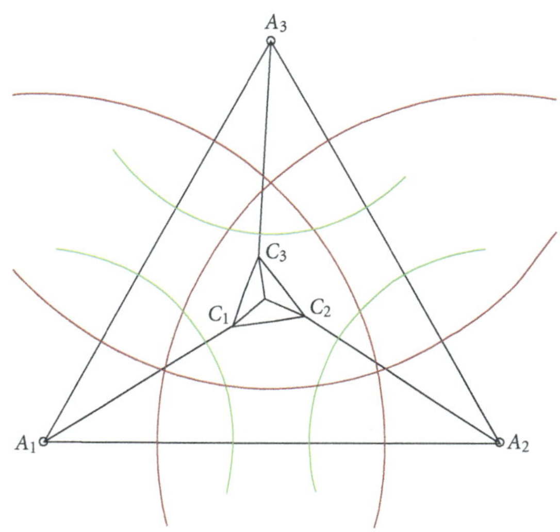

FIGURE 3: The possible workspace of 3RPR manipulator.

2.1.1. Kinematic Modeling of 3RPR Parallel Manipulator. The solutions are found using the distance norm constraint

$$L_i^2 = \left\| {}^B\underline{C_i} - {}^B\underline{A_i} \right\|^2 \quad i = 1, 2, 3. \tag{1}$$

In the forward pose kinematics for the given prismatic joint lengths $L = [L_1, L_2, L_3]^T$, the Cartesian pose coordinate $X = \begin{bmatrix} x & y & \phi \end{bmatrix}^T$ can be calculated. Equation (1) can be expressed in analytical form as

$$\begin{aligned} L_i^2 &= x^2 + y^2 + l^2 + a_{ix}^2 + a_{iy}^2 - 2a_{ix}x - 2a_{iy}y \\ &\quad - 2la_{ix}\cos(\gamma_i + \phi) - 2la_{iy}\sin(\gamma_i + \phi) \\ &\quad + 2lx\cos(\gamma_i + \phi) + 2ly\sin(\gamma_i + \phi), \end{aligned} \tag{2}$$

where ϕ is the orientation of the moving platform with reference x-axis, γ_i is angle between vector C_iP and x-axis of the moving platform, l is length of vector C_iP, and a_i is distance of the vertex A_i of the fixed platform from the origin of the reference frame. The revolute joint angle, which is attached to the fixed platform, is expressed as

$$\theta_i = a\tan 2\left({}^BC_{iy} - {}^BA_{iy}, {}^BC_{ix} - {}^BA_{ix}\right) \quad i = 1, 2, 3. \tag{3}$$

It is expressed in analytical form as

$$\tan\theta_i = \frac{y - a_{iy} + l\sin(\gamma_i + \phi)}{x - a_{ix} + l\cos(\gamma_i + \phi)}. \tag{4}$$

2.1.2. Actuation Schemes and Velocity Analysis. The major disadvantage of parallel kinematic chains is their limited workspace and in order to overcome this, recently some researchers have investigated hybrid serial-parallel manipulators that present a compromise between the high rigidity of parallel manipulators and the extended workspace of serial manipulators. However, the usefulness of this manipulator is also limited because the compliance of the manipulator is more, and so it leads to inaccuracy in controlling the manipulator. There is another trend to eliminate the singularities and increase the workspace by the redundant actuation with the use of more than one actuator in a leg. However, this redundant actuation is expensive due to the additional actuators and the complication in controlling the manipulator.

The actuation scheme is the set of active input joint rates of the manipulator in which only one joint from each leg is to be selected. Each actuation scheme has its own singularity free zones of the end-effecter. Since each scheme implies a different set of actuator motions, this leads to different movements of the end-effecter. A proper actuation scheme not only avoids singularities of the workspace but even enhances the stiffness and accuracy of the manipulator.

The 3RPR manipulator can be actuated either by revolute joint or by prismatic joint in each leg so to attain the required end-effecter output velocities $\dot{X} = \begin{bmatrix} \dot{x} & \dot{y} & \dot{\phi} \end{bmatrix}^T$. In order to change the end-effecter position, we have to actuate either the base revolute joint θ_i or the linear actuator L_i, that is, only one input from each leg so as to avoid redundancy since it has its own drawbacks. Only six joints $(L_1, L_2, L_3, \theta_1, \theta_2, \theta_3)$ are available for input out of the nine, and among these, we have to choose three joints, one from each leg. Hence there exist eight possible combinations and each such set is called an actuation scheme. These actuation schemes are presented in Table 1. The underlined letters show the input pairs, "R̲" for revolute with input angle θ_i and "P̲" for prismatic with input displacement L_i.

2.1.3. Jacobian Matrix. The velocity vector of a point is given in two different directions of a loop closure. Each loop closure consists of a fixed base, moving platform, and all links of a leg. The unactuated joint rates in each leg are eliminated by taking dot product of the velocity vector loop equation with a vector that is normal to all vectors of unactuated joint rates. Finally the resulting equations are assembled into a Jacobian matrix. In order to obtain the Jacobian matrix for the actuating joint L_i differentiate (2) with respect to time and is expressed as

$$\dot{L}_i = \frac{1}{L_i}\left[L_i\cos\theta_i \;\; L_i\sin\theta_i \;\; l\sin(\gamma_i + \phi)(a_{ix} - x) - l\cos(\gamma_i + \phi)\left(a_{iy} - y\right)\right]\dot{X} = j_{xLi}\dot{X} \quad \text{for } i = 1, 2, 3, \tag{5}$$

where

$$\dot{X} = \left[\dot{x} \;\; \dot{y} \;\; \dot{\phi}\right]^T. \tag{6}$$

The Jacobian for the actuating joint θ_i is obtained by differentiating (4) with respect to time and is expressed as

$$\begin{aligned} \dot{\theta}_i &= \frac{1}{L_i}\left[-\sin\theta_i \;\; \cos\theta_i \;\; l\cos(\theta_i - \gamma_i - \phi)\right] \\ &= j_{x\theta i}\dot{X} \quad \text{for } i = 1, 2, 3. \end{aligned} \tag{7}$$

Using the above equations (5) and (7), the Jacobian matrix for eight actuation schemes is expressed as

$$\text{Actuation scheme 1}: \quad J_1 = \left[j_{xL1} \;\; j_{xL2} \;\; j_{xL3}\right]^T,$$

$$\text{Actuation scheme 2}: \quad J_2 = \left[j_{x\theta 1} \;\; j_{x\theta 2} \;\; j_{x\theta 3}\right]^T,$$

$$\text{Actuation scheme 3}: \quad J_3 = \left[j_{xL1} \;\; j_{xL2} \;\; j_{x\theta 3}\right]^T,$$

$$\text{Actuation scheme 4}: \quad J_4 = \left[j_{xL1} \;\; j_{x\theta 2} \;\; j_{x\theta 3}\right]^T,$$

TABLE 1: Actuation schemes of 3RPR manipulator.

Actuation no.	Actuation scheme	Active joints
1	RPR1-RPR2-RPR3	L_1 L_2 L_3
2	RPR1-RPR2-RPR3	θ_1 θ_2 θ_3
3	RPR1-RPR2-RPR3	L_1 L_2 θ_3
4	RPR1-RPR2-RPR3	L_1 θ_2 θ_3
5	RPR1-RPR2-RPR3	L_1 θ_2 L_3
6	RPR1-RPR2-RPR3	θ_1 L_2 L_3
7	RPR1-RPR2-RPR3	θ_1 θ_2 L_3
8	RPR1-RPR2-RPR3	θ_1 L_2 θ_3

$$\begin{aligned}
&\text{Actuation scheme 5}: \quad J_5 = \begin{bmatrix} j_{xL1} & j_{x\theta 2} & j_{xL3} \end{bmatrix}^T, \\
&\text{Actuation scheme 6}: \quad J_6 = \begin{bmatrix} j_{x\theta 1} & j_{xL2} & j_{xL3} \end{bmatrix}^T, \\
&\text{Actuation scheme 7}: \quad J_7 = \begin{bmatrix} j_{x\theta 1} & j_{x\theta 2} & j_{xL3} \end{bmatrix}^T, \\
&\text{Actuation scheme 8}: \quad J_8 = \begin{bmatrix} j_{x\theta 1} & j_{xL2} & j_{x\theta 3} \end{bmatrix}^T.
\end{aligned} \tag{8}$$

For actuation scheme 1 (RPR1-RPR2-RPR3), the joint variables θ_i are passive joints, while the active joints are the prismatic joints (L_1, L_2, L_3). The Jacobian matrix mapping joint rates $\dot{q}_i = \begin{bmatrix} \dot{L}_1 & \dot{L}_2 & \dot{L}_3 \end{bmatrix}^T$ into Cartesian rates $\dot{X} = \begin{bmatrix} \dot{x} & \dot{y} & \dot{\phi} \end{bmatrix}^T$ is also expressed in two separate Jacobian matrices as

$$\begin{bmatrix} L_1 & 0 & 0 \\ 0 & L_2 & 0 \\ 0 & 0 & L_3 \end{bmatrix} \begin{bmatrix} \dot{L}_1 \\ \dot{L}_2 \\ \dot{L}_3 \end{bmatrix} = \begin{bmatrix} L_1 \cos\theta_1 & L_1 \sin\theta_1 & Z_1 \\ L_2 \cos\theta_2 & L_2 \sin\theta_2 & Z_2 \\ L_3 \cos\theta_3 & L_3 \sin\theta_3 & Z_3 \end{bmatrix} \begin{bmatrix} \dot{x} \\ \dot{y} \\ \dot{\phi} \end{bmatrix}, \tag{9}$$

where

$$Z_i = l \sin(\gamma_i + \phi)(a_{ix} - x) - l \cos(\gamma_i + \phi)\left(a_{iy} - y\right). \tag{10}$$

Which is analogous to the $J_q \dot{q} = J_x \dot{X}$, where J_q and J_x are inverse and direct Jacobian matrices, respectively, the overall Jacobian matrix $J = J_q^{-1} J_x$. For actuation scheme 2 (RPR1-RPR2-RPR3), the active joint rates $\dot{q} = \begin{bmatrix} \dot{\theta}_1 & \dot{\theta}_2 & \dot{\theta}_3 \end{bmatrix}^T$ into the Cartesian rates $\dot{X} = \begin{bmatrix} \dot{x} & \dot{y} & \dot{\phi} \end{bmatrix}^T$. The velocity relation is expressed in two separate matrix forms as

$$\begin{aligned}
&\begin{bmatrix} L_1 & 0 & 0 \\ 0 & L_2 & 0 \\ 0 & 0 & L_3 \end{bmatrix} \begin{bmatrix} \dot{\theta}_1 \\ \dot{\theta}_2 \\ \dot{\theta}_3 \end{bmatrix} \\
&= \begin{bmatrix} -\sin\theta_1 & \cos\theta_1 & l\cos(\theta_1 - \gamma_1 - \phi) \\ -\sin\theta_2 & \cos\theta_2 & l\cos(\theta_2 - \gamma_2 - \phi) \\ -\sin\theta_3 & \cos\theta_3 & l\cos(\theta_3 - \gamma_3 - \phi) \end{bmatrix} \begin{bmatrix} \dot{x} \\ \dot{y} \\ \dot{\phi} \end{bmatrix}.
\end{aligned} \tag{11}$$

2.2. 3PRR Manipulator Architecture. A 3-DOF planar parallel manipulator with three identical legs, which has been studied extensively by many researchers including Zhang et al. [16], is shown in Figure 4. Each kinematic chain is of PRR type and consists of one prismatic joint and two revolute joints. The manipulator is intended to position and orient the equilateral triangle-shaped platform $C_1C_2C_3$. The geometric centre of platform $C_1C_2C_3$ denoted by p is the operation point of the manipulator, with its position and orientation being (x_p, y_p, ϕ). Three intermediate links B_iC_i for $i = 1, 2, 3$ connect the sliders and the moving platform.

The base of the manipulator, also an equilateral triangle with vertices A_1, A_2, and A_3, is fixed to the ground. The joint coordinate associated with the ith actuator that actuates the prismatic joint is ρ_i for $i = 1, 2, 3$. The variable α_i is the angle at A_i between the x-axis of the fixed frame and ith linear guide $[\alpha_1, \alpha_2, \alpha_3] = \begin{bmatrix} 30^\circ & 150^\circ & 270^\circ \end{bmatrix}$. The revolute variable β_i is the angle made by the ith intermediate link B_iC_i with the x-axis of the fixed frame. The vector loop closure equation for the ith leg is written as

$$\overline{OP} + \overline{PC_i} = \overline{OA_i} + \overline{A_iB_i} + \overline{B_iC_i}. \tag{12}$$

Writing the above equation in component form,

$$\begin{aligned}
x_p &= x_{ai} + \rho_i \cos\alpha_i + l\cos\beta_i + e_i \cos(\gamma_i + \phi), \\
y_p &= y_{ai} + \rho_i \sin\alpha_i + l\sin\beta_i + e_i \sin(\gamma_i + \phi).
\end{aligned} \tag{13}$$

Using (11) and (12), we can establish a quadratic equation in ρ_i

$$\rho_i^2 + k_{2i}\rho_i + k_{1i} = 0 \quad \text{for } i = 1, 2, 3, \tag{14}$$

where

$$\begin{aligned}
k_{1i} = \Big(&x_p^2 + y_p^2 + x_{ai}^2 + y_{ai}^2 + e_i^2 - 2x_p x_{ai} - 2y_p y_{ai} \\
&- 2x_p e_i \cos(\gamma_i + \phi) - 2y_p e_i \sin(\gamma_i + \phi) \\
&+ 2x_{ai} e_i \cos(\gamma_i + \phi) + 2y_{ai} e_i \sin(\gamma_i + \phi) - l^2\Big),
\end{aligned} \tag{15}$$

$$\begin{aligned}
k_{2i} = \Big(&2x_{ai}\cos\alpha_i + 2y_{ai}\sin\alpha_i - 2x_p\cos\alpha_i \\
&- 2y_p \sin\alpha_i + 2e_i\cos(\alpha_i - \gamma_i - \phi)\Big).
\end{aligned} \tag{16}$$

It is clear that there are two possible solutions for each individual link, and the manipulator can take on a maximum of eight configurations for a given set of coordinates of the moving platform.

2.2.1. Velocity Analysis of Actuation Schemes. The 3RPR manipulator can be actuated either by revolute joint or by prismatic joint in each leg of the manipulator in order to attain the required end-effecter output velocities. Thus there are only three input joint rates out of six $(\rho_1, \rho_2, \rho_3, \beta_1, \beta_2, \beta_3)$. There exist eight possible combinations; each combinational set of actuators is called an actuation scheme and these are shown in Table 2. The underlined letters show the input pairs, "R" for revolute with input angle β_i and "P" for prismatic with input displacement ρ_i. The actuation scheme 1 is represented as PRR1-PRR2-PRR3, where the prismatic joint variables $\begin{pmatrix} \rho_1 & \rho_2 & \rho_3 \end{pmatrix}$ are active joints.

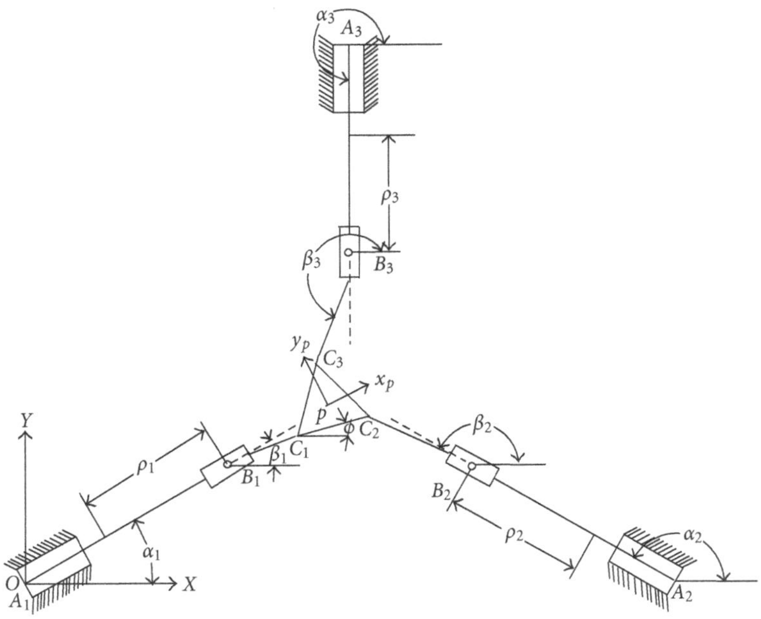

FIGURE 4: 3PRR planar manipulator.

TABLE 2: actuation schemes for 3PRR manipulator.

Actuation no.	Actuation scheme	Active joints
1	PRR1-PRR2-PRR3	$\rho_1\ \rho_2\ \rho_3$
2	PRR1-PRR2-PRR3	$\beta_1\ \beta_2\ \beta_3$
3	PRR1-PRR2-PRR3	$\rho_1\ \beta_2\ \beta_3$
4	PRR1-PRR2-PRR3	$\rho_1\ \rho_2\ \beta_3$
5	PRR1-PRR2-PRR3	$\beta_1\ \beta_2\ \rho_3$
6	PRR1-PRR2-PRR3	$\beta_1\ \rho_2\ \rho_3$
7	PRR1-PRR2-PRR3	$\beta_1\ \rho_2\ \beta_3$
8	PRR1-PRR2-PRR3	$\rho_1\ \beta_2\ \rho_3$

To obtainmanipulator the velocity equations for actuation scheme 1, we differentiate (11) with respect to time so that

$$\left(\dot{x}_p\bar{i} + \dot{y}_p\bar{j}\right) + \dot{\phi}\left(\bar{k}\times\bar{e}_i\right) = \dot{\rho}_i\bar{a}_i + \dot{\beta}_i\left(\bar{k}\times\bar{b}_i\right), \tag{17}$$

where $\bar{a}_i = (\cos\alpha_i, \sin\alpha_i)^T$, $\bar{b}_i$ is the positional vector from B_i to C_i, $\bar{e}_i$ is the positional vector from p to C_i, and $\bar{k}$ is unit vector orthogonal to both $\bar{i}$ and $\bar{j}$. The left-hand side of (9) can be interpreted as summation of the velocity of the platform at its mass centre and a velocity term due to the rotation of the platform. The right-hand side of (9) is the summation of the velocity of the slider and the velocity term due to rotation of the link. To remove the velocity term related to the rotational velocity of the link, $\dot{\beta}_i$ in (17), we apply vector dot product on both sides of (17) with $\bar{b}_i$

$$\left(\dot{x}_p\bar{i} + \dot{y}_p\bar{j}\right)\cdot\bar{b}_i + \left(\bar{e}_i\times\bar{b}_i\right)\cdot\bar{k} = \dot{\rho}_i\left(\bar{a}_i\cdot\bar{b}_i\right). \tag{18}$$

The above equation can be expressed as

$$\dot{\rho}_i = \frac{1}{\bar{a}_i\cdot\bar{b}_i}\left(b_{ix}\ \ b_{iy}\ \ e_{ix}b_{iy} - e_{iy}b_{ix}\right)\left(\dot{x}_p\ \ \dot{y}_p\ \ \dot{\phi}\right)^T = J_{pi}\dot{\overline{X}}_p. \tag{19}$$

Equation (19) presents the velocity relationship between the sliders and the platform with the Jacobian matrix J_P. Using (19) for $i = 1, 2, 3$, we can write Jacobian matrix for actuation scheme 1 (PRP1-PRR2-PRR3).

Performing vector cross product to (17) by $\bar{b}_i$,

$$\left(\dot{x}_p\bar{i} + \dot{y}_p\bar{j}\right)\times\bar{b}_i + \dot{\phi}\left(\bar{k}\times\bar{e}_i\right)\times\bar{b}_i = \dot{\rho}_i\left(\bar{a}_i\times\bar{b}_i\right) - \dot{\beta}_i^2\left|\bar{b}_i^2\right|\bar{k}. \tag{20}$$

From (19) and (20), the angular velocity of the link is given as

$$\begin{aligned}\dot{\beta}_i &= \frac{1}{\left|\bar{b}_i\right|^2}\Big\{\left(-b_{iy}\ \ b_{ix}\ \ e_{ix}b_{ix} + e_{iy}b_{iy}\right) \\ &\qquad - \left(b_{ix}a_{iy} - b_{iy}a_{ix}\right)J_{pi}\Big\}\dot{\overline{X}}_p \\ &= \left[J_{\beta i1}\ \ J_{\beta i2}\ \ J_{\beta i3}\right]\dot{\overline{X}}_p.\end{aligned} \tag{21}$$

Using (21) we have to get Jacobian matrix for the actuation scheme 2 represented as (PRR1-PRR2-PRR3). Here ρ_i is the passive joint variable and β_i is the active revolute joint variable for $i = 1, 2, 3$. The Jacobian matrix of other actuation schemes is derived using (19) and (21).

2.2.2. Dexterity Indices. Yoshikawa [17] proposed the concept of kinematics manipulability as

$$\omega = \sqrt{\det(JJ^T)}, \tag{22}$$

where J is the Jacobin matrix and depends on the instantaneous configuration of the manipulator defined by a joint vector q. This index is treated as a quantitative measure of the easiness to arbitrarily change the position and orientation at a certain posture. The significance of manipulability is the measure of the fast recovery ability from the escapable singular point for the redundant manipulator. Manipulability has two problems: scale and order deficiencies. Those problems prevent the fair comparison among the manipulators with different dimension and make it impossible to derive the physical meaning of the manipulability. Trying to eliminate the order dependency, Kim and Khosla [18] gave another way to calculate manipulability

$$\omega = \sqrt[m]{\det(JJ^T)}, \quad (23)$$

where m is the number of degrees of freedom of the manipulator.

Another usually used index is the condition number of the Jacobian matrix that is expressed as

$$\kappa = \|J\|\|J^{-1}\|. \quad (24)$$

Condition number signifies the error amplification factor between the joint rate errors in to task space errors. The condition number depends on the manipulator configuration, link lengths. It varies from one at isotropic configurations to infinity at singular configurations so this is also defined as the measure of degree of ill-conditioning of the Jacobian matrix. Both indices, condition number and manipulability, are local measures of performance at a particular pose of the end-effecter within the workspace region; using these indices, we cannot compare the performance of any two different manipulators for their usefulness. In order to evaluate the global behavior of a manipulator over the workspace, a global index is proposed by Gosselin and Angeles [19] as

$$\mathrm{GCI} = \frac{\int_W (1/\kappa) dW}{\int_W dW}, \quad (25)$$

where W is the workspace and κ is the condition number. In other sense, GCI is average value of $1/\kappa$ over workspace region; it represents the uniformity of dexterity over the entire workspace so this index makes sense for the optimal design of manipulator for which the average value performance is an important design factor.

3. Results and Discussion

3.1. Performance of Actuation Schemes for 3RPR Manipulator. Symmetric architectures are considered here since they are widely used in practice. The values chosen for the sides of the fixed link are $A_1A_2 = A_2A_3 = A_3A_1 = 12$ m, while $C_1C_2 = C_2C_3 = C_3C_1 = 2$ m are the sides of the moving platform. The leg length range $L_i = A_iB_i$ is chosen as $\begin{bmatrix} 5 & 9 \end{bmatrix}$ m.

The reachable workspace for all eight actuation schemes using the Monte Carlo algorithm is plotted and is shown in Figure 5. The Global Condition Index (GCI) for each actuation scheme within the maximum possible reachable workspace is computed, and the results are shown in Table 3. Actuation scheme 1 (R$\underline{P}$R1-R$\underline{P}$R2-R$\underline{P}$R3) has the highest GCI of 0.4351 as well as a reasonably large dexterous workspace as indicated by the Monte Carlo points (29328). Actuation scheme 2 ($\underline{R}$PR1-$\underline{R}$PR2-$\underline{R}$PR3), on the other hand, showed

Table 3: GCI of actuation schemes at $A_1A_2 = 12$ m, $C_1C_2 = 2$ m; $[L_{i\min} \;\; L_{i\max}] = [5 \;\; 9]$ m.

Actuation no.	Monte Carlo workspace points	GCI
1	29328	0.4351
2	28778	0.0074
3	29734	0.1121
4	28860	0.0661
5	30202	0.0792
6	29811	0.0530
7	24765	0.0077
8	29269	0.0712

very poor performance with a GCI of 0.0074 and also lowest dexterous workspace (27778). The potential of each actuation scheme is quantified by these two numbers. The shape and magnitude of the reachable workspace is varied from one actuation scheme to other actuation scheme. The reachable workspace for actuation scheme 1 (R$\underline{P}$R1-R$\underline{P}$R2-R$\underline{P}$R3) is registered for almost all orientations of the end-effecter as shown in Figure 5(a). The reachable workspace for actuation scheme2 and for actuation scheme3 is shown in Figures 5(b) and 5(c), respectively, where the tool platform is unable to move in certain orientations (ϕ). For actuation scheme 4 ($\underline{P}$RR1-$\underline{P}$RR2-P$\underline{R}$R3) a continuously increasing reachable workspace for wide range of orientations of moving platform is observed.

3.1.1. Effect of Geometric Parameters on GCI. The performance values shown above pertain to one set of dimensions, and obviously these need not necessarily be the best. Hence various dimensions have been tried out keeping in mind that generally linear actuators are available in standard sizes. Of all possible combinations, the optimum ones based on GCI have been determined and the results have been plotted as shown in Figure 6. The variables considered are maximum length of the linear actuator $L_{i\max}$ and the size of the end-effecter (l) for the given $L_{i\min} = 5$ m and the base platform $A_1A_2 = A_2A_3 = A_3A_1 = 12$ m. The peak values of GCI for each actuation scheme are shown in Table 4.

It can be seen that the global maximum of all GCIs occurs for actuation scheme 1 (R$\underline{P}$R1-R$\underline{P}$R2-R$\underline{P}$R3) with a value of 0.4487 at the joint variable $L_{i\max} = 9$ m, $l = 0.75$ m. Even though the GCI is greatly influenced by the geometric parameters in all the actuation schemes as shown in Figures 6(a) to 6(h), this value is changed from one actuation scheme to other actuation scheme for a given set of geometric parameters.

3.2. Workspace and Performance Evaluation of 3PRR. Precise calculation of workspace and its boundaries is important because they influence the dimensional design, manipulator's positioning in the work environment, and its dexterity within the reachable workspace. The workspace is limited by the boundary obtained through solving inverse kinematics, reachable extent of drives and joints, and

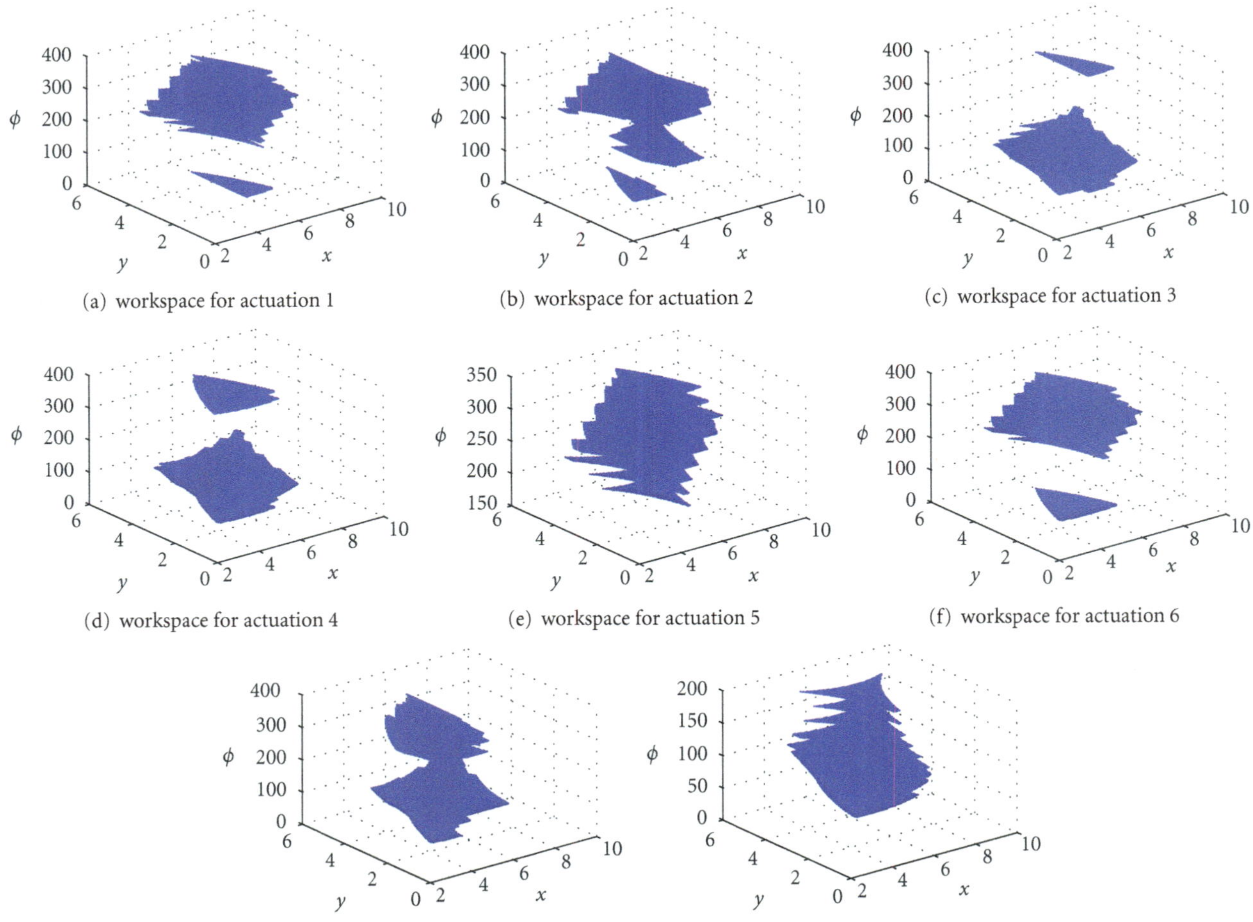

(a) workspace for actuation 1 (b) workspace for actuation 2 (c) workspace for actuation 3

(d) workspace for actuation 4 (e) workspace for actuation 5 (f) workspace for actuation 6

(g) workspace for actuation 7 (h) workspace for actuation 8

FIGURE 5: Workspace of actuation schemes for 3RPR manipulator.

TABLE 4: Design variables corresponding to maximum GCI for all eight possible actuation schemes for 3RPR parallel manipulator at $A_1A_2 = 12$ m, $C_1C_2 = 2$ m; $L_{i\min} = 5$ m.

Actuation scheme	1	2	3	4	5	6	7	8
$L_{i\max}$	9.00	7.00	11.00	11.00	7.00	9.00	7.00	7.00
l	0.75	0.75	1.25	1.25	1.25	0.75	0.75	1.25
GCI	0.4487	0.008	0.1143	0.0711	0.0778	0.0559	0.0092	0.0732

TABLE 5: GCI of actuation schemes at $A_1A_2 = 10$ m, $C_1P = 2$ m, $\rho_{\min i} = 2$ m; $\rho_{\max i} = 5$ m.

Actuation no.	Monte Carlo workspace points	GCI
1	2294	0.2438
2	218622	0.3241
3	29566	0.1724
4	4206	0.0820
5	45082	0.1437
6	8522	0.1170
7	34215	0.1699
8	12321	0.1045

the occurrence of singularities. A numerical example of the force-unconstrained poses of the 3PRR planar manipulator is presented. Both fixed and moving platforms are equilateral triangles, whose sides are $A_1A_2 = A_2A_3 = A_3A_1 = 10$ m for the fixed and $C_1P = C_2P = C_3P = 2$ m for the moving platform. The stroke of the prismatic joint is $2 \le \rho_i \le 5$ m for $i = 1, 2, 3$, the intermediate leg lengths $B_iC_i = 2$ m for $i = 1, 2, 3$. For the workspace calculation, Rastegar and Fardanesh [20] introduced the Monte Carlo random sampling numerical method to generate the workspace and the boundary surfaces. The Global Condition Index and the number of Monte Carlo points N_m are shown in Table 5 for the eight actuation schemes, and the workspace plots are shown in Figure 7. The size and shape of the workspace of the manipulator varies with the actuation. Among all the eight actuation

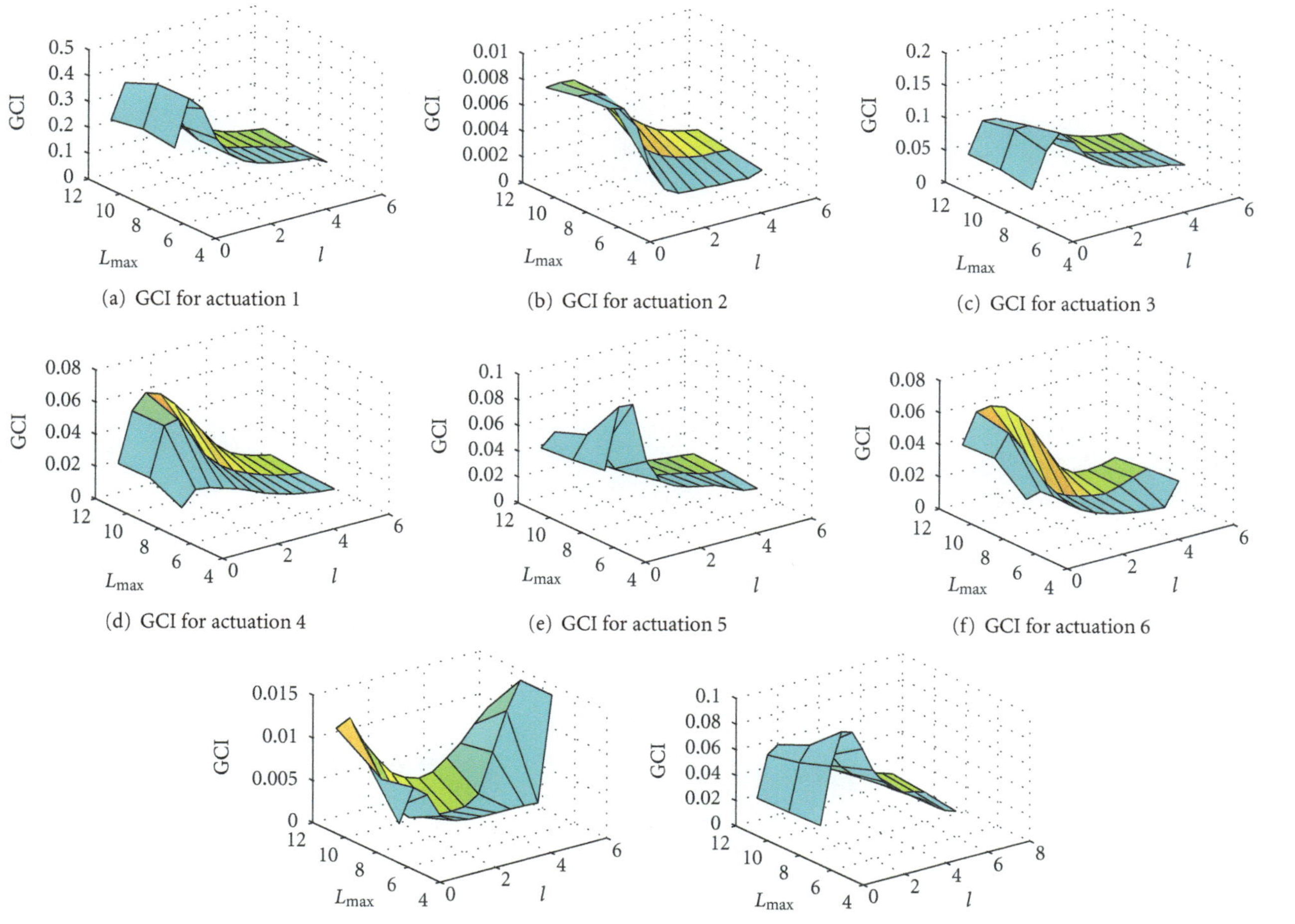

(a) GCI for actuation 1 (b) GCI for actuation 2 (c) GCI for actuation 3

(d) GCI for actuation 4 (e) GCI for actuation 5 (f) GCI for actuation 6

(g) GCI for actuation 7 (h) GCI for actuation 8

FIGURE 6: Effect of design variables on GCI of 3RPR manipulator.

TABLE 6: Optimized design variables for various actuation schemes at $\rho_{\min i} = 2$ m; $\rho_{\max i} = 5$ m.

Actuation scheme	Design vector $K = [l\ q\ \alpha_1\ \alpha_2\ \alpha_3]^T$					GCI
	l	q	α_1	α_2	α_3	
1	1.000	1.093	24.55°	150.02°	255.22°	0.4015
2	1.097	1.002	10.00°	120.00°	243.39°	0.5489
3	1.000	1.000	10.01°	122.15°	240.00°	0.1344
4	1.000	2.269	13.75°	120.00°	240.00°	0.1068
5	1.045	1.000	11.04°	120.21°	240.00°	0.2227
6	1.000	1.000	10.00°	120.00°	140.00°	0.1516
7	1.000	1.097	10.04°	120.00°	243.80°	0.2046
8	2.095	1.022	10.00°	127.86°	240.35°	0.1569

schemes, the actuation scheme 2 (PRR1-PRR2-PRR3) showed the best performance with a GCI of 0.3241 and the number of Monte Carlo workspace points 218622. The GCI itself represents the quality of the conditioned workspace, and the number of Monte Carlo points indicates the size of the reachable workspace. Actuation scheme 7 (PRR1-PRR2-PRR3) also shows good performance with GCI of 0.1699 along with larger workspace size next to the actuation scheme 2 among all eight actuation schemes for 3PRR manipulator. The reachable workspace for actuation scheme 2 (PRR1-PRR2-PRR3) is uniform and in regular shape for all the orientations of the moving platform as shown in Figure 7(b). The workspace for actuation scheme 2 (PRR1-PRR2-PRR3) as shown in Figure 7(b) is registered for only one orientation of the tool platform; the moving platform is unable to move in other orientations within the maximal reachable workspace due to the presence of complex singularities. Actuation scheme 4 (PRR1-PRR2-PRR3)

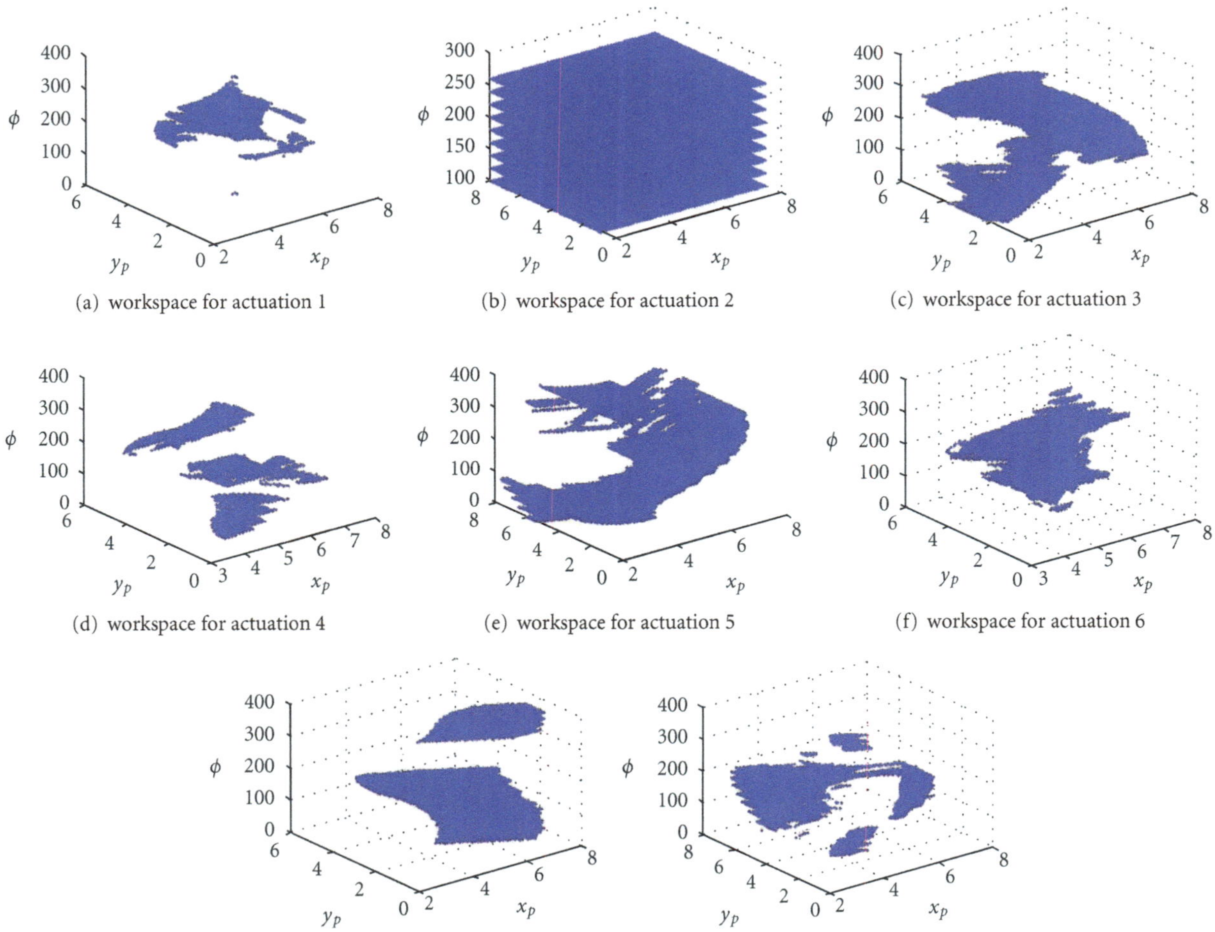

(a) workspace for actuation 1 (b) workspace for actuation 2 (c) workspace for actuation 3

(d) workspace for actuation 4 (e) workspace for actuation 5 (f) workspace for actuation 6

(g) workspace for actuation 7 (h) workspace for actuation 8

FIGURE 7: Workspaces for actuation schemes of 3PRR manipulator.

and actuation scheme 8 (PRR1-PRR2-PRR3) have got very limited reachable workspaces, and also the moving platform failed to move many orientations as shown in Figures 7(d) and 7(h), respectively.

3.2.1. Optimization of Geometric Structure for 3PRR Manipulator. Sequential quadratic programming (SQP) methods attempt to solve a nonlinear program directly rather than convert it to a sequence of unconstrained minimization problems. The basic idea is analogous to Newton's method for unconstrained minimization. At each step, a local model of the optimization problem is constructed and solved, yielding a step toward the solution of the original problem. A quadratic approximation to this function is now constructed and so along with linear constraints forms a quadratic programming problem that is, the minimization of a function quadratic in the variables, subject to linear and nonlinear constraints. The solution of the original optimization problem, say $\dot{x}$, is now obtained from an initial estimate and solving a sequence of updated quadratic programs. Sequential quadratic programming was used by Attolico et al. [21] and Xu and Wang [22]. It is a successful approach to solving constrained optimization problems. The vector of design parameters is $K = \begin{bmatrix} l & q & \alpha_1 & \alpha_2 & \alpha_3 \end{bmatrix}^T$.

The objective of this optimization algorithm is to determine optimal dimensions and optimal performance index for the eight actuations for the three legged parallel manipulators for a specified task. The optimization algorithm solves a nonlinear constrained optimization problem defined as follows:

$$\begin{aligned} &\text{Maximize} && \left(\prod_{i=1}^{n}\frac{1}{k_i}\right)^{1/n} \\ &\text{subjected to the constraints} && \beta_{\min i} \le \beta_i \le \beta_{\max i}, \\ &&& \text{(reach ability constraint)} \\ &&& \alpha_{\min i} \le \alpha_i \le \alpha_{\max i}, \\ &&& \rho_{\min i} \le \rho_i \le \rho_{\max i}, \\ &&& l + q \le \rho_{\max i}, \\ &&& l + q \ge \rho_{\min i} \\ &&& \text{(workspace constraint)} \\ &&& \text{for } i = 1, 2, 3. \end{aligned} \tag{26}$$

For the given $\beta_{\min i} = 0°$, $\beta_{\max i} = 360°$, $\alpha_{\min 1} = 0°$, $\alpha_{\max 1} = 60°$, $\alpha_{\min 2} = 120°$, $\alpha_{\max 2} = 170°$, $\alpha_{\min 3} = 220°$, $\alpha_{\max 3} = 280°$, $\rho_{\min i} = 2$ m; $\rho_{\max i} = 5$ m. The optimized results are shown in Table 6.

Actuation scheme 2 (P$\underline{R}$R1-P$\underline{R}$R2-P$\underline{R}$R3) possesses maximum reachable workspace size with maximum GCI of 0.5489, the optimal design vector corresponding to this actuation mode is $K = \begin{bmatrix} 1.097 & 1.002 & 10.00° & 120.00° & 243.39° \end{bmatrix}^T$.

4. Conclusions

Kinematic analysis of 3RPR and 3PRR parallel manipulators with symmetric properties is presented in this paper. In this problem, actuation scheme is identified as one of the design variables of the mechanism. It is shown that actuation scheme affects the performance of the manipulator and size and shape of the workspace significantly. The dimensional synthesis for optimization of Global Condition Index for each actuation scheme is presented.

Thus this paper offers a methodology to choose the ideal actuation scheme so as to attain the best performance of the manipulator with desired workspace; this is illustrated on two planar 3PRR and 3RPR manipulators as an example. The same technique can be applied to other planar PKMs.

References

[1] H. R. Mohammadi Daniali, P. J. Zsombor-Murray, and J. Angeles, "Singularity analysis of planar parallel manipulators," *Mechanism and Machine Theory*, vol. 30, no. 5, pp. 665–678, 1995.

[2] C. M. Gosselin, "Parallel computational algorithms for the kinematics and dynamics of planar and spatial parallel manipulators," *Journal of Dynamic Systems, Measurement and Control, Transactions of the ASME*, vol. 118, no. 1, pp. 22–28, 1996.

[3] J. P. Merlet, S. Lemieux, and C. M. Gosselin, "New architecture of planar three-degree-of-freedom parallel manipulator," in *Proceedings of the 13th IEEE International Conference on Robotics and Automation*, pp. 3738–3743, April 1996.

[4] Y. Jin, I. M. Chen, and G. Yang, "Structure synthesis and singularity analysis of a parallel manipulator based on selective actuation," in *Proceedings of the IEEE International Conference on Robotics and Automation*, pp. 4533–4538, New Orleans, La, USA, May 2004.

[5] V. Arakelian, S. Briot, and V. Glazunov, "Improvement of functional performance of spatial parallel manipulators using mechanisms of variable structure," in *Proceedings of the 12th IFTOMM World Congress*, Besancon, France, 2007.

[6] H. J. Su, P. Dietmaier, and J. M. McCarthy, "Trajectory planning for constrained parallel manipulators," *Journal of Mechanical Design, Transactions of the ASME*, vol. 125, no. 4, pp. 709–716, 2003.

[7] I. Ebrahimi, J. A. Carretero, and R. Boudreau, "Actuation scheme for a 6-dof kinematically redundant planar parallel manipulator," in *Proceedings of the 12th IFTOMM World Congress*, Besancon, France, 2007.

[8] R. Matone and B. Roth, "The effects of actuation schemes on the kinematic performance of manipulators," *Journal of Mechanical Design, Transactions of the ASME*, vol. 119, no. 2, pp. 212–217, 1997.

[9] K. Ramji, V. Ramachandra Raju, and S. Ramanababu, "Effect of actuator relocation on performance of industrial manipulators," *International Journal of Mechanics of Solids*, vol. 5, pp. 71–82, 2009.

[10] Y. Li and Q. Xu, "Design and optimization of an XYZ parallel micromanipulator with flexure hinges," *Journal of Intelligent and Robotic Systems*, vol. 55, no. 4-5, pp. 377–402, 2009.

[11] Q. Xu and Y. Li, "Design and analysis of a new singularity-free three-prismatic-revolute-cylindrical translational parallel manipulator," *Proceedings of the Institution of Mechanical Engineers Part C*, vol. 221, no. 5, pp. 565–577, 2007.

[12] Y. Li and Q. Xu, "Kinematic analysis and design of a new 3-DOF translational parallel manipulator," *Journal of Mechanical Design, Transactions of the ASME*, vol. 128, no. 4, pp. 729–737, 2006.

[13] N. Rakotomanga, D. Chablat, and S. Caro, "Kinetostatic performance of planar parallel mechanism with variable actuation," *Advances in Robot kinematics*, pp. 1–11, 2008.

[14] V. Kumar, "Characterization of workspaces of parallel manipulators," in *Proceedings of the ASME 21th Biennial Mechanisms Conference*, pp. 321–329, Chicago, Ill, USA, 1990.

[15] G. R. Pennock and D. J. Kassner, "Workspace of a general geometry planar three-degree-of-freedom platform-type manipulator," *Journal of Mechanical Design, Transactions Of the ASME*, vol. 115, no. 2, pp. 269–276, 1993.

[16] X. Zhang, J. K. Mills, and W. L. Cleghorn, "Dynamic modeling and experimental validation of a 3-PRR parallel manipulator with flexible intermediate links," *Journal of Intelligent and Robotic Systems*, vol. 50, no. 4, pp. 323–340, 2007.

[17] T. Yoshikawa, "Manipulability of robotic mechanisms," *International Journal of Robotics Research*, vol. 4, no. 2, pp. 439–446, 1984.

[18] J. O. Kim and P. Khosla, "Dexterity measures for design and control of manipulators," in *Proceedings of the IEEE/RSJ International Workshop on Intelligent Robot and Systems (IRSO '91)*, pp. 758–763, November 1991.

[19] C. Gosselin and J. Angeles, "Global performance index for the kinematic optimization of robotic manipulators," *Journal of Mechanisms, Transmissions, and Automation in Design*, vol. 113, no. 3, pp. 220–226, 1991.

[20] J. Rastegar and B. Fardanesh, "Manipulation workspace analysis using the Monte Carlo Method," *Mechanism and Machine Theory*, vol. 25, no. 2, pp. 233–239, 1990.

[21] M. Attolico, P. Masarati, and P. Mantegazza, "Trajectory optimization and real-time solution for robotics applications," in *Proceedings of the ECCOMAS Thematic Conference: Multibody Dynamics*, Madrid, Spain, 2005.

[22] J. X. Xu and W. Wang, "Two optimization algorithm for solving robotics inverse kinematics with redundancy," in *Proceedings of the IEEE International Conference on Control and Automation (ICCA '07)*, pp. 3021–3028, June 2007.

17

A Mathematical Model for the Analysis of Elastic Field in a Stiffened Cantilever of Laminated Composite

N. M. L. Huq[1] and A. M. Afsar[2]

[1] *Department of Mechanical Engineering, Dhaka University of Engineering and Technology (DUET), Gazipur 1700, Bangladesh*
[2] *Department of Mechanical Engineering, Bangladesh University of Engineering and Technology (BUET), Dhaka 1000, Bangladesh*

Correspondence should be addressed to A. M. Afsar, mdafsarali1967@yahoo.com

Academic Editor: Seung Bok Choi

A mathematical model is developed for the analytical solution to elastic filed in a deep stiffened cantilever beam of laminated composite under mixed boundary conditions. The two displacement parameters associated with the two-dimensional elasticity problems are defined in terms of a single displacement potential function such that one of the equilibrium equations is satisfied automatically. This reduces the problem to the solution of a single fourth-order partial differential equation, which is solved in terms of Fourier series. To present some numerical results, cantilever beams of glass/epoxy cross-ply and angle-ply laminated composites are considered and different components of stress and displacement at different sections of the beam are presented. The effects of laminate stacking sequence and ply-angle of the cross-ply and angle-ply composite beams, respectively, on the elastic field are analyzed. The numerical results justify that the present mathematical model is simple whereas capable to generate exact results of elastic field in a cantilever beam even at the critical regions of supports and loadings.

1. Introduction

It is now well known that composite materials are superior to conventional monolithic materials in many respects. In particular, these materials possess superior mechanical properties, such as specific strength and stiffness, superior physical property, such as specific density, and superior thermal property, such as minimized coefficient of thermal expansion. As a result, these materials are being widely used in modern structural elements to ensure better characteristics. However, prior to their applications, these structural elements should be properly studied under actual boundary conditions in order to understand, quantify, and improve their characteristics. The elastic characteristic is one of such important issues that should be paid attention to understand the stress and displacement fields. In most of the cases, the boundary conditions experienced by engineering structural elements are of the mixed type. Thus, it is imperative to develop a suitable mathematical model that is capable of dealing with mixed boundary conditions as well as other boundary conditions. Obviously, experimental approaches are sometimes not feasible for complex boundary conditions and complex geometries of elements, in addition to the cost involved with experiments. Numerical approaches, on the other hand, are not always adequate to generate accurate results, particularly at and near the boundaries of structural elements. In an attempt to get rid of these problems, the present study is aimed to develop an analytical approach for the analysis of elastic field in a cantilever beam of laminated composite materials.

Airy stress function approach [1] appears to be convenient to obtain analytical solution only for some idealized two-dimensional elasticity problems when boundary conditions are prescribed in terms of stresses only. It appears to be inadequate for the problems when the boundary conditions are prescribed in terms of displacement or strain. Displacement parameter approach [2] formulates the two-dimensional elasticity problems in terms of two displacement parameters u_x and u_y. Thus, this approach can be used for the solution of problems associated with displacement boundary conditions. However, this approach involves two major difficulties. Firstly, formulations of two-dimensional elasticity problems in terms of displacement parameters yield two simultaneous second-order partial differential equations which are quite difficult to solve analytically. Secondly, the mixed boundary conditions appears to be tedious to deal with by this approach.

Recently, a relatively new approach, called displacement potential approach [3], has been introduced for the solution of two-dimensional elasticity problems of an orthotropic composite lamina. This approach was used to solve a number of practical problems of structural elements of composite materials both analytically [4–7] and numerically [8]. This approach has gained importance in the field of two-dimensional elasticity problems because of its outstanding advantage that it can deal with any modes of boundary conditions prescribed in terms of either stress or displacement or any combination of these, that is, mixed boundary conditions. In this approach, both the displacement parameters u_x and u_y are defined in terms of a single function ψ, called displacement potential function, such that one of the equilibrium equations is satisfied automatically. The other equilibrium equation is reduced to a fourth-order partial differential equation for the solution of ψ. However, as this approach was developed for an orthotropic composite lamina only, it cannot be applied to laminated composites.

As most of the structural elements are designed with laminated composites, the above displacement potential approach was later extended by Huq [9] to cover the problems of laminated composites. Based on this extended displacement potential approach [9], in this study, a mathematical model is developed for the analysis of elastic field in a deep stiffened cantilever beam of laminated composite. A stiffened cantilever beam is obtained by using stiffeners at its boundaries. It is noted that stiffeners are usually used at the boundaries of structural elements to increase their stiffness, which in turn, reduce the degree of deformation. The developed model is demonstrated for two cantilever beams of glass/epoxy cross-ply and angle-ply laminated composites. Numerical results of different components of stress and displacement at different sections of the beam are presented graphically. Also, the effects of laminate stacking sequence in the case of cross-ply composite beam and ply-angle in the case of angle-ply composite beam on the elastic field are investigated in details.

2. Displacement Potential Formulation for Laminated Composites

For a symmetric laminate subjected to an in-plane loading, the bending-extension coupling stiffness matrix vanishes [10] and the mid-plane strains become equal to the global strains. Furthermore, for a cross-ply laminate and an angle-ply laminate with even number of plies, the shear-extension coupling terms of the extensional stiffness matrix are banished. For such laminates, the average stress-strain relations under plane stress in global coordinate system can be formulated as [10, 11]

$$\begin{bmatrix}\sigma_x\\ \sigma_y\\ \sigma_{xy}\end{bmatrix} = \frac{1}{h}\begin{bmatrix}A_{11} & A_{12} & 0\\ A_{12} & A_{22} & 0\\ 0 & 0 & A_{66}\end{bmatrix}\begin{bmatrix}\varepsilon_x\\ \varepsilon_y\\ \gamma_{xy}\end{bmatrix}, \tag{1}$$

where σ_x and σ_y are the components of normal stress in the x- and y-directions, respectively, σ_{xy} is the shear stress component, ε_x and ε_y are the normal strain components in the x- and y-directions, respectively, γ_{xy} is the shear strain component, and the elements of stiffness matrix $[A]$ are given by [10, 11]

$$A_{11} = \sum_{k=1}^{n} \left[Q_{11}c^4 + Q_{22}s^4 + 2(Q_{12} + 2Q_{66})s^2c^2\right]_k (h_k - h_{k-1}), \tag{2}$$

$$A_{12} = \sum_{k=1}^{n} \left[(Q_{11} + Q_{22} - 4Q_{66})s^2c^2 + Q_{12}(c^4 + s^4)\right]_k \times (h_k - h_{k-1}), \tag{3}$$

$$A_{22} = \sum_{k=1}^{n} \left[Q_{11}s^4 + Q_{22}c^4 + 2(Q_{12} + 2Q_{66})s^2c^2\right]_k (h_k - h_{k-1}), \tag{4}$$

$$A_{66} = \sum_{k=1}^{n} \left[(Q_{11} + Q_{22} - 2Q_{12} - 2Q_{66})s^2c^2 + Q_{66}(s^4 + c^4)\right]_k \times (h_k - h_{k-1}). \tag{5}$$

Here, $Q_{11} = E_1(1 - \nu_{21}\nu_{12})^{-1}$, $Q_{12} = \nu_{12}E_2(1 - \nu_{21}\nu_{12})^{-1}$, $Q_{22} = E_2(1 - \nu_{21}\nu_{12})^{-1}$, $Q_{66} = G_{12}$, $c = \cos\theta$, $s = \sin\theta$, $(h_k - h_{k-1})$ is the thickness of the k th ply of the laminate, h is the total thickness of the laminate, θ is the angle between the x-axis and the fiber direction of a lamina in the laminate, E_1 and E_2 are the Young's modulus in the longitudinal and the transverse directions, respectively, ν_{12} and ν_{21} are the major and minor Poison's ratios, respectively, and G_{12} is the in-plane shear modulus of a lamina in the laminate.

Substitution of (1) and making use of strain-displacement relations into two-dimensional equilibrium equations without body forces yield

$$A_{11}\frac{\partial^2 u_x}{\partial x^2} + (A_{12} + A_{66})\frac{\partial^2 u_y}{\partial x \partial y} + A_{66}\frac{\partial^2 u_x}{\partial y^2} = 0, \tag{6}$$

$$A_{22}\frac{\partial^2 u_y}{\partial y^2} + (A_{12} + A_{66})\frac{\partial^2 u_x}{\partial x \partial y} + A_{66}\frac{\partial^2 u_y}{\partial x^2} = 0. \tag{7}$$

Although the above two equations are theoretically sufficient to solve for two displacement parameters u_x and u_y, it is in fact very tedious to solve two second-order partial differential simultaneous equations analytically. To overcome this difficulty, the displacement potential approach is used to reduce the number of equations which is outlined in the following section.

2.1. Definition of Displacement Potential Function. If the two unknown displacement parameters u_x and u_y in (6) and (7) can be expressed in terms of a single function $\psi(x, y)$, the number of unknowns will be reduced to one. It implies that only one equation will then be sufficient for the determination of this unknown function $\psi(x, y)$. However, both the expressions in (6) and (7) must be satisfied by the function $\psi(x, y)$. Therefore, the two displacement parameters u_x and u_y are defined in terms of the single function $\psi(x, y)$ in such a way that one of the equilibrium equations in (6) and (7) is always satisfied automatically. To achieve this goal, the displacement potential function $\psi(x, y)$ is defined as [9]

$$u_x = \frac{\partial^2 \psi}{\partial x \partial y}, \tag{8}$$

$$u_y = -\frac{A_{11}}{A_{12} + A_{66}} \frac{\partial^2 \psi}{\partial x^2} - \frac{A_{66}}{A_{12} + A_{66}} \frac{\partial^2 \psi}{\partial y^2}. \tag{9}$$

The above definition of $\psi(x, y)$ satisfies the first equilibrium equation in (6) and (7) and the second equilibrium equation is reduced to

$$\frac{\partial^4 \psi}{\partial x^4} + \left[\frac{A_{22}}{A_{66}} - \frac{A_{12}^2}{A_{11} A_{66}} - \frac{2A_{12}}{A_{11}}\right] \frac{\partial^4 \psi}{\partial x^2 \partial y^2} + \frac{A_{22}}{A_{11}} \frac{\partial^4 \psi}{\partial y^4} = 0. \tag{10}$$

Now, one needs to satisfy only (10). By making use of (1), (8), (9), and strain-displacement relations, the components of stress can be expressed in terms of the same displacement potential function $\psi(x, y)$ as

$$\sigma_x = \frac{A_{66}}{h(A_{12} + A_{66})} \left[A_{11} \frac{\partial^3 \psi}{\partial x^2 \partial y} - A_{12} \frac{\partial^3 \psi}{\partial y^3}\right], \tag{11}$$

$$\sigma_y = \frac{1}{h(A_{12} + A_{66})} \times \left[(A_{12}^2 + A_{12} A_{66} - A_{11} A_{22}) \frac{\partial^3 \psi}{\partial x^2 \partial y} - A_{22} A_{66} \frac{\partial^3 \psi}{\partial y^3}\right], \tag{12}$$

$$\sigma_{xy} = \frac{A_{66}}{h(A_{12} + A_{66})} \left[A_{12} \frac{\partial^3 \psi}{\partial x \partial y^2} - A_{11} \frac{\partial^3 \psi}{\partial x^3}\right]. \tag{13}$$

Equation (10) is a fourth-order partial differential equation for a single unknown function $\psi(x, y)$. Once this equation is solved for $\psi(x, y)$, the components of displacement and stress can readily be determined from (8), (9), (11), (12), and (13), respectively.

3. Cantilever Beam

3.1. Modelling of the Problem. The displacement potential formulations developed in the preceding sections are used to modelling a cantilever beam problem for the solution of elastic field due to a tip load. Figure 1 shows a thin deep cantilever beam of laminated composite with length b and height a. A Cartesian coordinate system x-y is considered with the origin located at the left lower corner of the beam and the x- and y-axes aligned with the longitudinal and lateral edges of the beam, respectively. The left lateral edge is rigidly fixed and the two longitudinal edges parallel to the x-axis are stiffened. As mentioned earlier, the application of stiffeners to structural elements is a common practice to improve the stiffness which, in turn, reduces the degree of deformation. The stiffeners are mathematically modelled by imposing zero deformation along their length [4–7]. The right lateral edge is subjected to a parabolic shear load $\sigma_{xy}^0 = 4P(y^2 - ay)/a^2$, where P is the maximum value of the stress at the center of the edge. It is noted that a load at the tip of a cantilever can be simulated as a parabolic shear load [1]. As the thickness of the cantilever is small in comparison with other dimensions, the analysis is carried out under plane stress condition. The relevant boundary conditions are

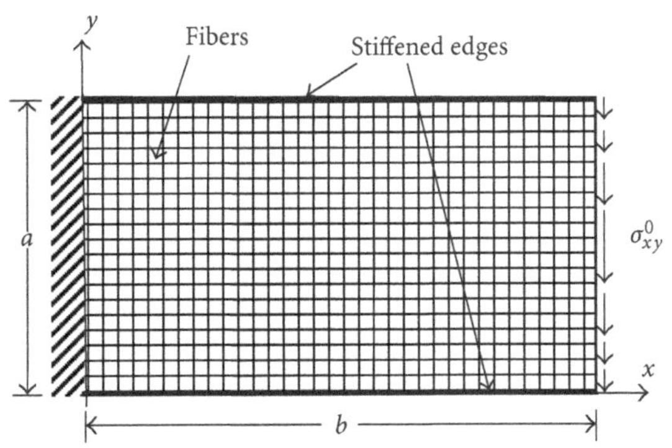

Figure 1: A cantilever beam of laminated composite under a parabolic shear load.

(i) $u_x(x, 0) = u_x(x, a) = 0$; $0 \le x \le b$ (upper and lower stiffened edges),

(ii) $\sigma_y(x, 0) = \sigma_y(x, a) = 0$; $0 \le x \le b$ (upper and lower stiffened edges),

(iii) $u_x(0, y) = u_y(0, y) = 0$; $0 \le y \le a$ (left fixed edge),

(iv) $\sigma_x(b, y) = 0, \sigma_{xy}(b, y) = \sigma_{xy}^0$; $0 \le y \le a$ (right lateral edge).

3.2. Solution of the Problem. To obtain the solution of the above cantilever beam problem, (10) should be solved for $\psi(x, y)$ that must satisfy all the above boundary conditions. Here we assume the solution of (10) in the form of Fourier series as

$$\psi(x, y) = \sum_{m=1}^{\infty} X_m(x) \cos \alpha y + M x^3, \tag{14}$$

where $X_m(x)$ is the function of x only, $\alpha = m\pi/a$, and M is an arbitrary constant. Note that any other expression for $\psi(x, y)$ may also be assumed for the solution of (10). However, a simple expression with some additional benefits may be desirable. A close examination of (8), (12), and (14) shows that the boundary conditions (i) and (ii) on the stiffened edges are automatically satisfied by the expression of $\psi(x, y)$ in (14), which are the additional benefits of our assumption. Now, the combination of (10) and (14) yields a fourth-order ordinary differential equation with constant coefficients as

$$X_m'''' - BX_m''\alpha^2 + CX_m\alpha^4 = 0, \tag{15}$$

where $B = (A_{22}/A_{66}) - (A_{12}^2/A_{11}A_{66}) - (2A_{12}/A_{11})$ and $C = A_{22}/A_{11}$. The general solution of (15) is

$$X_m = A_m e^{n_1 x} + B_m e^{n_2 x} + C_m e^{n_3 x} + D_m e^{n_4 x}, \tag{16}$$

where

$$n_1, n_2 = \alpha\sqrt{\frac{\left(B \pm \sqrt{B^2 - 4C}\right)}{2}}, \tag{17}$$

$$n_3, n_4 = -\alpha\sqrt{\frac{\left(B \pm \sqrt{B^2 - 4C}\right)}{2}}, \tag{18}$$

and A_m, B_m, C_m, and D_m are arbitrary constants. Now, by making use of (14) and (16) into (8), (9), (11), (12), and (13), the components of displacement and stress can be expressed as

$$u_x(x, y) = \sum_{m=1}^{\infty} [-(n_1 A_m e^{n_1 x} + n_2 B_m e^{n_2 x} + n_3 C_m e^{n_3 x} + n_4 D_m e^{n_4 x})\alpha \sin \alpha y], \tag{19}$$

$$u_y(x, y) = \frac{1}{A_{12} + A_{66}} \times \sum_{m=1}^{\infty} [\{A_{66}(A_m e^{n_1 x} + B_m e^{n_2 x} + C_m e^{n_3 x} + D_m e^{n_4 x})\alpha^2 - A_{11}(n_1^2 A_m e^{n_1 x} + n_2^2 B_m e^{n_2 x} + n_3^2 C_m e^{n_3 x} + n_4^2 D_m e^{n_4 x})\} \times \cos \alpha y - 6MxA_{11}], \tag{20}$$

$$\sigma_x(x, y) = -\frac{A_{66}}{h(A_{12} + A_{66})} \times \sum_{m=1}^{\infty} [A_{12}(A_m e^{n_1 x} + B_m e^{n_2 x} + C_m e^{n_3 x} + D_m e^{n_4 x})\alpha^3 + A_{11}(n_1^2 A_m e^{n_1 x} + n_2^2 B_m e^{n_2 x} + n_3^2 C_m e^{n_3 x} + n_4^2 D_m e^{n_4 x})\alpha] \sin \alpha y, \tag{21}$$

$$\sigma_y(x, y) = -\frac{1}{h(A_{12} + A_{66})} \times \sum_{m=1}^{\infty} [A_{22}A_{66}(A_m e^{n_1 x} + B_m e^{n_2 x} + C_m e^{n_3 x} + D_m e^{n_4 x})\alpha^3 + (A_{12}^2 + A_{12}A_{66} - A_{11}A_{22}) \times (n_1^2 A_m e^{n_1 x} + n_2^2 B_m e^{n_2 x} + n_3^2 C_m e^{n_3 x} + n_4^2 D_m e^{n_4 x})\alpha] \sin \alpha y, \tag{22}$$

$$\sigma_{xy}(x, y) = -\frac{A_{66}}{h(A_{12} + A_{66})} \times \sum_{m=1}^{\infty} [A_{12}(n_1 A_m e^{n_1 x} + n_2 B_m e^{n_2 x} + n_3 C_m e^{n_3 x} + n_4 D_m e^{n_4 x})\alpha^2 + A_{11}(n_1^3 A_m e^{n_1 x} + n_2^3 B_m e^{n_2 x} + n_3^3 C_m e^{n_3 x} + n_4^3 D_m e^{n_4 x})] \cos \alpha y - \frac{6MA_{11}A_{66}}{h(A_{12} + A_{66})}. \tag{23}$$

As mentioned earlier, the boundary conditions (i) and (ii) are satisfied automatically. Thus, the boundary conditions (iii) and (iv) are remaining to be satisfied. Before application of the boundary conditions (iii) and (iv), the parabolic shear load applied to the edge $x = b$ is expressed in terms of Fourier series as

$$\sigma_{xy}^0(b, y) = E_0 + \sum_{m=1}^{\infty} E_m \cos \alpha y, \tag{24}$$

where $E_0 = -2P/3$ and $E_m = 8P[1 + \cos(m\pi)]/m^2\pi^2$. Now, applying the boundary conditions (iii) and (iv) into (19), (20), (21), and (23), one obtains the constant M as

$$M = \frac{hP(A_{12} + A_{66})}{9A_{11}A_{66}}, \tag{25}$$

and the following four simultaneous equations for the determination of the constants A_m, B_m, C_m, and D_m:

$$n_1 A_m + n_2 B_m + n_3 C_m + n_4 D_m = 0, \quad 0 < y < a, \tag{26}$$

$$(A_{11}n_1^2 - A_{66}\alpha^2)A_m + (A_{11}n_2^2 - A_{66}\alpha^2)B_m + (A_{11}n_3^2 - A_{66}\alpha^2)C_m + (A_{11}n_4^2 - A_{66}\alpha^2)D_m = 0, \tag{27}$$

$$(A_{11}n_1^2\alpha+A_{12}\alpha^3)A_m e^{n_1 b}+(A_{11}n_2^2\alpha+A_{12}\alpha^3)B_m e^{n_2 b} + (A_{11}n_3^2\alpha + A_{12}\alpha^3)C_m e^{n_3 b} + (A_{11}n_4^2\alpha + A_{12}\alpha^3)D_m e^{n_4 b} = 0, \tag{28}$$

$$(A_{11}n_1^3+A_{12}n_1\alpha^2)A_m e^{n_1 b}+(A_{11}n_2^3+A_{12}n_2\alpha^2)B_m e^{n_2 b} + (A_{11}n_3^3 + A_{12}n_3\alpha^2)C_m e^{n_3 b} + (A_{11}n_4^3 + A_{12}n_4\alpha^2)D_m e^{n_4 b} = \frac{-E_m h(A_{12} + A_{66})}{A_{66}}. \tag{29}$$

The constants A_m, B_m, C_m, and D_m obtained from the above equations are substituted into (19), (20), (21), (22), and (23) to obtain the explicit expressions of displacement and stress which are valid to the entire region of the cantilever beam.

4. Results and Discussion

The modelling of the present cantilever beam problem is demonstrated for glass/epoxy laminated composites. The properties of an orthotropic unidirectional glass/epoxy composite lamina are shown in Table 1. These properties are used in calculating the laminate properties of interest. All the results presented in this section correspond to the aspect ratio of the beam $b/a = 3.0$. The maximum value of the applied load P is taken as 1000 MPa and the thickness of each ply is considered to be 1.0 mm. The convergence of the series solution is ensured by comparing the results for different values of the number of terms m in the series. It is found that the results converge very well for $m = 5$ only. However, for further accuracy, 10 terms in the series have been used to generate all the results presented hereafter.

Figure 2 shows the normalized longitudinal displacement u_x/b at different sections of the cantilever beam as a function of normalised position y/a. The beam consists of a cross-ply laminate with laminate stacking sequence [0/90/0]. Unless otherwise stated, all the results from this point forward correspond to the beam of cross-ply laminate [0/90/0]. As seen, the displacement is zero at the section $x/b = 0$ and at the top $(y/a = 1.0)$ and bottom $(y/a = 0)$ edges of the cantilever. This satisfies the physical boundary conditions of the problem. The maximum displacement occurs at the right lateral edge $(x/b = 1.0)$ and it gradually decreases towards the fixed support $(x/b = 0)$. Shown in Figure 3 is the normalized lateral displacement (u_y/a) at different sections of the beam. The lateral displacement is also exactly zero at the fixed support $(x/b = 0)$ which satisfies the boundary conditions of the problem. The magnitude of this displacement is also the maximum at the right lateral edge and gradually diminishes towards the fixed support. Further, the variation of the displacement with y/a is almost zero at all the sections except the right lateral edge $x/b = 1.0$. However, the magnitude of the displacement varies significantly from one section to another section, that is, with the variation of x/b.

TABLE 1: Mechanical properties of a glass/epoxy unidirectional lamina.

E_1 (MPa)	E_2 (MPa)	G_{12} (MPa)	ν_{12}	ν_{21}
38.60×10^3	8.27×10^3	4.14×10^3	0.26	0.06

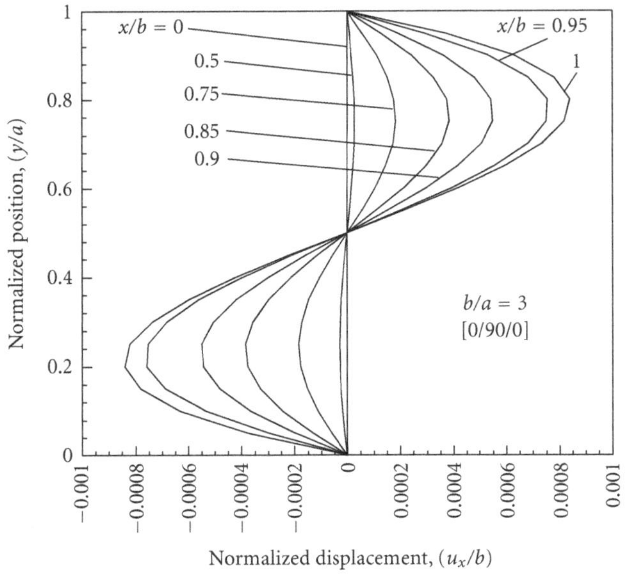

FIGURE 2: Longitudinal displacement at different sections of the cross-ply composite beam.

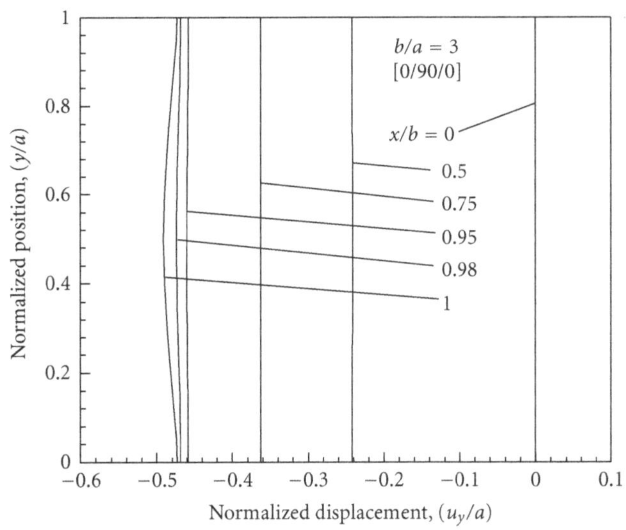

FIGURE 3: Lateral displacement at different sections of the cross-ply composite beam.

The distribution of normalized longitudinal stress component (σ_x/P) as a function of normalized position (y/a) is illustrated in Figure 4. At the right lateral edge $(x/b = 1.0)$, the stress is zero that satisfies the boundary condition of the problem. Further, the longitudinal stress is zero at the fixed support and at the top and bottom stiffened edges

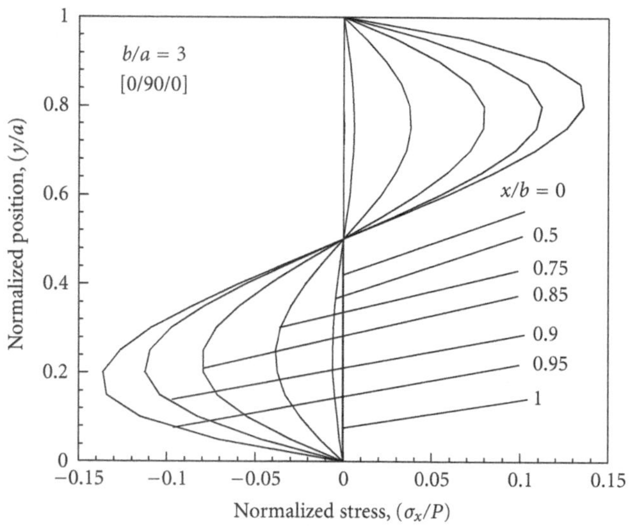

FIGURE 4: Longitudinal stress at different sections of the cross-ply composite beam.

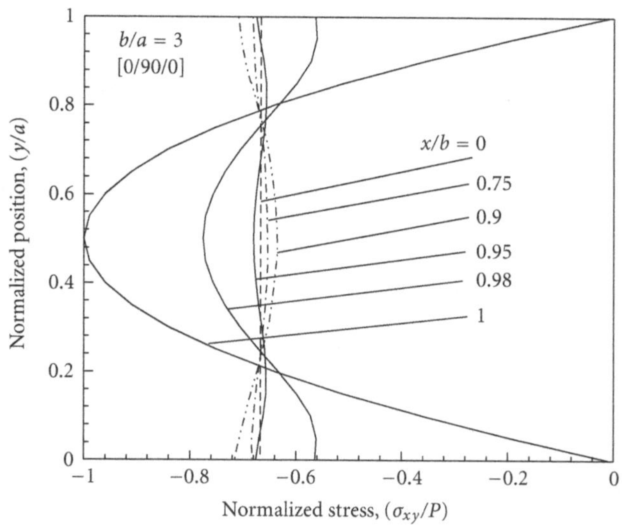

FIGURE 6: Shear stress at different sections of the cross-ply composite beam.

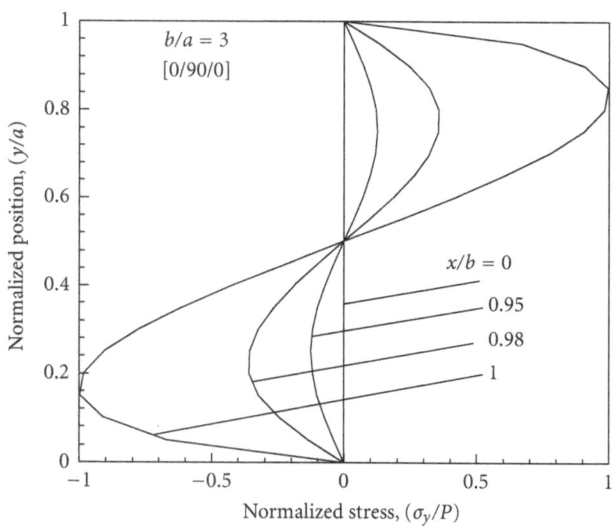

FIGURE 5: Lateral stress at different sections of the cross-ply composite beam.

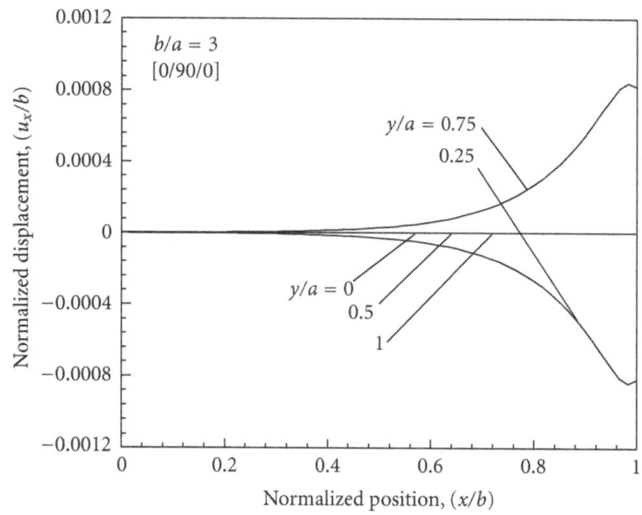

FIGURE 7: Longitudinal displacement at different sections of the cross-ply composite beam.

of the cantilever. The magnitude of the stress is maximum near the right lateral edge ($x/b = 0.95$) of the cantilever and it gradually decreases towards the fixed support. The distribution of this stress component shows that the neutral axis coincides with the geometric mid-plane $y/a = 0.5$, where stress is zero for any section (any value of x/b). The normalized lateral stress component (σ_y/P) is displayed in Figure 5. The distribution of this stress component is similar to that of the longitudinal stress component except that this stress is the maximum at the right lateral edge. Figure 6 depicts the distribution of normalized shear stress component (σ_{xy}/P). At the right lateral edge ($x/b = 1.0$), the distribution of the shear stress exactly matches with the applied load. For the sections near the right lateral edge (e.g., $x/b = 0.98$), the magnitude of the shear stress is the maximum at the mid-plane ($y/a = 0.5$) which decreases towards the top and bottom edges of the beam. This scenario changes for the sections away from the right lateral edge of the beam. More specifically, the magnitude of this stress component is the minimum at the mid-plane ($y/a = 0.5$) which increases towards the top and bottom edges of the beam as seen for the sections $x/b = 0.9$ and 0.75.

The distribution of displacement and stress along the longitudinal direction of the beam is illustrated in Figures 7 to 11. The longitudinal displacement u_x/b is zero along the entire mid-plane ($y/a = 0.5$), which is the neutral axis of the beam. Further, the displacement is zero along the top ($y/a = 1.0$) and bottom ($y/a = 0$) edges of the beam, which conforms to the boundary condition of the present stiffened beam problem. However, at other longitudinal sections between the neutral plane and top or

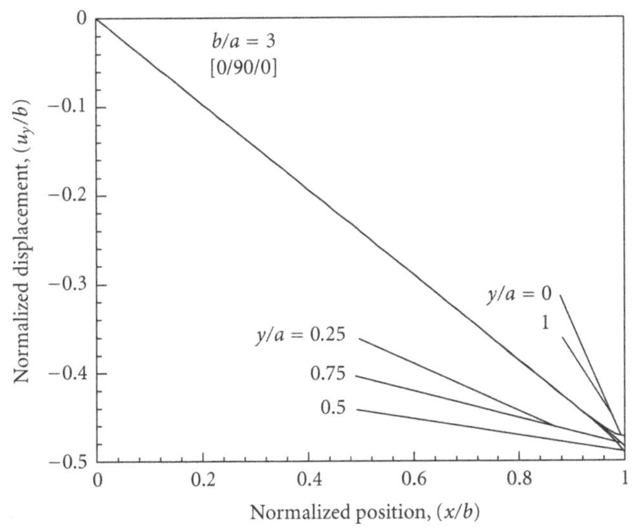

FIGURE 8: Lateral displacement at different sections of the cross-ply composite beam.

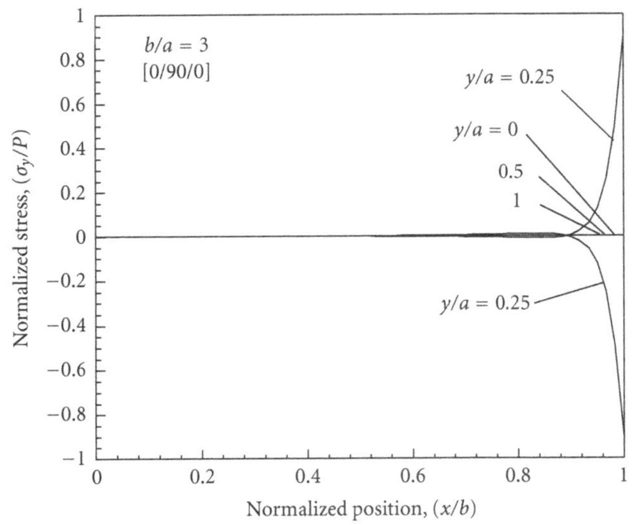

FIGURE 10: Lateral stress at different sections of the cross-ply composite beam.

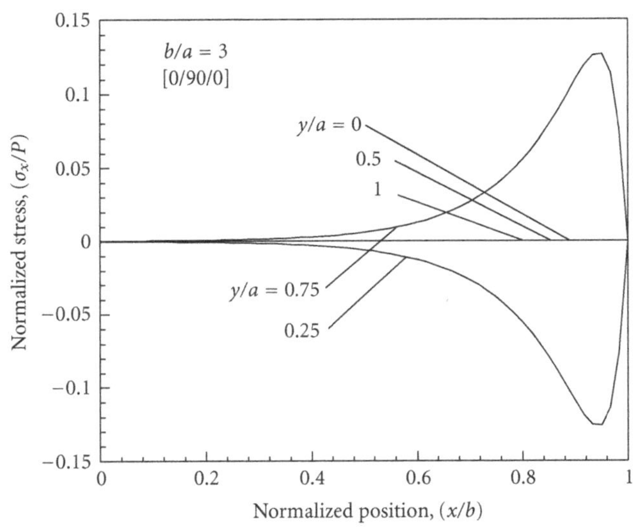

FIGURE 9: Longitudinal stress at different sections of the cross-ply composite beam.

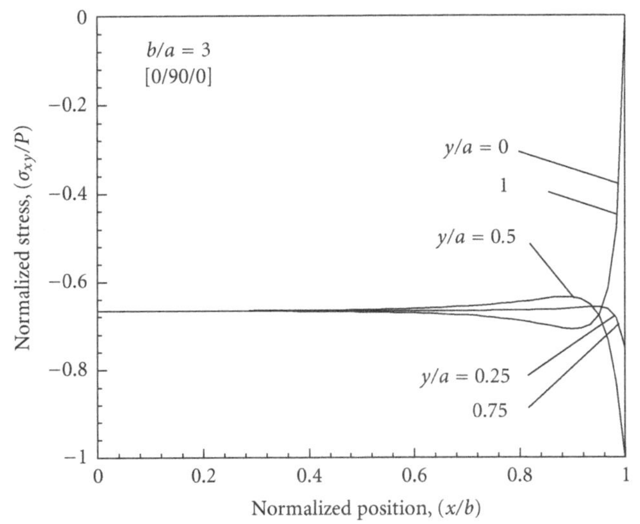

FIGURE 11: Shear stress at different sections of the cross-ply composite beam.

bottom edge of the beam, this displacement component varies with x/b as shown by the curves corresponding to $y/a = 0.25$ and 0.75. The variation is substantial at the right half of the beam, particularly at and near the right edge of the beam. Further, the distribution of the displacement is symmetric with respect to the neutral axis of the beam. The lateral displacement component u_y/b as a function of x/b at different longitudinal sections is illustrated in Figure 8. The distribution of this displacement component is the same for any section except the small region of $x/b > 0.96$ where a small variation of the displacement from one longitudinal section to another can be observed.

The longitudinal stress distribution along the longitudinal direction of the beam as shown in Figure 9 is similar to that of the longitudinal displacement except that the stress is zero for any value of y/a at the right lateral edge $x/b = 1.0$. The lateral stress σ_y/P varies from one section to another section (except the sections $y/a = 0$, 0.5, and 1.0) only for $x/b > 0.92$ as can be seen from Figure 10. The distribution of the shear stress component along the longitudinal direction is shown in Figure 11. Unlike longitudinal and lateral stress components, the shear stress is nonzero at any section of the beam. From the fixed end to $x/b = 0.6$, the stress does not vary with x/b. A close observation shows that the stress distribution at the right lateral edge ($x/b = 1.0$) conforms to the parabolic applied load at that edge. At $y/a = 0$ and 1.0, the stress is zero and its value is maximum and equal to the applied load P at $y/a = 0.5$.

To investigate the effect of laminate stacking sequence of a cross-ply laminate consisting of three plies on the elastic field, four different stacking sequences, [0/0/0],

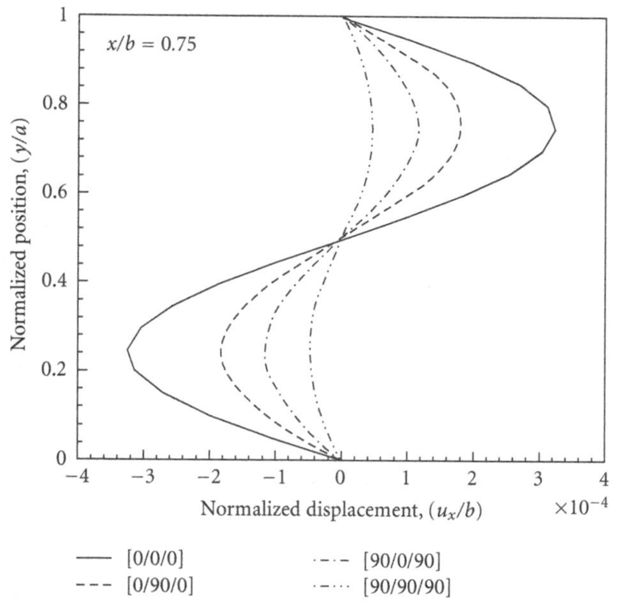

FIGURE 12: Effect of laminate stacking sequence on longitudinal displacement.

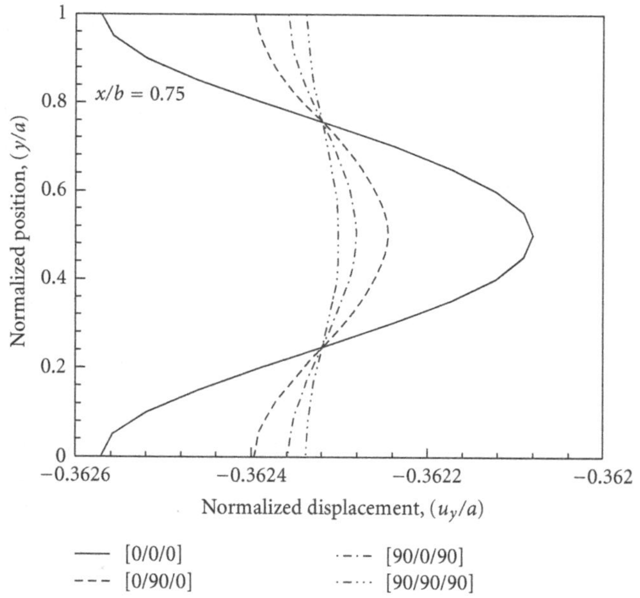

FIGURE 13: Effect of laminate stacking sequence on lateral displacement.

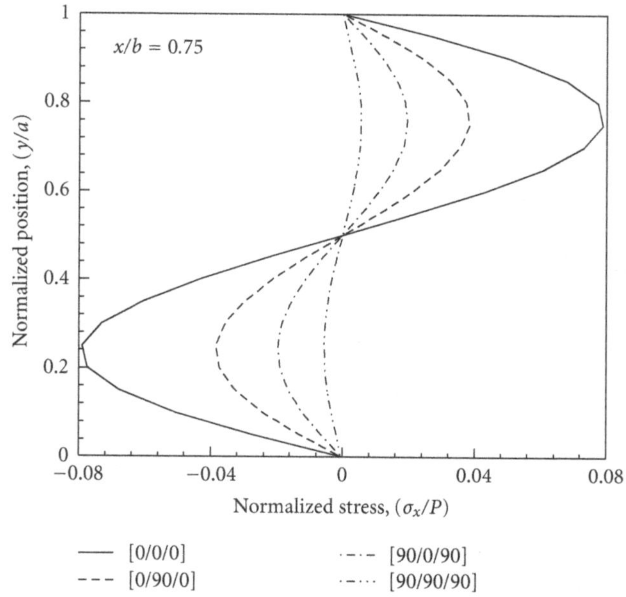

FIGURE 14: Effect of laminate stacking sequence on longitudinal stress.

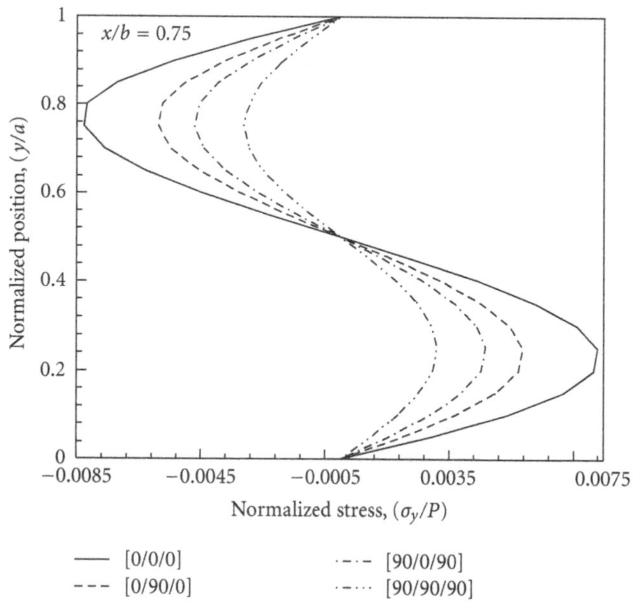

FIGURE 15: Effect of laminate stacking sequence on lateral stress.

[0/90/0], [90/0/90], and [90/90/90], are considered. Only the results of the vertical section $x/b = 0.75$ are shown in Figures 12 to 16. Figure 12 shows that the peak values of longitudinal displacement occur at $y/a = 0.25$ and 0.75 for any stacking sequence. These peak values are found to decrease as the number of 90 degree plies increases. The value of lateral displacement is maximum at the bottom ($y/a = 0$) and top ($y/a = 1$) surfaces of the beam when all the plies have 0 degree orientation as can be seen from Figure 13. The maximum value decreases as the number of 90 degree plies increases. Further, the minimum value of lateral displacement is obtained at $y/a = 0.5$ when all the plies have 0 degree orientation. As the number of 90 degree plies increases, the value of lateral displacement increases at $y/a = 0.5$. The longitudinal stress shown in Figure 14 exhibits the similar characteristics as the longitudinal displacement displayed in Figure 12. Shown in Figure 15 is the lateral stress as a function of laminate stacking sequence. This stress component has similar characteristics as the longitudinal stress of Figure 14 except that the positive and negative stresses interchange their position between $y/a =$ 0.25 and 0.75. The effect of laminate stacking sequence on

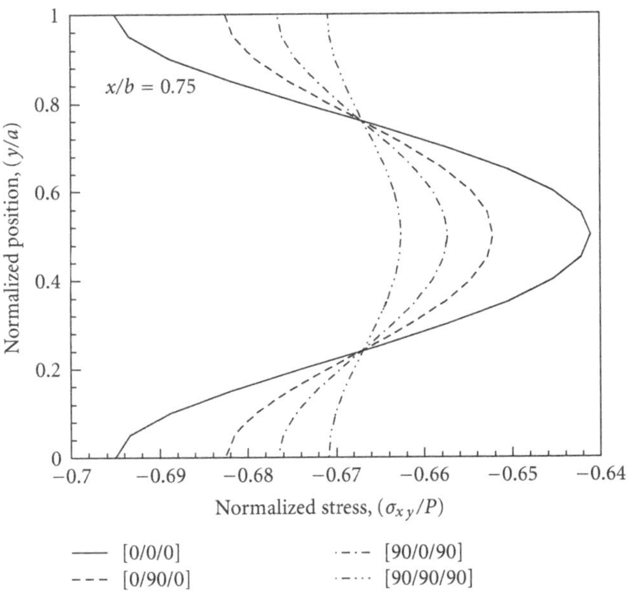

FIGURE 16: Effect of laminate stacking sequence on shear stress.

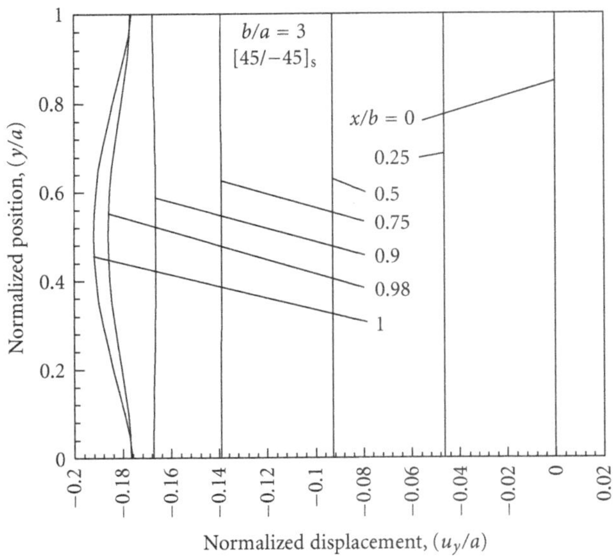

FIGURE 18: Lateral displacement at different sections of the angle-ply composite beam.

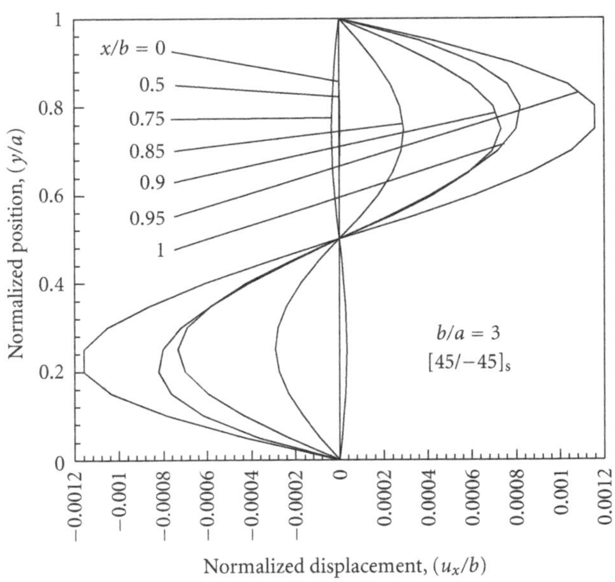

FIGURE 17: Longitudinal displacement at different sections of the angle-ply composite beam.

the shear stress component is illustrated in Figure 16. The characteristics of these stresses are similar to those of lateral displacement shown in Figure 13.

In the preceding paragraphs, the results presented correspond to the cantilever beam of cross-ply laminate. Now some results will be presented and discussed for the beam of $[45/-45]_s$ angle-ply laminate.

Figure 17 displays the distribution of normalized longitudinal displacement component u_x/b as a function of y/a at different lateral sections of the beam. It shows a bit different characteristics than the case of the cross-ply laminated beam at and near the right lateral edge. For the angle-ply laminated beam, the longitudinal displacement is not maximum at the right lateral edge $x/b = 1.0$. Rather, the maximum value of the displacement occurs at a section (e.g., $x/b = 0.95$) near the right lateral edge $x/b = 1.0$. For the sections in the region $x/b < 0.95$, the characteristics of the displacement are similar to the case of the cross-ply beam, that is, the value of the displacement decreases towards the fixed edge. The normalized lateral displacement shown in Figure 18 exhibits exactly similar characteristics to that of the cross-ply cantilever beam.

The normalized longitudinal stress σ_x/P as a function of y/a is exhibited in Figure 19. At the right lateral edge $x/b = 1.0$, the stress is zero that increases towards the fixed support until the value of $x/b = 0.95$. After this, the magnitude of this stress decreases as a section further moves towards the fixed support ($x/b = 0$). The distribution of the normalized lateral stress σ_y/P as a function of y/a at different sections of the beam as shown in Figure 20 is similar to that of the cross-ply cantilever beam, that is, the magnitude of the stress is the maximum at the right lateral edge $x/b = 1.0$, which gradually decreases to zero as the fixed support is approached. Shown in Figure 21 is the distribution of normalized shear stress σ_{xy}/P that shows the characteristics similar to that of the cross-ply cantilever beam. Note that, like longitudinal and lateral stress components, the shear stress is not zero at the fixed support $x/b = 0$ for both the cases of cross-ply and angle-ply beams.

To investigate the effect of ply-angle on the elastic field in the case of the angle-ply cantilever beam, three different points $y/a = 0.25$, 0.50, and 0.75 on the lateral section $x/b = 0.75$ are considered for which results are computed as a function of ply-angle θ varying from 0 to 90 degrees. Figure 22 illustrates the variation of normalized longitudinal displacement with the variation of ply-angle θ which shows that the displacement at the mid-longitudinal plane, $y/a = 0.5$, is zero and remains uninfluenced by the ply-angle θ.

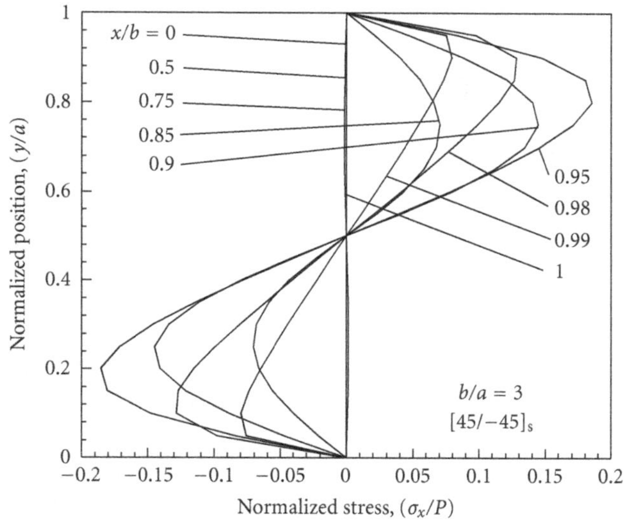

FIGURE 19: Longitudinal stress at different sections of the angle-ply composite beam.

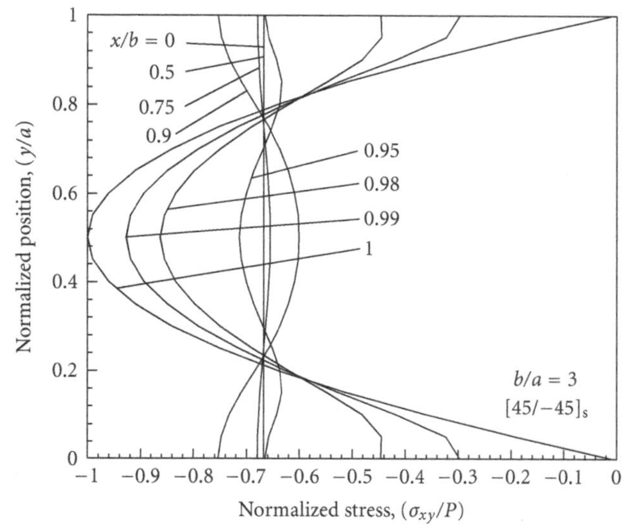

FIGURE 21: Shear stress at different sections of the angle-ply composite beam.

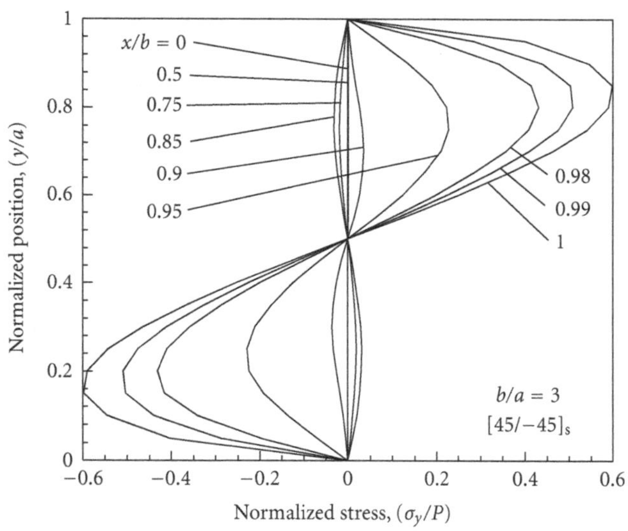

FIGURE 20: Lateral stress at different sections of the angle-ply composite beam.

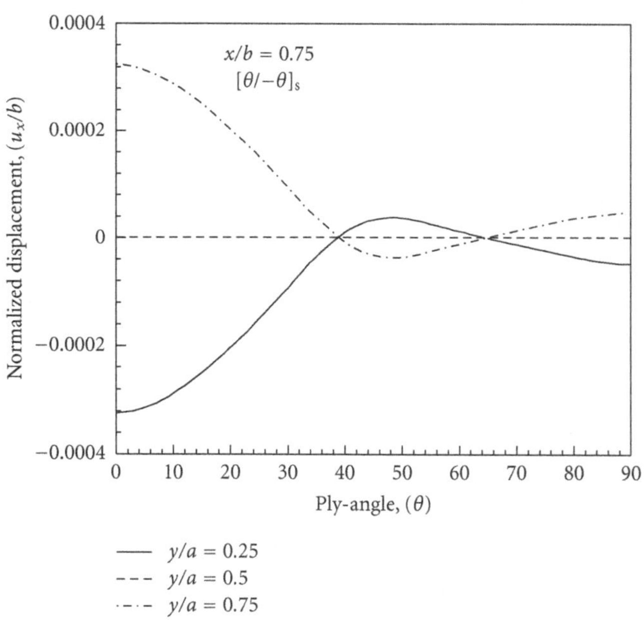

FIGURE 22: Effect of ply-angle on longitudinal displacement.

However, near the top ($y/a = 0.75$) and bottom ($y/a = 0.25$) surfaces of the beam, the longitudinal displacement varies significantly as the ply-angle varies from 0 to about 40 degrees. With further increase of the ply-angle, the variation of the displacement is insignificant. Also, the displacement at any point along the lateral section $x/b = 0.75$ is zero for the ply-angle 40 and 60 degrees. Further, it is noted that the value of the displacement is the maximum for $y/a = 0.25$ and 0.75 when the ply-angle is zero, which implies the unidirectional composite. This conforms to the results corresponding to the [0/0/0] curve as shown in Figure 12. Thus, zero degree is the critical ply-angle in terms of longitudinal displacement. The normalized lateral displacement is shown in Figure 23 that gives the information that the magnitude and variation of this displacement component with the variation of ply-angle is the same at any point along a lateral section (e.g., $x/b = 0.75$). The magnitude of the lateral displacement is the minimum at $\theta = 45$ degrees and its value increases as the ply-angle deviates in either direction from 45 degrees.

The normalized longitudinal stress component as shown in Figure 24 has the similar characteristics to that of the longitudinal displacement component shown in Figure 22. However, the lateral stress component as shown in Figure 25 has a completely different characteristic than lateral displacement. The lateral stress at the mid-point ($y/a = 0.5$) of

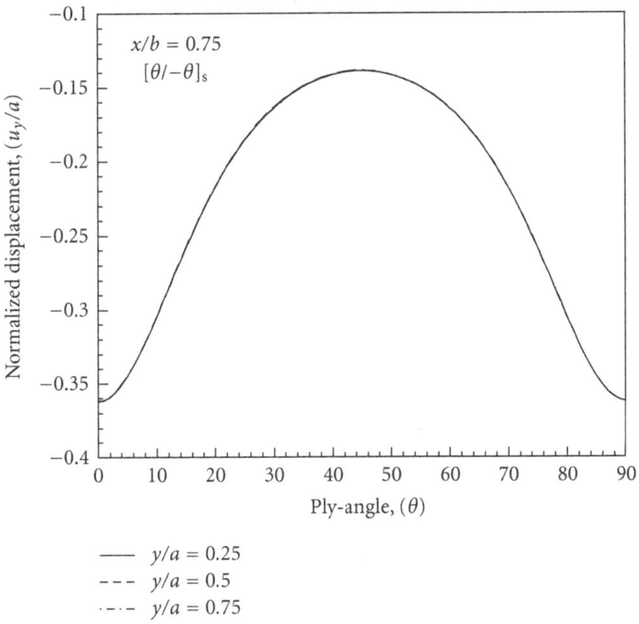

Figure 23: Effect of ply-angle on lateral displacement.

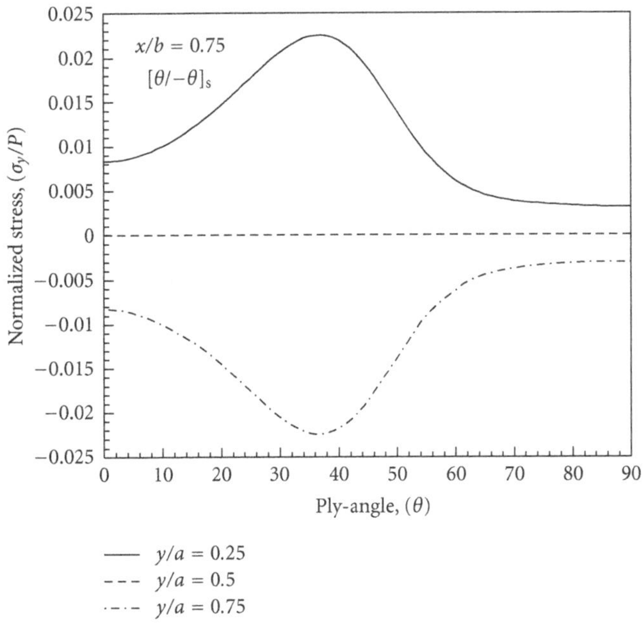

Figure 25: Effect of ply-angle on lateral stress.

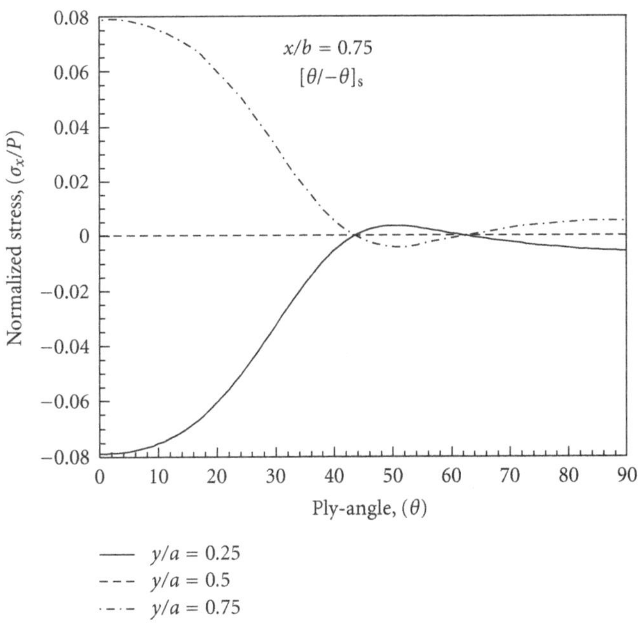

Figure 24: Effect of ply-angle on longitudinal stress.

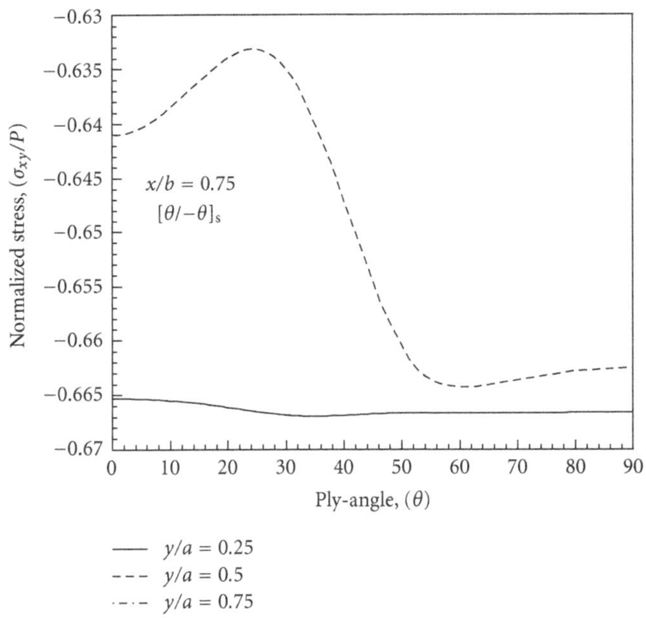

Figure 26: Effect of ply-angle on shear stress.

the lateral section $x/b = 0.75$ is uninfluenced by the ply-angle which always remains zero. However, near the top and bottom surfaces of the beam, the magnitude of the stress increases as the ply-angle varies from 0 to about 35 degrees. Then it falls as the ply-angle further increases. In terms of the lateral stress, $\theta = 37$ degree is the critical ply-angle for which the value of the lateral stress near the top $(y/a = 0.75)$ and bottom $(y/a = 0.25)$ surfaces is the maximum.

Shown in Figure 26 is the effect of ply-angle on the shear stress component. The shear stress exhibits the reverse trend to that of the longitudinal and lateral stress components. In this case, the stress at the mid-point $(y/a = 0.5)$ of the vertical section $x/b = 0.75$ has a significant variation with the variation of the ply-angle. On the contrary, the stress near the top and bottom surfaces of the beam has a little variation with the ply-angle. In addition, the shear stress at the points of equal distance from the mid-point of the vertical section is exactly the same. Note that, in terms of shear stress, $\theta = 34$ degree is the critical value of the ply-angle for which the stress is the maximum.

5. Conclusion

A stiffened cantilever beam of symmetric laminated composite is mathematically modelled in terms of a single displacement potential function for analytical solution to elastic field in the beam under mixed boundary conditions. The model is suitable for any boundary conditions prescribed in terms of either stress or constraints or any combination of them. The numerical results obtained for a glass/epoxy laminated composite beam show that all the boundary conditions are satisfied exactly. This justifies that the present model is reliable as well as simple in obtaining analytical solution to elastic field even in the critical regions of supports and loadings of structural elements of laminated composites under mixed boundary conditions. From the numerical results of the stiffened cantilever beam, the following major conclusions can be made.

(1) The analysis of the cross-ply cantilever beam shows that the beam with all the 90-degree plies is much better than that with all the zero-degree plies in terms of all the components of stress and displacement.

(2) For the cantilever made of angle-ply laminated composite, ply-angle is an important parameter that has a significant effect on the elastic field. The same value of the ply-angle does not correspond to the critical value for all the components of stress and displacement. Therefore, to design with a cantilever beam of angle-ply laminate, it is necessary to find out an optimum value of the ply-angle for which all the components of stress and displacement can be minimized to a desired level.

References

[1] S. Timoshenko and V. N. Goodier, *Theory of Elasticity*, McGraw-Hill, New York, NY, USA, 1979.

[2] M. W. Uddin, *Finite Difference Solution of Two-Dimensional Elastic Problems with Mixed Boundary Conditions [M.S. thesis]*, Deptartment of Mechanic Engineering, Carleton University, Ottawa, Canada, 1966.

[3] S. K. D. Nath, *Displacement Potential Approach to Solution of Elasticity Problems of Orthotropic Composite Structures [Ph.D. dissertation]*, Deptartment of Mechanic Engineering, Bangladesh University of Engineering and Technology, Dhaka, Bangladesh, 2007.

[4] S. K. D. Nath and A. M. Afsar, "Analysis of the effect of fiber orientation on the elastic field in a stiffened orthotropic panel under uniform tension using displacement potential approach," *Mechanics of Advanced Materials and Structures*, vol. 16, no. 4, pp. 300–307, 2009.

[5] S. K. D. Nath, S. R. Ahmed, and A. M. Afsar, "Displacement potential solution of short stiffened flat composite bars under axial loading," *International Journal of Applied Mechanics and Engineering*, vol. 11, no. 3, pp. 557–575, 2006.

[6] S. K. D. Nath, A. M. Afsar, and S. R. Ahmed, "Displacement potential approach to the solution of stiffened orthotropic composite panels under uniaxial tensile load," *Proceedings of the Institution of Mechanical Engineers G*, vol. 221, no. 5, pp. 869–881, 2007.

[7] S. K. D. Nath, A. M. Afsar, and S. R. Ahmed, "Displacement potential solution of a deep stiffened cantilever beam of orthotropic composite material," *Journal of Strain Analysis for Engineering Design*, vol. 42, no. 7, pp. 529–541, 2007.

[8] A. M. Afsar, S. K. D. Nath, S. R. Ahmed, and J. L. Song, "Displacement potential based finite difference solution to elastic field in a cantilever beam of orthotropic composite," *Mechanics of Advanced Materials and Structures*, vol. 15, no. 5, pp. 386–399, 2008.

[9] N. M. L. Huq, *Analysis of Elastic Field in Structural Elements of Laminated Composites by Displacement Potential Approach [Ph.D. thesis]*, Deptartment of Mechanic Engineering, Bangladesh University of Engineering and Technology, Dhaka, Bangladesh, 2007.

[10] A. K. Kao, *Mechanics of Composite Materials*, chapter 4-5, Taylor and Francis, New York, NY, USA, 2006.

[11] R. M. Jones, *Mechanics of Composite Materials*, chapter 4, Scripta, Washington, DC, USA, 1st edition, 1975.

Performance Evaluation of a Bench-Top Precision Glass Molding Machine

Peter Wachtel,[1] Peiman Mosaddegh,[2] Benn Gleason,[1] J. David Musgraves,[1] and Kathleen Richardson[1]

[1] *School of Materials Science & Engineering, COMSET, Clemson, SC 29634, USA*
[2] *Department of Mechanical Engineering, Clemson University, Clemson, SC 29634, USA*

Correspondence should be addressed to Peter Wachtel; wachtel@g.clemson.edu

Academic Editor: Essam Eldin Khalil

A Dyna Technologies Inc. GP-5000HT precision glass molding machine has been found to be a capable tool for bridging the gap between research-level instruments and the higher volume production machines typically used in industry, providing a means to apply the results of rigorous instrumentation analysis performed in the lab to industrial PGM applications. The GP-5000HT's thermal and mechanical functionality is explained and characterized through the measurement baseline functionality and the associated error. These baseline measurements were used to determine the center thickness repeatability of pressed glass parts, which is the main metric used in industrial pressing settings. The baselines and the repeatability tests both confirmed the need for three warm-up pressing cycles before the press reaches a thermal steady state. The baselines used for pressing a 2 mm glass piece to a 1 mm target center thickness yielded an average center thickness of 1.001 mm and a standard deviation of thickness of 0.0055 mm for glass samples pressed over 3 consecutive days. The baseline tests were then used to deconvolve the sources of error of final pressed piece center thickness.

1. Introduction

The continual drive for advancing optical systems has lead the optics industry towards extensive development of aspheric glass lenses. In many cases, the demand for aspherical optics has forced industry to evaluate a more cost effective process for the manufacturing of aspheric glass lenses over the labor and time intensive approach of conventional grinding and polishing. An emerging technology to handle the demand of aspherical lenses, called precision glass molding (PGM), is one in which the glass workpiece is heated and then molded into the desired lens geometry. Such processing requires significant understanding of mold and workpiece attributes, machine process parameters, and interactions at temperatures (viscosities) necessary to generate workpiece flow. A detailed understanding of the thermal and mechanical response of a given glass and mold material at elevated temperatures is not always available and is often realized by time-consuming iterative processes on expensive production tools. A bench-top research platform to satisfy the needs of the user to control specific process parameters such as temperature, force, and position through the entire molding cycle in order to tailor the molding cycle to the desired glass has been commercially developed and evaluated. This paper discusses the use of such a tool, Dyna Technologies Inc.'s (DTI's) GP-5000HT, in support of material and process parameter optimization for the PGM process of visible optical glasses. The attributes and limitations of the as-purchased configuration are discussed along with the improvements made towards enhancing tool performance repeatability.

Moore Nanotechnology Systems, LLC [1] and Toshiba Machine Co. [2] sell commercial lens molding tools worldwide, and both have designed their machines to be compatible with medium to large volume production of optics, typically 5,000 and 30,000 pieces a year per machine, respectively. These machines are very similar in their flexibility of pressing parameters. They allow for force or position control. Depending on the motor, they can apply around

25 kN of force. Current industrial machines are offered with one of the following: resistive, induction, or infrared heaters. Infrared heaters have been found to be the most common. The temperature control is typically dictated by standard or manufacturer proportional, integral, and derivative (PID) feedback values [1–3].

Scientific research not aimed at realizing production volumes has a need for a smaller, bench-top sized version of such tools with flexibility in machine design, optical component geometries, molding atmospheres, heating geometries, and process parameters. The GP-5000HT precision glass molding machine from DTI [4] was designed and further modified at Clemson University for this purpose. The machine has functionality comparable to commercially available products but also the flexibility for laboratory testing, research and prototyping of various lens/mold designs, mold coatings, and lifetimes, and assessment of resulting part surface quality. Since its arrival at Clemson in the spring of 2010, the GP-5000HT has had two modifications made to it, to enhance the research functions of this machine. The first modification was the implementation of a high-temperature force transducer (Futek LCM200) to compensate the temperature effect on the force measurement. The second modification was the introduction of mold tooling inserts (the pieces which come into direct contact with the glass during molding) to allow the mold tooling shape to be easily changed with mold design changes or change in glass type. The present paper discusses results on only one glass type with a designated set of inserts, to illustrate tool function and performance, therefore, eliminating cross contamination of multiple glass types and enabling a further mold material/glass family interaction study. It should be noted that the modifications and associated changes employed have allowed the modified DTI tool to be used for studies conducted on both visible (oxide) optical glasses and infrared transparent (nonoxide, chalcogenide) glasses [5–7].

The main objective of the present study is to summarize the DTI tool's functionality and measure the accuracy of the new design's control over thickness of pressed or molded lenses. To this end, we have chosen part center thickness, CT, as an attribute which is critical in lens manufacturing as our metric. Measurement of CT in this context relates to the ability of the tool to repeatedly heat, press, and cool a lens blank (in this case a plano-plano part) under identical process parameters to assess resulting part uniformity. While not representative of the full capabilities of the DTI tool, CT assessment has allowed us to obtain statistically significant data on the influence of tool systems, in carrying out a simple press cycle on a commercially relevant optical material. To this end, mechanical, thermal, and pressed part repeatability tests were conducted using two polished planar tungsten carbide (WC) surfaces (mold inserts, Kennametal Inc., Latrobe, PA, USA) and L-BAL35 (Ohara Corp., Branchburg, NJ, USA) as the glass workpiece sample.

2. Machine Functionality and Baselines

The ultimate performance of the DTI press (referred to as the tool) is defined by the structural and thermal attributes of the system and the ability of those components to return to a baseline condition in between pressing cycles. Thus, its ultimate functionality is a combination of the tool's structural and thermal response under defined molding cycle conditions and parameters. Knowledge of the baseline values of these attributes and how they vary/change with varying cycle times and "offline" cooling periods between presses dictates the cycle time possible for the tool. The GP-5000HT has a frame to house the heating system, mechanical system, and instrumentation. The frame is composed of two rectangular plates and four posts. The four posts close the structural loop of the machine by passing through the corners of the lower plate and connecting to the corners of the upper plate, so it has a closed frame structural loop. The components of the frame and other mechanical parts of the machine can be found in Figure 1. The machine systems described in this study are controlled by the limits and inputs of the press's instrumentation system. Simple thermal and mechanical baseline measurements were established to characterize the functionality of the press and determine possible sources of error during the glass molding process.

2.1. Instrumentation. The instruments used for assessing the press's performance are composed of sensors, measurement devices, and a controller. The sensors are used to send feedback to the controller enabling it to follow the command; they also act as safety features, preventing any damage to the heating system, mechanical system, or the user of the press.

The measurement devices for the tool include the thermocouples, pressure transducer, flow meter, load cell, and linear encoder used to govern the molding process. Holes aligned axially with the molds are used for inserting thermocouples to measure temperature and are located in the upper and lower molds up to 2 mm away from the back side of the mold tooling surfaces. This in turn places the thermocouples about 6 mm from the molded glass surface. The readings obtained from these thermocouples are used to estimate the temperature of the glass sample during the molding cycle. The outputs from these thermocouples are used as feedback for the heating elements. Two other thermocouples serve as safety sensors: one embedded in the wall of the molding chamber records the temperature within the molding chamber, while another is in the exhaust tube to monitor the exhaust gas temperature. The pressure transducer monitors the static pressure of one of the pneumatic lines coming into the molding chamber. The load cell is aligned axially with the molds, and it is connected in series to the actuator to instantaneously measure the force applied to the glass sample during molding. The linear encoder reads the displacement of the upper mold in respect to a home position. The home position is the furthest position from the bottom mold.

The signals from these sensors are sent to a Unitronics Vision570 Programming Logic Controller (PLC), which also serves as the human user interface (HUI) on the front of the press. There is Unitronics software that also allows the operator to run the press directly from a separate computer. The heating system has a separate software called EZ-Zone Configurator and is used to adjust the proportional, integral,

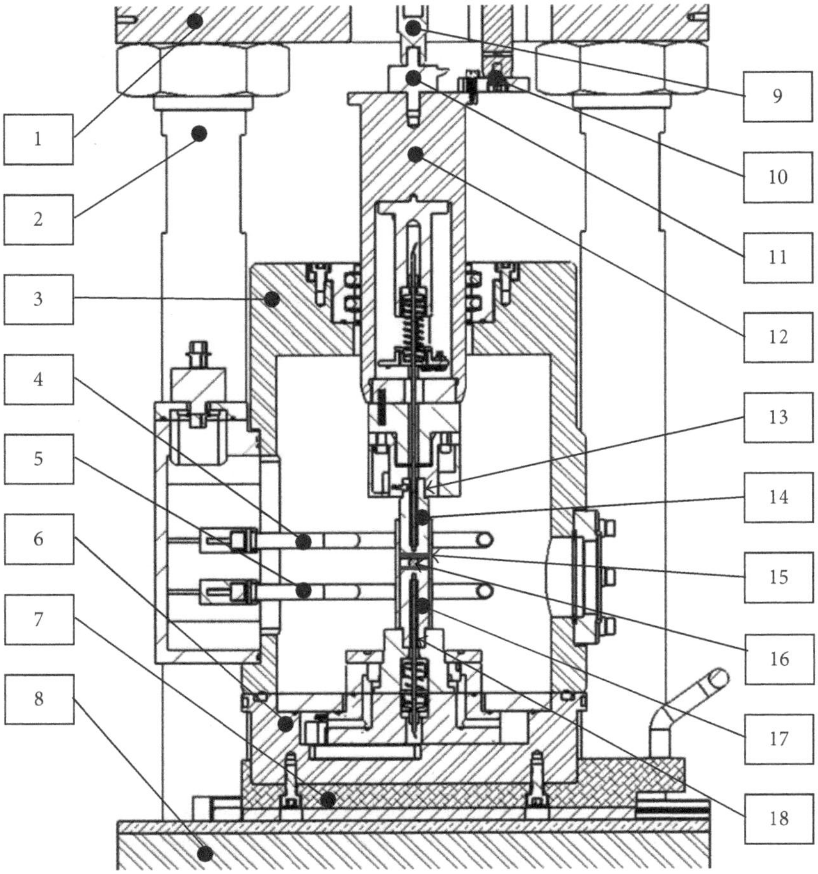

(1) Upper frame plate
(2) Corner support post
(3) Upper mold chamber
(4) Upper infrared heater
(5) Lower infrared heater
(6) Lower mold chamber
(7) Slide
(8) Lower frame plate
(9) Electric actuator
(10) Linear encoder
(11) Load cell/force transducer
(12) Post
(13) Upper thermocouple
(14) Upper mold
(15) Alignment sleeve
(16) Glass sample
(17) Lower mold
(18) Lower thermocouple
— Mold tooling inserts

FIGURE 1: Cross-section of the GP-5000HT revealing its critical components. The pneumatic air cylinders that lift the upper mold chamber along with the control boxes, which contain the electrical equipment for operating the GP-5000HT, have been excluded.

and derivative (PID) values; this software needs to be run from a separate computer because the PLC and the EZ-Zone software cannot both be connected at the same time as the inputs to heating system will interfere. The PLC stores memory on either the hard memory of the PLC or an SD card.

2.2. Mechanical Systems. The mechanical system is composed of an electric actuator and two pneumatic cylinders attached to the upper frame plate. The pneumatic cylinders are used to raise and lower the upper molding chamber. The upper portion of the molding chamber is connected to the ends of the pneumatic cylinders, and the lower portion is attached to slide rails on the lower frame plate. This allows the lower portion of the molding chamber to slide in and out from under the upper molding chamber. A back stop (set screw) on the slide rails aligns the lower molding chamber with the upper. The back stop is fitted with a sensor that activates when the lower molding chamber is resting against it allowing the upper molding chamber to be moved. The electric actuator is also linked to this sensor but can be disabled in the manual control mode.

The electric actuator provides the movement and force along the axis of the molds needed for molding the glass. The actuator is mounted to the upper frame plate and is connected in series with the load cell, the support post (post), upper mold, and upper mold tooling, respectively. The support post goes through a seal housing that is on the upper molding chamber so that the molding chamber is not disrupted by ambient air. The mold inserts/tooling are the only objects in the system that are in contact with the glass during the pressing cycle. The upper and lower mold tooling is approximately 4 mm in thickness and sit between their corresponding mold face and the glass. The mold tooling is used indirectly to monitor mold/glass interactions by measuring the change in tooling surface roughness as a function of cycle number, which after multiple cycles changes as nanoscale deposits of glass are formed or mold tooling material pulls out of the mold tooling. The use of multiple small tooling inserts can decrease contamination and allows increased flexibility of molded shapes. The lower mold and mold tooling remain stationary throughout the molding process and are mechanically connected to the lower molding chamber. A sleeve around the lower mold and mold tooling is utilized to align the upper and lower molds which are made of tungsten carbide (WC). The WC is used because of its robust mechanical and thermal properties at higher temperatures [8]. Alternative mold/insert materials can be employed as needed for the desired glass type/heating method/temperature range; however, the present work only describes finding for WC with L-BAL35 glass.

A preliminary set of baseline measurements were performed at room temperature in order to characterize the repeatability of the mechanical structural response and how these components settle under force over a number of pressing cycles. These baseline measurements define the position and the repeatability of the bottom mold position at room temperature because that position is used as a reference to determine the glass workpiece thickness during the molding cycle. A number of experiments were performed in order to determine the exact position where the molds are in contact. From one molding cycle to the next, there are four steps to loading the sample. The four steps involve putting the alignment sleeve in place, putting the bottom insert in place, inserting the workpiece sample, putting the top mold insert in place, and sliding the bottom mold housing directly under the upper mold housing. Experiments were conducted with each of these steps to understand sources of error in the position read-out. The bottom mold position repeatability experiments were conducted by bringing the upper tooling down until it reached an 88 N force, and the position was recorded; this load level was selected because it is a low force that would be able to overcome internal friction of the system during the pressing cycle. This procedure was repeated 10 times for the purpose of statistical analysis and the resulting bottom position recorded. Due to bottom position being a physical starting point for the measurement of the system, the range of the bottom position will directly correlate to the error of the final pressed part CT. Table 1 lists the total range in the bottom position measured for each experiment and a percentage of that range given by the standard deviation of the experiment. Another metric monitored during this test was the change in system friction by recording the load under movement before the upper mold came in contact with the bottom or the sample. This is important because a change in force would mean that the upper mold tooling was coming in contact with the sleeve and would then create extra stresses in the system, false force readings, bending moments, and wedging of the sample. Change of system friction can be seen by a change in the system's static force when the mold tooling was above the sleeve and in the sleeve.

Table 1 shows that the deviation in the bottom position remains smaller than 1 μm, which is the GP5000HT's LVDT position sensitivity, until experiment 4. The deviation increases in experiment 4 and increases even more in experiment 5. Experiment 5's procedure is identical to that used for inserting the glass molding preform during an industrial pressing cycle. This result indicates that a part of the deviation in final pressed part center thickness comes from the sample placement in the molding cavity and from the total movement of the bottom mold housing. Another observation from this test is that the static force did not increase, which means the friction of the mold tooling movement did not change with the addition of the pieces required for molding.

2.3. Heating System. The press has two different types of heat sources that can be employed, depending on the pressing process under consideration; one set uses resistive heaters and the other infrared (IR) heaters. The resistive heater setup uses resistive coils that surround and are in contact with the molding sleeve. Using the resistive heaters, the press is designed to reach temperatures of 600°C. The IR heater setup uses two omega-shaped IR lamps that circle and do not touch the sleeve and are connected to the upper mold chamber. The IR lamps have a reflector on the outer half to help reflect more light towards the sleeve. In addition to the reflectors on the lamps, reflective foil is placed by DTI on inner surfaces of the chamber to improve the fraction of light reflected back to the sleeve and keep chamber parts from overheating.

The heating system uses two separate heating elements, an upper and a lower, to increase temperature uniformity. Uniform temperature control becomes important if the glass being pressed has a steep viscosity curve (where the viscosity changes drastically as a function of temperature) because a temperature gradient could cause unfavorable, nonuniform glass flow. The temperature uniformity is especially important for PGM process during the pressing and the cooling sections of the pressing cycle, and not as important in the heating and soaking sections, as the glass and the system are still reaching equilibrium and less than ten percent of the total pressed part deformation should be taking place (see Figure 7 for a schematic of the PGM force and temperature profiles showing these sections).

The feedback loop controlling the heaters is based on two thermocouples embedded in the upper and lower molds roughly 6 mm from the molding surface, and the thermocouple output feeds back for the corresponding upper or lower heater. The heaters are controlled through one of two approaches, an open-loop control and a closed-loop control. The closed loop is controlled by a proportional, integral, and derivative (PID) feedback loop in the PLC and is activated by the feedback from the thermocouples. The PID values can be adjusted to increase the variability of the temperature control with separate software to control the system. Different temperature settings in a program will be optimized with different PID values and can be determined with an autotune function. The autotune has three different settings: over,

Table 1: Results of the bottom position repeatability experiments.

	Experiment 1	Experiment 2	Experiment 3	Experiment 4	Experiment 5
With sleeve	No	Yes	Yes	Yes	Yes
With mold inserts	No	No	Yes	Yes	Yes
With sample	No	No	No	Yes	Yes
With full movement	No	No	No	No	Yes
Total bottom position range (μm)	1	0	0	5	6
Bottom position standard deviation (μm)	0.5	0	0	1.7	1.8
Change in static force (N)	N.A.	0	0	0	0

under, and critical, which allow for slightly different tunability for the heating of the system to temperature; the three autotune settings react differently at different temperatures, and the response needs to be experimentally determined for each new thermal and material profile used. For even more advanced control of the heaters, the PID values can be changed manually. The IR and resistive heaters respond differently to the autotune function as shown in Figure 2. Figure 2 shows that the IR lamps heat up faster than the resistive heaters. The IR heaters are the heating devices the machine is designed for as the instrument is designed to operate at temperatures up to 800°C. The IR heaters also hold temperature better than the resistive heaters as shown in the reduced modulation seen in the temperature variation with time data depicted in the inset of Figure 2.

In the second method of control, the open loop, the ability to manually set the heater power is accessible; however, there is no feedback from the thermocouple. The open-loop control is valuable as a tool for adjusting the gain or power going to the heaters. This ability is important to permit heater power optimization and in turn can allow higher operating temperatures and higher heating rates. If higher operating temperatures and faster heating rates are not necessary for the desired pressing application, the gain can be adjusted down accordingly to preserve the life of the IR lamps. With the current gain settings on the controller at about 90% of the allowed gain for the heaters, the 100%-power heating rate can be realized as is shown in Figure 3. A simple exponential decay has been used to fit the data as can be assumed based on heat transfer considerations [9].

Optimization of the PID values has led to a temperature control during the pressing section of the molding cycle of ±1°C of the commanded temperature for the two different heating elements. The IR lamps attained this level of control directly from an autotune; however, the resistive heater setup needed further adjustment after the autotune by manually changing the PID values in order to reach that level of control.

The same PID values used for the heating cycle are also used for the cooling cycle. Inert gas, typically unfiltered, reagent-grade nitrogen, is used to cool the press from operating temperature to 150°C. This temperature, if below the workpiece's glass transition temperature, is deemed a safe temperature to open the system housing/sleeve. The cooling limitations are based on the fastest cooling provided by maximum nitrogen flow in the system at a given temperature. Figure 3 also depicts the measured cooling rate data with

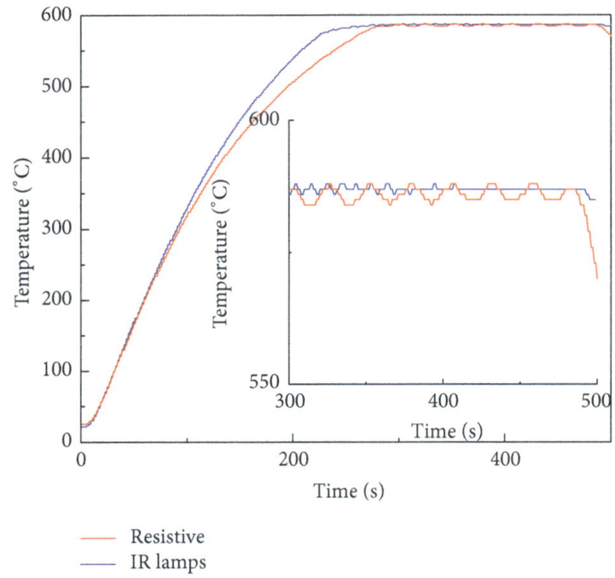

Figure 2: The difference in resistive and IR heaters during the heating and pressing stages of a molding cycle after the autotune.

curve fit for the maximum gas flow. The curve fit to the data is simple exponential demonstrating basic heat transfer with Newton's law of cooling. Newton's law of cooling does not fit the data exactly because the system components are at different temperatures at the start of the cooling cycle and the total cooling rate will be a distribution of cooling curves.

The IR heaters have demonstrated superior thermal stability; they will be the heating elements used for the remainder of the present study. The heating system causes thermal expansion of the press components when at molding temperatures, which can have a dramatic impact on the resulting pressed piece thickness if not properly accounted for. An evaluation of the system thermal expansion is performed with the experimental settings of experiment 5 from Table 1. The sample used was fused silica, with a glass transition temperature above 900°C, it does not compress or flow at the temperatures studied, and the thermal expansion of the silica is negligible compared to that of the rest of the system. The tests were conducted by finding the bottom position and holding an 88 N force at room temperature as defined in the baseline section above and then bringing the furnace up to temperature while recording the change in

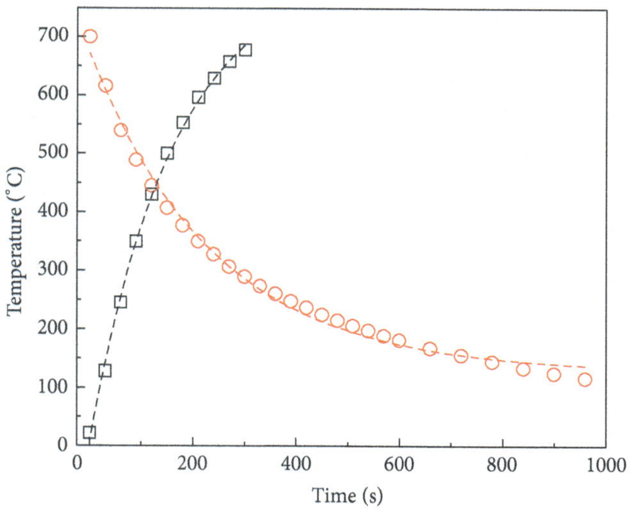

FIGURE 3: Data from temperature response, as measured by the average of the mold tooling thermocouples, from the IR lamps on full power fitted with a simple exponential decay to the data (squares, heating); and the maximum cooling of the system with a simple exponential decay fit (circles).

bottom position to assess and quantify the overall system thermal expansion. An example of the results of a single test is shown in Figures 4(a) and 4(b), the thermal expansion was measured at specific times at different temperatures across multiple consecutive cycles.

It should be noted that this test does not give a method for finding the bottom position at temperature. This is because of the large thermal mass of the press which prevents the system from returning to room temperature before the next pressing cycle. The time for the press to reach room temperature can be over eight hours; this is not a practical time scale for producing lenses. To understand the effect of the tool's thermal mass on the systems thermal expansion, an extended thermal expansion study has been made by recording the bottom position at the end of the pressing cycle for seven consecutive cycles. These results are shown in Figure 4(b). Knowing the position accurately is important in precise control of the PGM process because transitions between stages in the molding cycle (depicted later in Figure 7) are position limited. For example, only when the press reaches the final, user defined press position, it ends the pressing stage of the molding cycle and begins the cooling stage. Analysis of the final press position as a function of molding time and temperature is done by ramping the heating elements at the maximum ramp rate to 300, 500, or 700°C and holding for 120, 600, or 900 seconds. These temperatures were chosen as temperatures close to those of chalcogenide lower molding temperatures and oxide higher molding temperatures, effectively representing the full planned temperature range of the press. The 120-second and 600-second soaking times were chosen because they are the typical limits of time at temperature for industry; the 900-second time was chosen as a time that allows the system to get closer to true thermal equilibrium. Each combination of the time/temperature parameters was run for seven consecutive molding cycles. The extremity results, the highest and lowest temperatures, with the shortest and longest hold times, of this test are shown in Figure 4(b). These results demonstrate that the thermal mass of the system does not reach an equilibrium state until the third or fourth consecutive molding cycle, therefore, demonstrating the need for warm-up cycles to bring the system into a "steady-state" following initial power on first thing in the day. The steady state region, where the thermal expansion is essentially constant with subsequent pressing cycles, begins at cycle number four for each of the different expansion runs tested. This figure shows that the final pressing position is greatly affected by the thermal expansion of the system, which at higher molding temperatures can be greater than 1,200 μm. Figure 4(b) also shows that the final position set point will be a function of the time at temperature, due to the large thermal mass of the system. This experiment demonstrates why it is not currently possible in industry to predict the exact final press position without doing iterative pressing cycles and changing the final position, a trial-and-error type solution that our work seeks to minimize. The results of the measurement of the deviation in bottom position for the varying temperature and time regimes shown in Figure 4(b) were analyzed to quantify the magnitude of this variation, and the resulting deviation data are summarized in Table 2.

Table 2 indicates that the error in the final press position is neither a function of temperature nor of the duration of time at temperature. Therefore, the error could be from the nonequilibrium thermal state of the system; the heater power is still pulsing when trying to hold a molding temperature. These thermal expansion measurements show that the average error in the press position when using a molding temperature above 300°C is up to $\pm 3\,\mu$m. The final positions measured in these tests were used to generate a predictive final position set point expression as a function of temperature and time at temperature. All fits for the baseline thermal expansion were calculated using a first-order exponential decay function following Newton's first law of cooling; the first-order exponential is needed because the system does not fully reach equilibrium. The baseline function is comprised of a first-order exponential decay function as a function of time where each fitting parameter is then an exponential decay function of temperature. The full thermal expansion baseline function to predict the final press position (FPP_E) in μm is given below:

$$\mathrm{FPP}_E = \left[A_1 e^{(-T_m/B_1)} + C_1\right]\left[e^{(-t/(A_2 e^{(-T_m/B_2)}+C_2))}\right] + \left[A_3 e^{(-T_m/B_3)} + C_3\right], \quad (1)$$

where T_m is the molding temperature in °C and t is the time in seconds. The fitting parameters are given in Table 3.

Due to the exponential nature of the fit, care needs to taken in the fitting process to not disregard significant figures. The model fit with two less significant figures overestimates the fit by 10 μm at high temperature and time. The FPP_E fitting parameters in (1) at a molding temperature of 566°C

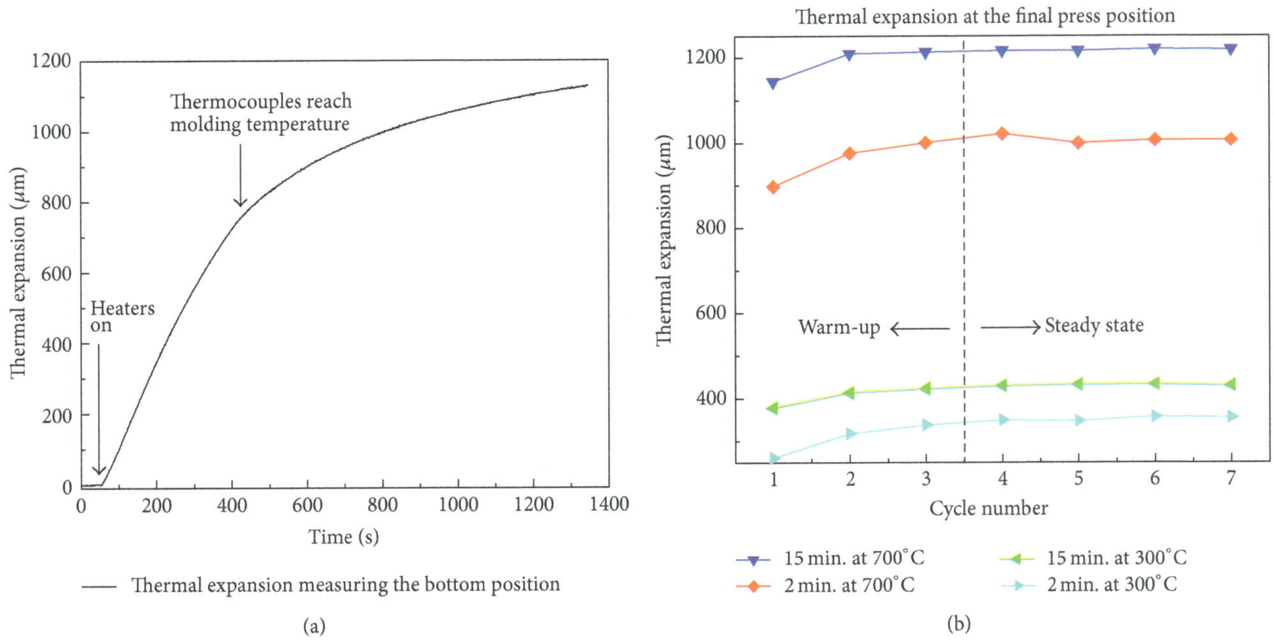

FIGURE 4: (a) A typical thermal expansion of the system, measuring the bottom position change in the press as a function of temperature. (b) The trend of the bottom position measurement at defined times and temperature for seven consecutive cycles.

TABLE 2: Deviations in the final press position for the test shown in Figure 4(b).

Temperature (°C)		300			500			700	
Time (s)	120	600	900	120	600	900	120	600	900
Thermal Expansion (μm)	277 ± 2	372 ± 4	376 ± 2	534 ± 3	663 ± 2	694 ± 4	848 ± 1	1018 ± 1	1090 ± 3

TABLE 3: Fitting parameters for the FPP_E in (1).

	A	B	C
1	−0.07425	−85.05	−225.47
2	61.90	−246.84	−65.13
3	243.00	−394.21	−145.04

with a soak time of 300 seconds were tested for model validation. The model predicts an expansion of 711 μm, and the experimental value was measured to be 710 ± 2 μm, matching the predicted model value fit within the experimental values and system error. It should be noted that the thermal mass affected by a given molding cycle changes with temperature until the molding temperature is above 300°C due to the nonsymmetric geometry of the system. This effect causes the FPP_E baseline to be valid only above 300°C. More evidence of the nonsymmetric nature of the tool's thermal mass impacting results below a 300°C molding temperature can be seen in the structural stiffness measurements of the machine discussed below.

2.4. Structural Stiffness of the Machine. The theoretical stiffness of the machine including upper plate, corner posts, and lower plate is designed and stated by the manufacturer to be 351 MN/m meaning that the theoretical deflection is expected to be approximately 2.85 micrometers at a force of 1 kN.

To experimentally evaluate the machine structural stiffness at room temperature and elevated temperatures, the linear drift of the force transducer was measured using a shunt calibrator to both calibrate and monitor the force during testing. The shunt is attached to the force transducers circuitry where it is not affected by heat, and the effect of the force transducers internal circuitry was then monitored as a function of molding temperature. The shunt calibration showed a linear output from the force transducer as a function of increasing temperature up to 800°C, indicating that the transducer was within its working temperature range during the subsequent tests. The linear drift of force as a function of molding temperature determined from the temperature dependence of the shunt test is roughly given by

$$F_a = F_r - (0.045 T_m), \quad (2)$$

where F_a is the actual force in Newtons when F_r is the machine's recorded force in Newtons and T_m is equal to the molding temperature in degrees Celsius. The correction factor of 0.045, determined from the shunt test, has units of Newtons per degree Celsius and indicates that the forces change by 4.5 N per 100°C.

With the force correction as a function of temperature known, systems structural stiffness as a function of temperature can be measured. The structural stiffness of the instrument is determined by measuring the press position as a function of applied force. The structural stiffness experiments were performed at room temperature, 300, 500, and 700°C in order to cover the usable temperature range of the press. The tests were conducted by bringing the press to the target temperature and soaking for a minimum of one hour, then bringing the mold tooling in contact with a silica sample, and applying a predetermined force. The bottom position was measured at an 89 N force before and after all of the experiments to ensure that the system was at equilibrium. The test was conducted by allowing the press to control the movement of the mold tooling while reaching a set force up to 890 N and also by manually moving the actuator and measuring the force the press records once the actuator is stationary. The two methods of obtaining data were to determine if the machine could record a dynamic effect to the structural stiffness. If the output from the DTI for the position or the force was at different rates or one lagged behind another, the structural stiffness would be skewed higher or lower depending on which one lagged behind. If there was a lag, the data would also seem to be a function of actuator speed. The data between the two recording methods did not show any significant difference in results, and the data was combined to create a larger data set for determining the structural stiffness. A graphical example of a single data set is shown in Figure 5(a).

The raw position output data has been zeroed at furthest point of the actuator from the bottom mold, so as the top mold comes closer to the bottom mold, the raw data shows an increase in position. It is clear that the molds come into contact at a position of 15.439 mm. Figure 5(b) shows the calculation of the structural stiffness of the instrument through the application of a linear fit to the change in force as a function of position (the red line). As expected, as the position is increased (i.e., the molds are forced into contact) the force measured by the system increases linearly, and the slope of this line is the structural stiffness.

The first step in analyzing the raw data was to isolate the values after the first contact was made with the mold tooling and when force is initially applied. To complete this, all of the data with a force of 22 N or less was discarded. Furthermore, a linear fit of the data will minimize the deviation of the data in the y variable only, and due to the large scatter in the x variable values, the fit does not minimize the total deviation of the data. To correct this directional discrepancy for a linear fit, the force values over every 22 N were averaged, and the fit was performed using the averaged points. An example of the 22 N bin average plot and the linear fit is seen in Figure 5(b). By creating 22 N bins and taking the average, the deviation in the position is minimized, and the fit will minimize the deviation in the force. As seen in Figure 5, there is a very clear linear trend, and the fit has an R^2 value of 0.8 which is quite good for such a large data set.

This process of data reduction was used for all of the temperatures studied. The slope of the linear fit in Figure 5(b) is the structural stiffness, and the slopes across all of the temperatures can then be plotted as a function of temperature. Figure 6 shows the structural stiffness of the GP5000-HT as a function of temperature.

The measured structural stiffness at room temperature is 125 (MN/m) and is almost one third the theoretical value of the machine's frame stiffness of 351 (MN/m). This drop in measured stiffness can be attributed to the many mechanical connections and joints in the axis of the molds and the structural loop connecting to the frame.

Figure 6 shows that the structural stiffness decreases with increasing temperature, which is the expected trend. From room temperature to the highest operating temperature, the structural stiffness of the machine decreases by 32%. At the actuator's maximum load of 2.2 kN, the compression of the system measured by the LVDT can vary by 10 μm. The stiffness is not a linear function of temperature in this system because the heat transfer and the thermal mass are not symmetric and change as function of temperature. Although the structural stiffness does not have a perfectly linear dependence across all temperatures, due to the change in thermal mass up to 300°C, a first-order approximation of a linear fit may be used in the molding temperatures from 300°C to 700°C. This linear fit of the structural stiffness as a function of temperature takes the following form:

$$\mathrm{SS}(T) = \left(-88.9 \times 10^{3}\right) T + 142.5 \times 10^{6}, \tag{3}$$

where SS is the structural stiffness in N/m and T is the molding temperature in degrees Celsius. The slope constant has units of N/m · °C, and the intercept constant has units of N/m. Equation (2) is used to help determine the position as a function of force in

$$P_a = P_r - \frac{F_a}{\left(\left(-88.9 \times 10^3\right) T + 142.5 \times 10^3\right)}, \tag{4}$$

where P_a is the actual position and P_r is the recorded position of the LVDT measured in millimeters from the top home position. The force F_a is taken from (2). Scatter from the raw data as seen in Figure 4(a) was used for (5). Equation (5) shows an estimated relationship of the force and the error of position:

$$P_{\mathrm{err}} = \pm\left(3 + 0.002 F_a\right), \tag{5}$$

where P_{err} is the error of the position values and the constant 3 has units of microns while the constant 0.002 has units of microns per N. The temperature does not seem to have a statistically significant effect on the error of the position or the scatter in the stiffness data as a function of force. All of these information are important to understand prior to pressing in order to achieve the precision needed for the PGM for optical elements. The information can be used to determine an approximate baseline for an experiment and can simultaneously show where possible sample center thickness error could be arising. These findings ultimately will save time and money, as finding the optimal pressing positions will be less of an iterative process.

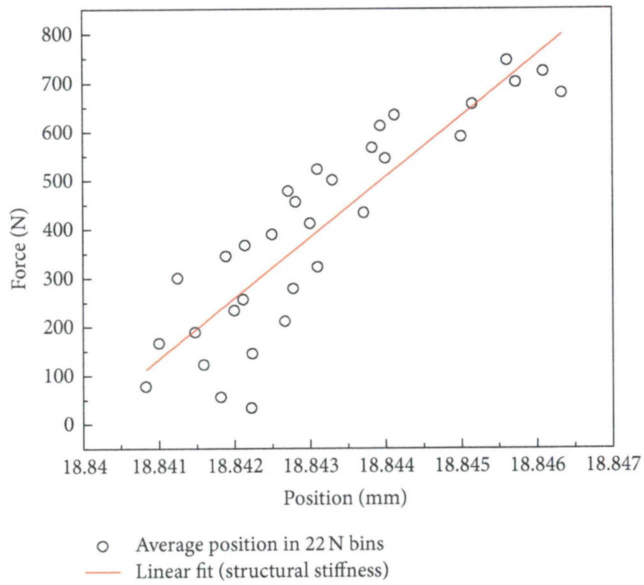

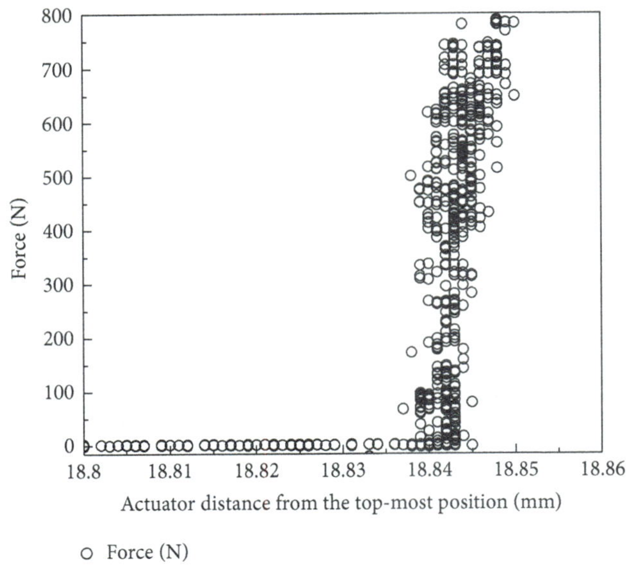

FIGURE 5: (a) An example of the raw data taken at room temperature for the structural stiffness experiments. (b) Analysis of the room temperature raw data for determining the structural stiffness using 22 N bin averages of position with a linear fit.

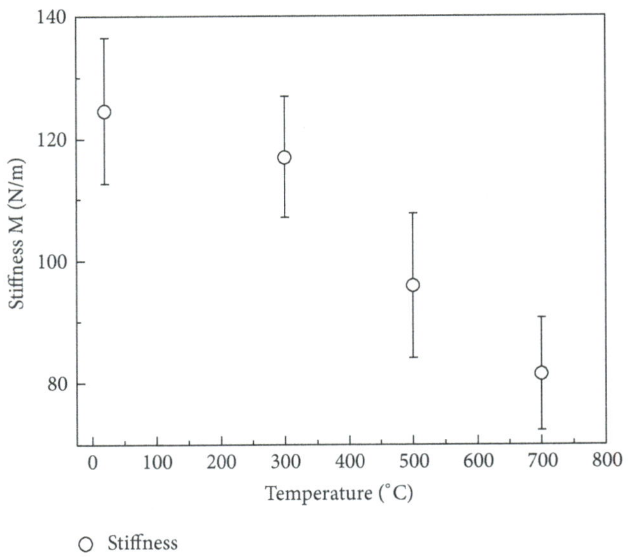

FIGURE 6: Structural stiffness of the GP500-HT as a function of temperature.

2.5. Molding Cycle. The molding cycle is comprised of several stages: purging, heating, soaking, molding, and cooling. The measurement devices embedded in the press drive the transition between these stages in the molding process. Before a cycle starts, the user programs the setpoints for a molding program. These setpoints are the triggers that cause the press to move to the next stage in the cycle. The user has two options for running a given pressing cycle: auto mode runs through a full cycle when the start button is pressed, and manual mode gives the user control over when the different stages begin.

Figure 7 shows a typical molding cycle with temperature and force as a function of time. The stages of the pressing cycle are labeled.

Once the molding cycle is started, the upper molding chamber lowers and seals the molding chamber. The press is then evacuated using a vacuum pump and then backfilled with inert gas (nitrogen) three times in order to prevent oxidation of the molds and other components in the molding chamber. The evacuation pressure, back fill pressure, and number of cycles are user specified inputs.

Once purged, the soaking stage begins, and the actuator moves into the sleeve quickly to a user defined height above the position it will come into contact with the glass. The actuator continues to move from that point slowly until it comes into contact with the glass sample and applies a user-defined force. Once the soaking force is reached, the heaters heat either at full output until 95% of the target temperature where the PID values take over or at a user-defined ramp rate. Once the thermocouples near the molding surface reach the commanded temperature, the soak timer begins. The soak time can be as short as 1 second or extend for hours. The soak allows the glass sample to equilibrate at a uniform temperature to ensure the glass has a uniform viscosity. When the soak time has elapsed, the pressing stage begins. The pressing stage is force driven but position limited: the actuator begins to press the glass until a specified force is reached, and the actuator holds the pressing force until the final position is reached. The final press position is determined from the room temperature (RT) bottom position (P_{RT}):

$$\text{FPP} = P_{RT} + \text{CT}_{\text{Final}} + \text{Baseline}, \quad (6)$$

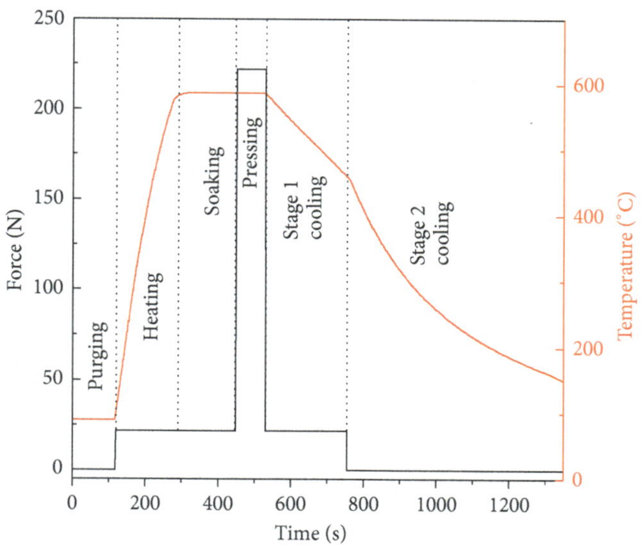

FIGURE 7: The molding cycle used for the repeatability tests in this study with temperature and force as a function of time.

where CT_{Final} is the desired final press part center thickness and the Baseline is determined by combining (1), (2), and (4). The full master equation for the FPP is

$$\begin{aligned} \text{FPP} = {} & P_{RT} - CT_{Final} \\ & - \frac{1}{1000}\left[\left(A_1 e^{(-T_m/B_1)} + C_1\right) e^{(-t/(A_2 e^{(-T_m/B_2)} + C_2))} \right. \\ & \left. + \left(A_3 e^{(-T_m/B_3)} + C_3\right)\right] \\ & + \frac{F_r - (0.045T_m)}{((-88.9T_m) + 142.5 \times 10^3)}. \end{aligned} \quad (7)$$

The final pressing position is also a safety feature: regardless of the pressing stage or cycle, when the LVDT output reaches the final position, the cooling stage begins. The press can be cooled at the maximum cooling rate or using up to three separate cooling sections with control over the cooling rate and applied force. This cooling ramp control gives more control over the amount of stress remaining in the glass. The whole pressing cycle is completed when the press reaches a safe user-defined temperature; at that time, the upper mold chamber is raised, the lower chamber is slid out, and the upper tooling is removed. The molded glass sample can then be removed for analysis.

Auto-mode allows for full cycles to be complete under the set points, but to have more versatility on the segments of the molding cycle, the manual mode is needed. Manual mode allows the user to start and stop cycles at any time and gives full freedom of the actuator, the cooling, and can be used to find set points. Manual mode does not allow for 3 sections in the cooling stage but does give the flexibility of individually using the upper, lower, and chamber gas cooling. Manual mode also has an option to monitor the heaters power and temperature control during any stage of the molding cycle.

3. Repeatability Results

The goal of the repeatability measurements is to determine the ability of the press to produce similar results on a day-to-day basis and also on a per cycle basis. The target function for both of the data sets is the center thickness of the sample after it has been pressed.

3.1. Methodology. The samples used during the pressing cycles are square-cut plane parallel plate (PPP) windows of L-BAL35 glass from Ohara Inc., as this is a glass currently being used for molding in industrial settings. The glass's physical properties needed for PGM such as viscosity, glass transition temperature, viscoelastic response, and structural relaxation parameters are known for L-BAL35 [10]. Dimensionally, each PPP window square measures approximately 5 mm on a side, with a thickness of approximately 2 mm. The cycle parameter set points for this pressing cycle are listed in Table 4.

While in auto mode, the press heats the sample to 590°C and molds the sample with a force of 222 N. The sample is cooled in two stages, with different rates of cooling. A graphical representation of the molding cycle used for the repeatability is shown in Figure 7.

The data collected on the repeatability can be broken into two sets. The first data set includes only the data from the first cycle of the day. This data set is representative of the day-to-day variations exhibited by the instrument. The second set of tests includes data from all cycles, including the first-cycle-of-the-day data already presented in the first data set. The second data set mimics a typical production cycle used in industry [3]. Twelve consecutive pressing cycles were taken on three different days. The sample thickness was measured in 10 places to determine a mean value and standard deviation of the center thickness for each individual sample. In the 12 consecutive cycles there are two distinct phases in the press' repeatability. The first phase, or the "warm up" phase, includes the first 2–4 runs (shown in Figures 4(b) and 8), where the press experiences thermal expansion between pressing cycles. The second phase, referred to as the "steady state," includes all cycles after the warm-up phase, where the thermal expansion is constant between pressing cycles. The boundary line between the two phases is somewhat arbitrary but is distinguished by a significant drop in the standard deviation of the sample thickness across the three days.

3.2. Results. The individual measurements of each sample show that they have a highly uniform thickness, varying only a few microns with a maximum standard deviation of 2 microns as listed in Table 5.

Table 5 also shows the total range and the standard deviation of the first cycle and all of the steady state samples. A large difference exists in repeatability capabilities from the first cycle to the steady state; an order of magnitude drop of the standard deviation and the total range from the first cycle to the steady state cycles is shown. The steady state average center thickness is 1.001 mm, showing that the press can reach desired center thicknesses.

TABLE 4: Pressing parameters used in determining system repeatability.

	Cycle parameters
Purge limits (psi)	−12/2
Soak force (N)	22
Pressing temperature (°C)	590
Soak time (min)	2
Pressing force (N)	222
Final position (μm from P_{RT})	1717
Cooling section 1 temperature (°C)	590
Cooling section 1 ramp rate (°C/min)	35
Cooling section 1 force (N)	22
Cooling section 2 temperature (°C)	450
Cooling section 2 ramp rate (°C/min)	None
Cooling section 2 force (N)	4
Chamber open temperature (°C)	150

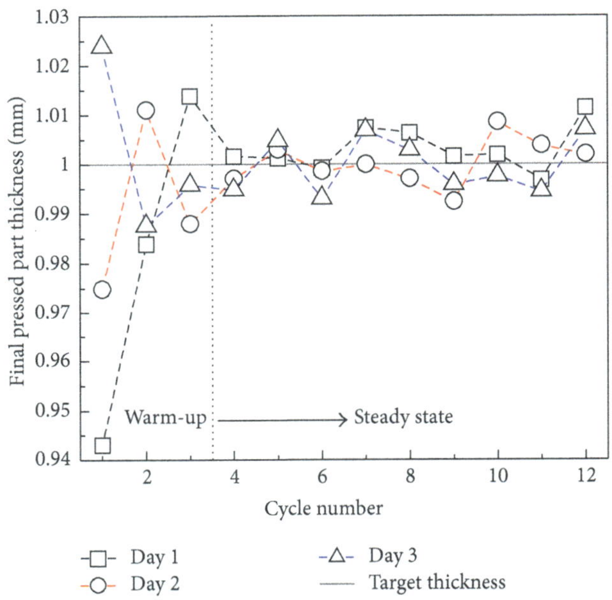

FIGURE 8: The center thicknesses of the final pressed part pieces through the cycle number over multiple days in 12 consecutive pressing cycles.

TABLE 5: The standard deviation and the total range of the thickness of an individual sample, the first molding cycle of the day, and the steady state.

	Individual samples	First cycle	Steady state
Standard deviation (μm)	2	81	5.5
Total range (μm)	6	41	21

Within the whole data set, the effect of thermal expansion on the target function can be analyzed. These results show several machine tool attributes that would be important to a lens manufacturer: the uniformity of the pressed part thicknesses and how quickly (how many pressing cycles) before this uniformity can be obtained. As the press transitions from room temperature through the warm-up phase, the average thickness of the resulting samples continually changes. However, once the press has reached its "steady-state phase," the thicknesses of the pressed pieces remain essentially constant. Comparing the deviation from the target thickness of each numbered cycle, from 1 to 12, over 3 consecutive days of cycles, a definite trend is evident, and is shown in Figure 8. As the number of cycles increases past the third cycle of that day, the standard deviation of the sample thickness decreases.

Figure 8 shows the actual center thicknesses of the pieces run over three different days. There is a drastic difference in the error of the center thicknesses after 3 cycles, once again showing the warm-up period of the press. The standard deviation of the thickness of the individual samples is within the data points.

Errors in the pressed part can be traced to three factors in the experiment: the cycle-to-cycle mechanical movement, the thermal mass cycling, and the actual part being pressed. The error of the final pressed part can be shown to come from three aspects of molding: mechanical, thermal, and the press material. The cycle-to-cycle mechanical movement shows a 1.8 μm error. The error recorded from the thermal mass cycling was as high as 4 μm. With 1.8 μm error coming just from the mechanical movement of the system, 2.2 μm error comes from that actual thermal cycling. By introducing the sample and the full molding cycle, the error is 5 μm. This means the error from the sample and the entire molding cycle contribution is a 1 μm addition to standard deviation of the total error.

4. Conclusions

A commercially available, bench-top precision glass molding machine is a valuable tool for research in the demanding and growing optoelectronic industry and scientific field. Based on experience in our facility, it has been found to be a tool that can help bridge the difficult and important gap between scientific research and industrial manufacturing. The GP-5000HT is a glass molding machine with functionality similar to a full-production molding machine but is also versatile enough to be used for scientific research purposes.

A steady-state baseline was shown to be needed due to pressing cycle setpoints being position limited, and the systems position was found to be a function of the temperature, time at temperature, and force; this finding demonstrates one of the reasons why molding is still an iterative process. A master baseline was created for this press to account for the system position change. The baseline proved to be predictive and valid for a standalone test of temperature, time, and force. Furthermore, the master baseline was found to be valid for the repeatability tests.

Preliminary repeatability results utilizing pressed part thickness of plano-plano samples have shown an increase in thickness consistency across a number of days as the number of molding cycles increases. There is still a large uncertainty to the first cycle results and that there is a definite "warm-up" period, though industry routinely runs "dry" cycles, without glass, over some defined period. The tolerances of the final thickness become much tighter after the system reaches thermal equilibrium, typically after 3 molding cycles. During the steady state cycles done over multiple days, the total range of thicknesses is 21 μm with a standard deviation of 5 μm. The systematic study of the system has also shown that about 36% of the error comes from the mechanical movement from cycle to cycle, 44% comes from repeatability of thermal mass from cycle to cycle, and 20% comes from the introduction of the sample in the system.

This study has shown that the DTI GP5000-HT is capable of routine future tests (daily, weekly, etc.) with target boundaries and final piece tolerances that can also be developed from this set to ensure the press is performing within acceptable limits of operation.

References

[1] Nanotechsys.com, *New Hampshire*, Moore Nanotechnologies LLC., 2010, http://www.nanotechsys.com/machines/nanotech-gpm-glass-press-molding/.

[2] Toshiba-machine.co.jp, Toshiba Machine Co. LTD., Tokyo, Japan, 2010, http://www.toshiba-machine.co.jp/english/product/high/contents/molding.html.

[3] Private communication with industrial partners.

[4] Dti-fla.com, Dyna Technologies, Inc., Sanford, Fla, USA, 2010, http://dti-fla.com/glassmolding.html.

[5] K. D. Fischbach, K. Georgiadis, F. Wang et al., "Investigation of the effects of process parameters on the glass-to-mold sticking force during precision glass molding," *Surface and Coatings Technology*, vol. 205, no. 2, pp. 312–319, 2010.

[6] D. H. Cha, H. J. Kim, H. S. Park et al., "Effect of temperature on the molding of chalcogenide glass lenses for infrared imaging applications," *Applied Optics*, vol. 49, no. 9, pp. 1607–1613, 2010.

[7] S. Gaylord, B. Ananthasayanam, B. Tincher et al., "Thermal and structural property characterization of commercially moldable glasses," *Journal of the American Ceramic Society*, vol. 93, no. 8, pp. 2207–2214, 2010.

[8] P. Mosaddegh, *Friction measurement in precision glass molding [Dissertation]*, Clemson University, Clemson, Mich, USA, 2010.

[9] Y. A. Cengel, *Heat Transfer—A Practical Approach*, McGraw Hill, New York, NY, USA, 2003.

[10] S. Gaylord, *Thermal and structural properties of candidate moldable glass types [Dissertation]*, Clemson University, Clemson, Mich, USA, 2008.

19

A Rough Set Approach of Mechanical Fault Diagnosis for Five-Plunger Pump

Jiangping Wang

School of Mechanical Engineering, Xi'an Shiyou University, Xi'an, Shaanxi 710065, China

Correspondence should be addressed to Jiangping Wang; jpwang@xsyu.edu.cn

Academic Editor: Duc Truong Pham

Five-plunger pumps are widely used in oil field to recover petroleum due to their reliability and relatively low cost. Petroleum production is, to a great extent, dependent upon the running condition of the pumps. Closely monitoring the condition of the pumps and carrying out timely system diagnosis whenever a fault symptom is detected would help to reduce the production downtime and improve overall productivity. In this paper, a rough set approach of mechanical fault diagnosis is proposed to identify the five-plunger pump faults. The details of the approach, together with the basic concepts of the rough sets theory, are presented. The rough classifier is a set of decision rules derived from lower and upper approximations of decision classes. The definitions of these approximations are based on the indiscernibility relation in the set of objects. The spectrum features of vibration signals are abstracted as the attributes of the learning samples. The minimum decision rule set is used to classify technical states of the considered object. The diagnostic investigation is done on data from a five-plunger pump in outdoor conditions on a real industrial object. Results show that the approach can effectively identify the different operating states of the pump.

1. Introduction

In many fields of nature science, society science, engineering, and so forth, there are many uncertain and imperfect data needed to be processed. The data collected from the real systems usually include not only useful information but also noisy data. That is, they are inaccurate, even incomplete. It is not practical to just eliminate or dodge this uncertainty by using some hypothesis mathematically. On the contrary, over the past decades many theories and techniques have been developed to deal with incompleteness and uncertainty in information analysis scientifically. Among them, the most successful ones are fuzzy set theories [1, 2] and the Dempster-Shafer evidence reasoning [3, 4]. However, additional information and prior knowledge are needed when they are used to solve some problems. For example, the membership functions are determined by experience and judgement of the experts in certain fields, and the basic probability assignment and statistical probability are sometimes not easy to be obtained, which means that the treatment of the uncertain information of the running conditions of the pump, to some extent, is subjective. In 1982, Pawlak, a Poland researcher, presented the rough sets theory [5] which is a mathematical tool depicting the imperfectness and uncertainty. It can analyze the imprecise, inconsistent, and incomplete information effectively. It can also process the data and also reason and mine the knowledge implied in the data to reveal the potential laws. This theory has strong practicability. Having been studied and developed for more than 20 years, it has great progress both in theoretical and practical application. Rough sets have been used in the areas of inductive reasoning, number logic analysis and redact, building predictive model, decision support, process control, machine learning, database mining, automatic classification, pattern recognition, and so forth.

Five-plunger pump is widely used in oil field to recover petroleum due to its reliability and relatively low cost. It is a typical reciprocating machinery, consisting of many rotating and back-forth motion parts. Diagnosis and isolation of the faults of a five-plunger pump is a challenging problem, because the pump's dynamic responses, generated by a wide range of possible impulsive sources, are very complex. Common techniques used for a pump or reciprocating machinery fault detection include time and frequency domain analyses for either vibration signals [6, 7] or acoustic emission (AE) signals [8]. In recent years, newly developed signal process methods and artificial diagnosis approaches have been paid more attention to. Wavelet transform [1] and local wave method [9] are utilized to extract diagnosis features from the vibration acceleration measured at the hydraulic-end or the dynamic-end of a pump. Expert system using adaptive order tracking technique and artificial neural networks [10], combined method based on wavelet transformation, fuzzy logic and neuron networks [1], wavelet packet analysis and BP and RBF networks [11, 12], local wave time-frequency spectrum and support vector machine (SVM) [9], and evidence theory for fusing the results from several single-source and multi-feature [3, 13] have been studied for fault diagnosis in a pump or reciprocating machinery.

However, in condition monitoring and fault diagnosis of the five-plunger pump, the raw knowledge collected from the pump may contain uncertainty, which can be imprecise or incomplete. Imprecise information refers mainly to raw knowledge that is fuzzy, conflicting, or contradictory. For example, the opinion about the performance of a pump perceived by two engineers could be different. This would introduce inconsistency to the raw knowledge concerning the performance of the pump. On the other hand, incomplete information refers mainly to missing data in the raw knowledge which can be caused by the unavailability of equipment or oversight of operators. This imprecise and incomplete nature of raw knowledge is the greatest obstacle to automatic rule extraction. The ability of acquiring decision rules from empirical data or the environment is an important requirement for both natural and artificial organisms. In artificial intelligence, decision rules can be acquired by performing inductive learning. Many developed techniques such as decision tree learning, neural network learning, and genetic algorithm-based learning could be introduced to carry out such a task.

In this paper, a novel approach based on the rough sets theory to fault diagnosis, which aims at identifying the basic events occurring in the running condition of the pump, is proposed. As the proposed approach involves subjective judgements by domain experts, which may be inconsistent or sometimes contradictory, rough sets theory is adopted. It uses the information offered by the testing data themselves and does not need the prior knowledge. Also it can represent and process the imperfect information. And on the premise of retaining the principal information of the pump running conditions, it can reduce the fault features, find a simpler representation of knowledge in the form of decision table, and drive diagnostic rules only from the facts presented in the data.

2. Basic Notions of the Rough Set Theory

In this section, only a short reminder of the basic notions of the rough sets theory is given. These concepts should be useful for us to understand the analysis of the considered diagnostic problem performed in the next sections. For a more exhaustive presentation of the rough sets theory, readers can refer to several researches [5, 14, 15] devoted to this theory.

2.1. Equivalence Relation and Lower/Upper Approximation. The concept of "knowledge" has different meanings in different categories. In rough sets theory, the "knowledge" can be referred to as the ability of classification. According to their characteristic differences, the ability to distinguish the objects on the basis of imprecise information about them is considered some kind of "knowledge." In other words, in the process of classification, the objects, whose characteristic information is not too different, can be classified into same class. The relation between them is named the indiscernibility relation, also called equivalence relation. Imprecise information causes indiscernibility of objects. The indiscernibility relation is very important in rough set theory. It deeply reveals the granular structure of the knowledge and is the basis of other concepts. Here the knowledge can be considered a family of equivalence relations. It partitions U, which is a set called the universe, into a series of equivalence classes. Rough sets theory extends the classical sets theory. It implants the knowledge of classification to the sets. Whether one object α belongs to a set X, depending on the knowledge, can be sorted to three cases: (1) the object α belongs to X certainly, (2) α does not belong to X certainly, and (3) α may or may not belong to X. Sets are partitioned according to the knowledge related to U, thus, therefore, are relative, not absolute.

Let U be the universe, and let R be an equivalence relation on U, that is, the knowledge related to U. The pair $A = (U, R)$ is called an approximation space. R will be called indiscernibility relation, if $x, y \in R$, x, y will be indistinguishable in A.

Equivalence classes of the relation R will also be called elementary sets (or atoms) in A. The set of all atoms in A will be denoted by U/R. Every finite union of elementary sets in A will be called a composed set in A. The family of all composed sets in A will be denoted as $\mathrm{Com}(A)$. The family of all composed sets is closed under intersection, union, and complement of sets.

Assuming that x is an object in U and $R(x)$ represents set of the objects which have strong relationship with x, all the objects in $R(x)$ have similar attributes to x. Briefly, $R(x)$ is an equivalence relation decided by x. If one certain set of equivalence relation R is not able to be represented by union of the elementary sets, it is called rough set.

Rough sets theory processes the uncertainty on the basis of concept of lower/upper approximation. Let X be a certain subset of U. The union of all elementary sets contained in X consists of the lower approximation of $R(X)$, in symbol $R_*(X)$. The union of nonempty set of all elementary sets intersection with X consists of the upper approximation of $R(X)$, in symbol $R^*(X)$. The mathematic representation is

$$R_*(X) = \{X \in U \mid R(X) \subseteq X\},$$
$$R^*(X) = \{X \in U \mid R(X) \cap X \neq \Phi\}, \tag{1}$$

where Φ is an empty set.

As a matter of fact, $R_*(X)$ is the greatest set composed by all elementary sets which surely belong to X, judged by the available knowledge. It is also called positive domain, records $\mathrm{pos}(X) = R_*(X)$, while negative domain of X, records $\mathrm{Neg}(X)$, is a set composed by all subsets which do not surely

TABLE 1: Information table for diagnosis.

U	C_1	C_2	C_3	D
x_1	High	High	Low	Yes
x_2	High	Normal	High	Yes
x_3	High	High	High	Yes
x_4	High	Normal	Low	No
x_5	Normal	Normal	High	No
x_6	Normal	High	Very high	Yes
x_7	Normal	High	High	Yes
x_8	Normal	Normal	Low	No

belong to X judged by the available knowledge. Obviously, $\text{Neg}(X) = U - R_*(X)$. $R^*(X)$ the least composed set may be containing X. The set $\text{Bnd}(X) = R^*(X) - R_*(X)$ is called the boundary of X. If $\text{Bnd}(X)$ is an empty set it is a crisp set with respect to R, while if $\text{Bnd}(X)$ is nonempty it will be a rough set with respect to R. Lower (upper) approximations are discernibility, which depicts the approximation property of a set having a vague boundary.

2.2. Information Table and Its Reduction. The data analysed by rough sets theory concern a set of machine running states described by a set of multivalued features or symptoms. They are called condition attributes. In the considered diagnostic problems, the data about the objects (technical states) are represented in a structure called information system.

Suppose that an information system S can be described as an ordered quadruple

$$S = (U, Q, V, \rho), \quad (2)$$

where U is the universe of S—elements of U are called objects; Q is the union of condition, denoted by C, and decision attributes, denoted by D (i.e., $Q = C \cup D$); $V(= \cup_{q \in Q})$ is a set of values of attributes—V_q will be called the domain of q, and $\rho: U \times Q \rightarrow V$ is a description function, such that $\rho(x, q) \in V_q$ for every $q \in Q$ and $x \in U$.

As an example, Table 1 shows the details of eight observations. Each of the observation can be associated with a set of attributes, namely, C_1, C_2, and C_3 which represent noise level, vibration level and machine temperature, respectively. Noise level is evaluated by the RMS value of the noise magnitude measured with sound-level meter. Vibration level is estimated by the peak amplitude of vibration acceleration collected with accelerometer. Machine temperature is tested by thermoelectric sensor.

Let Q be a finite set representing the observed symptoms of a machine malfunction on the universe of the performance of a machine, U, with

$U = \{x_1, x_2, \ldots, x_8\}$,

Q = {Noise level (C_1), Vibration level (C_2), Machine temperature (C_3), Machine malfunction (D)},

V_{C_1} = {High, Normal}, V_{C_2} = {High, Normal}, V_{C_3} = {Low, High, Very high}, D = {Yes, No}.

And information function ρ is given by the table (Table 1) called information table.

The rows of the information table are objects being considered, and the columns are the attributes for each of the objects. There are two kinds of attributes in the information table, namely, the condition (C_1, C_2, C_3) and decision (D) attributes. These attributes will be employed to depict the equivalence relations between the objects. And $U/C_1 = \{\{x_1, x_2, x_3, x_4\}, \{x_5, x_6, x_7, x_8\}\}$ is a discretion of U by the attribute C_1. Any subset in U/C_1 is called an equivalence relation of C_1. That is, the objects x_1, x_2, x_3, and x_4 are in indiscernibility relation with respect to C_1. The other subset in U/C_1 is also indiscernible each other. While $U/\{C_1, C_2\}$ shows the intersection of the subsets of U/C_1 and U/C_2 ($U/C_2 = \{\{x_1, x_3, x_6, x_7,\}, \{x_2, x_4, x_5, x_8\}\}$). Thus, the elementary subsets are $\{x_1, x_3\}$, $\{x_2, x_4\}$, $\{x_6, x_7\}$, and $\{x_5, x_8\}$ determined by attributes C_1 and C_2. The diagnosis problem is described as generating a decision algorithm, denoted by DS (U/D), from the information system (S) that relates elements of U/C to that of U/D. The attributes can be combined to form decision rules.

For the classification and diagnosis, all the condition attributes in S are not necessary. Some of them, however, are superfluous. The result of the decision will not be influenced if the superfluous condition attributes are removed. An intersection of the minimal subset of the attributes is called the core. A subset $P \subset Q$ is a reduction of Q in S if Q-P is superfluous in Q and P is independent of S; the corresponding system $S' = (U, P, V, \rho')$ is called reduced system (ρ' is the restriction of ρ to set $U \times P$).

Moreover, the reduced information system can be identified with the information table. From the information table a decision algorithm can be derived. The decision algorithm consists of a set of decision rules which are logical statements (if…then…). As an example, using the attributes displayed in row 1 in Table 1, the following decision rule can be induced.

> *If the noise level and vibration level are high, and the machine temperature is low, then the machine is in the condition of malfunction.*

Such an approach can be realised using the bespoke basic concepts of rough sets theory as follows.

Step 1. Set up an information (attribute-decision) table according to the specific domain knowledge.

Step 2. Compute the reduction of the condition attributes with respect to the decision attributes.

Step 3. Compute the core value of each decision rule and eliminate redundant rules.

Step 4. Generate the minimal rule set.

Step 5. Determine the logical rules according to the minimal reduction.

In the example shown in Table 1, the elementary sets defined by attributes C_2 and C_3 may be $\{x_1\}$, $\{x_2, x_3, x_6, x_7\}$, $\{x_4\}$, $\{x_5\}$, and $\{x_8\}$, while the subsets of the indiscernibility defined by C_1, C_2, and C_3 are similar to that defined by attributes C_2 and C_3. So it can be said that the attribute C_1 is

TABLE 2: Reduced decision table for classifying.

U	C_2	C_3	D
x_1	High	Low	Yes
x_2	Normal	High	Yes
x_3	High	Very high	Yes
x_4	Normal	Low	No
x_5	Normal	High	No
x_6	High	Very high	Yes
x_7	High	High	Yes
x_8	Normal	Low	No

TABLE 3: Lower approximation.

U	C_2	C_3	D
x_1	High	Low	Yes
x_3, x_6	High	Very high	Yes
x_4, x_8	Normal	Low	No
x_7	High	High	Yes

TABLE 4: Upper approximation.

U	C_2	C_3	D
x_1	High	Low	Yes
x_2	Normal	High	Yes
x_3, x_6	High	Very high	Yes
x_4, x_8	Normal	Low	No
x_5	Normal	High	No
x_7	High	High	Yes

superfluous and can be eliminated. A minimum set of condition attributes in which no superfluous attributes are included then can be obtained and used for diagnosis correctly, as shown in Table 2. The decision "machine malfunction", however, is not dependent on the attributes C_2 and C_3 completely because the decision for class x_5 conflicts with class x_2. The condition attributes are the similar one by one in this case, so the decisions are not distinguished by the attributes. For these situations, the lower/upper approximation in rough sets theory can be employed to handle the conflictions. If we consider nondescript class, we will get the lower/upper approximation as shown in Tables 3 and 4, respectively.

The rules determined by the lower approximation of the decision are inevitably valid, while the rules from the upper approximation may be interpreted deterministically or nondeterministically. From the Table 3, the certain rules are as follows.

Rule 1. If vibration level is high, and the machine temperature is low, then the machine is in the condition of malfunction.

Rule 2. If vibration level is high, and the machine temperature is very high, then the machine is in the condition of malfunction.

Rule 3. If vibration level is normal, and the machine temperature is low, then the machine is not in the condition of malfunction.

Rule 4. If vibration level is high, and the machine temperature is high, then the machine is in the condition of malfunction.

And the probable rules obtained from Table 4 are as follows

Rule 1. If vibration level is high, and the machine temperature is low, then the machine is in the condition of malfunction.

Rule 2. If vibration level is normal, and the machine temperature is high, then the machine may be or may not be in the condition of malfunction.

Rule 3. If vibration level is high, and the machine temperature is very high, then the machine is in the condition of malfunction.

Rule 4. If vibration level is normal, and the machine temperature is low, then the machine is not in the condition of malfunction.

Rule 5. If vibration level is high, and the machine temperature is high, then the machine is in the condition of malfunction.

3. Case Study for Five-Plunger Pump Fault Diagnosis

The basic concept to apply the rough set theory to diagnose the pump faults is that the real information of characteristics for a verity of fault sorts serves as the condition attributes C, and the types of the fault module samples will be the decision attributes D in information system. We can set up the information table according to the real situation of the fault samples and then analyse the set of decision attributes, whether they are determined by the condition attributes, namely, whether there exists the dependence between C and D, shortly $C \Rightarrow D$. If the relation $C \Rightarrow D$ exists it means that the decision set can be defined by category of condition attribute set. We can then calculate the condition set C whether the decision attributes are independent, remove a condition attribute according to whether it will damage the compatibility of the information table and abstract the diagnostic rules to diagnose the faults.

3.1. Construction of Condition Attributes for the Pump Fault Diagnosis. Vibration signals are extensively used in mechanical diagnostics. The reason for the wide use of vibration analysis may be due to the better understanding of vibration mechanism of machine operation and that the change in vibration signal can be easily attributed to the dynamic characteristics of the machine and its fault conditions. Of course, the five-plunger pumps generate vibration during their normal operation. Moreover, the vibration process is a

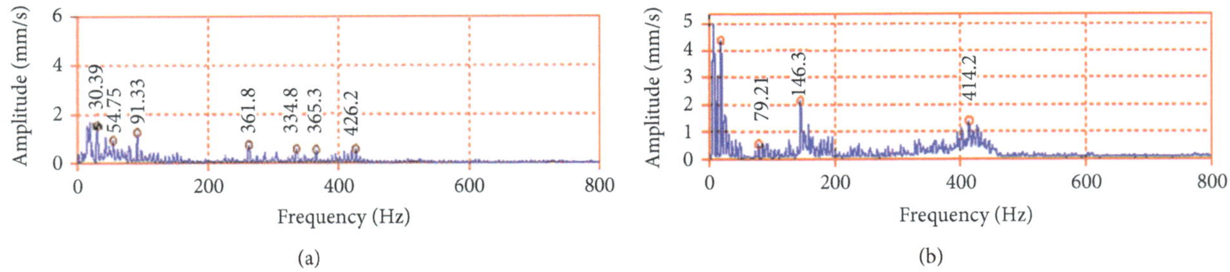

FIGURE 1: Frequency spectra from different vibration signals for different conditions. (a) Normal condition and (b) comprehensive faults both on the second and third inlet valves.

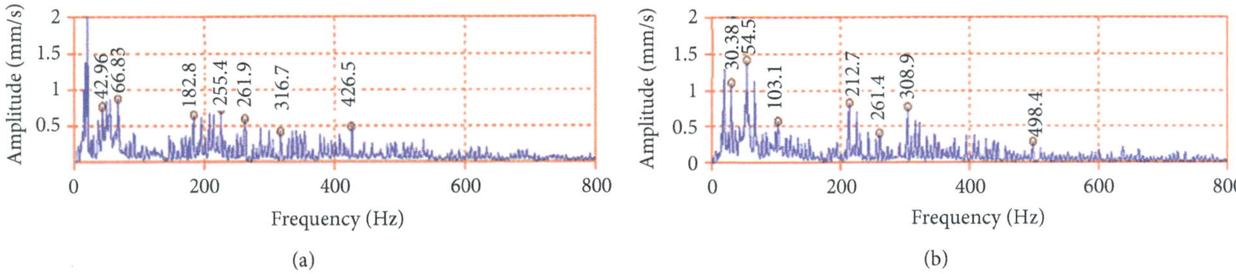

FIGURE 2: Two frequency spectrum representing well-developed faults of different nature that looks similar. (a) and (b) Damage on the second plunger and forth exhaust valve, respectively.

good carrier of the pump and condition-related information, and we are measuring this as vibration signals and transforming it (by FFT operation), to obtain symptoms of condition.

The testing work was carried out on an in-use pump in Tuha oil field in China. Under the consideration of the potential damage to the five-plunger pump and economic cost the faults were simulated mainly in the hydraulic-end of the pump during the experiment. Vibration signals were collected for both normal condition, and several different abnormal conditions. The accelerometers were mounted on the housing at the hydraulic-end of the pump. The crank shaft of the pump was running at approximately 370 rpm (6.17 Hz). The sampling rate was 2000 Hz. Figures 1(a) and 1(b) show the magnitude plot of the FFT of the vibration data (the frequency spectra of vibration signals).

Figure 1 is an example of two typical frequency spectra derived from a dynamic vibration signal collected from the hydraulic-end. This figure shows a fault spectrum and a nonfault spectrum. It is clear that in this case (as well as many others) differences between spectra are quite straightforward. Designing automatic diagnostic tools for these cases is not a problem. Difficulties arise when the differences between signals (and spectra) representing different faults are much more subtle. Figure 2 is an example of two spectra representing well-developed faults of different nature that look similar. In this case, there is a considerable amount of uncertainties and a rigorous and finely tuned automatic diagnostic tool is required.

Not only are individual samples of frequency spectra representing different faults often rather similar, but also the variability within a group of spectra that represent the same fault is often significant. Figures 3(a) and 3(b) show two frequency spectra that represent the same sample fault. The challenge is to provide early detection capability as well as distinction between fault types with a low risk of false alarms.

In the fault diagnosis of the pump, the frequency spectra of the vibration signals contain complicated information of the running condition. To ensure the reliability of the rough set theory for the fault diagnosis of the pump, it is important to extract the condition attributes from the frequency spectra of the vibration signals.

The vibration energies of the horizontal direction in the hydraulic-end of the pump mainly concentrated on the 3 frequency bands: 0–60 Hz, 90–190 Hz, and 300–500 Hz. In the rest of the frequency range 500–800 Hz, there is little of energy distribution with almost no changes in level. The impulse amplitude modulation due to the load variation through the crank shaft revolution (6.17 Hz) is evident. As shown in the above figures, the range where the energy occupies in the frequency spectra is wide and the impulse peaks of frequency spectra are numerous and appear as clusters. These demonstrate the hydraulic pulsation caused by reciprocating motions of the plungers of the pump, the impact impulses excited by the open and shut-off of inlet and exhaust valves in the hydraulic-end of the pump, and the periodic impact vibration caused by the motions of the rotational and reciprocating parts. Due to the subtle changes of the fine frequency structure in vibration signal spectra, direct application of spectra would result in a difficult interpretation and the main events would be difficult to isolate. However, we know that, when the fault of the pump occurs, the changes of the vibration of excitation resource not only lead to changes in each class fault pattern but also result in changes in the overall energy content of the signal which

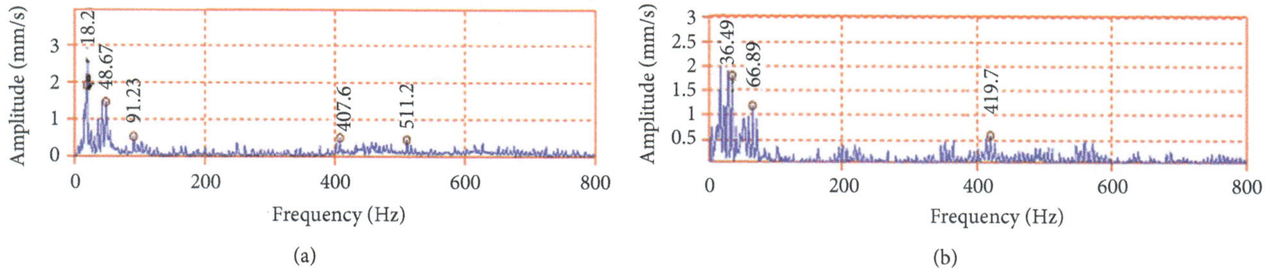

FIGURE 3: Two frequency spectra that represent the same sample fault; (a) and (b) failure both on the second inlet valve.

will lead to an increase or decrease in the general spectrum level. So we can utilize these changes to extract features of the frequency components from the spectra and then employ the fuzzy diagnostic method to identify the conditions of the pump. The values of the constituent parts of the frequency band can represent the energy level of the vibration signal in the frequency domain.

Eight features are selected from the frequency spectra obtained from the vibration signals tested on the horizontal direction of the hydraulic-end of the pump. These features are defined as follows.

(1) f_1: 5 kHz peak energy, to perform the peak energy analysis of the vibration signal, a three-step method is proposed, consisting of (a) using a band-pass filter of high cut-off frequency, for example, 5 kHz, to extract the lower and higher components from the acquired signal in time domain by removing the other events, (b) demodulating the reserved time signal, (c) transforming the demodulated signal to the frequency domain obtaining the 5 kHz spectrum, and then selecting the highest magnitude as the 5 kHz peak energy.

(2) f_2: 5 kHz total peak energy.

(3) f_3: the peak magnitude in the frequency range of 0–36 Hz which is mainly excited by the unbalance reciprocating motions of the plungers. The revolution of the crank shaft is 317 rpm (6.17 Hz), so the frequency of the plunger motion is about 31 Hz (5×6.17).

(4) f_4: the peak magnitude in frequency range of 37–66 Hz. This frequency range represents the vibration frequency of the open and setting-down of inlet and exhaust valves and the changes of the direction of the impulse forces applied by the changes of the motions of the plungers.

(5) f_5: the peak magnitude in range of low-middle frequency section of 67–100 Hz.

(6) f_6: the peak magnitude in range of middle frequency section of 101–250 Hz.

(7) f_7: the peak magnitude in range of middle-high frequency section of 251–500 Hz.

(8) f_8: the peak magnitude in range of high frequency section of 501–800 Hz.

The features 5 to 8 are all of the harmonics of the frequency components of motions of the crank and plungers. These eight eigenvalues compose the crisp universal set $F = \{f_1, f_2, f_3, f_4, f_5, f_6, f_7, f_8\}$ of the condition features. Table 5 shows the eight eigenvalues for normal running condition and different types of faults simulated in the hydraulic end of the pump. Four major types of mechanical faults, namely, *failure on the second inlet valve, failure on the second plunger assembly, damage on the forth exhaust valve, and comprehensive faults both on the second and third inlet valves*, are considered here. Among them, failure on plunger assembly is the most common fault that contributes nearly 56% of the pump faults. Inlet valve and exhaust valve are easily damaged components which account for about 11% and 13% of the pump faults, respectively. There are 14 sample sets representing 5 different types of the conditions given in Table 5. These 5 different types of conditions denote the decision set of fault causes $D = \{d_i\}\,(i = 1, 2, 3, 4, 5)$. The decision attributes can be described as

$$D = \begin{cases} d_1 = 1 & \text{Normal} \\ d_2 = 2 & \text{2nd inlet} \\ d_3 = 3 & \text{2nd plunger} \\ d_4 = 4 & \text{4th exhaust} \\ d_5 = 5 & \text{2nd 3rd inlet.} \end{cases} \quad (3)$$

3.2. Discretization of Continuous-Valued Attributes for the Pump Fault Diagnosis. It can be seen from Table 5 that the condition attributes are continuous variables. They take on numerical values and have a linearly ordered range of values. When dealing with the data of the attributes in the pump diagnosis in the context of rough sets methodology, only the discrete-valued attributes can be used to select the group of condition attributes [15, 16]. So the range of a continuous-valued attribute should be partitioned into subranges. Such a process is known as discretization. However, discretization should be done in such a way to provide useful classification information with respect to the classes within an attribute's range. That is, the purpose of the discretization of continuous-valued attributes is to translate the continuous values into some qualitative characters which are compatible with the information table. This translation involves a division of the original domain into some subintervals and an

TABLE 5: Simulated data of the different fault conditions of a five-plunger pump.

Samples	Condition features								Technical states
	f_1	f_2	f_3	f_4	f_5	f_6	f_7	f_8	
x_1	0.1449	0.3789	0.7	1.22	0.53	1.07	0.9	0.12	*Normal*
x_2	0.1078	0.237	0.8	1.3	0.5	1.2	0.9	0.15	*Normal*
x_3	0.0818	0.1932	0.85	1.6	0.45	0.87	0.72	0.15	*2nd inlet*
x_4	0.0779	0.1903	0.8	1.7	0.5	0.85	0.85	0.15	*2nd inlet*
x_5	0.0856	0.196	0.9	1.5	0.4	0.9	0.6	0.15	*2nd inlet*
x_6	0.2035	0.5404	0.5	1.25	0.45	1.55	1.2	0.24	*2nd plunger*
x_7	0.2003	0.5418	0.4	1.4	0.5	1.6	1.0	0.18	*2nd plunger*
x_8	0.2066	0.539	0.6	1.1	0.4	1.5	1.4	0.3	*2nd plunger*
x_9	0.0778	0.1689	1.17	1.5	0.4	1.1	0.92	0.22	*4th exhaust*
x_{10}	0.0643	0.1406	1.3	1.3	0.4	1.1	1.0	0.25	*4th exhaust*
x_{11}	0.0912	0.1972	1.05	1.7	0.4	1.1	0.85	0.2	*4th exhaust*
x_{12}	0.234	0.58	0.54	1.67	0.3	1.8	1.25	0.27	*2nd 3rd inlet*
x_{13}	0.244	0.5503	0.55	1.75	0.4	1.8	1.0	0.2	*2nd 3rd inlet*
x_{14}	0.224	0.5503	0.55	1.6	0.2	1.8	1.5	0.35	*2nd 3rd inlet*

In the last column of Table 5, *the normal, 2nd inlet, 2nd plunger, 4th exhaust and 2nd 3rd inlet* represent the normal running condition, failure on the second inlet valve, failure on the second plunger assembly, damage on the forth exhaust valve, and comprehensive faults both on the second and third inlet valves respectively.

assignment of qualitative codes to these subintervals. Definition of boundary values of the subintervals can influence considerably the quality of the classifications. Basically, a threshold value, T_i ($i = 1, 2, \ldots, N$, N is the maximum group number of the condition attributes), for a continuous-valued attribute, c_i, is first determined. The set $c_i < T_i$ (the value of attribute c_i which is less than T_i) is assigned to a certain code, while $c_i \geq T_i$ (the value of attribute c_i that is greater than or equal to T_i) is assigned to the other code. Such a threshold value, T_i, is called a quantization parameter. More specifically, given an information system, S, which is composed of M training examples, the discretization of a continuous-valued attribute, c_i, can be based on maximum covariance between the classes for the selection of a quantization parameter, to partition the domain range of c_i into two subdomains. Using a criterion of the consistency of classification it is possible to judge the consistence between the information systems before and after discretization. The details of the algorithm are as follows.

(1) Initiate the number of classes, $m = 2$.

(2) Set the initial number of condition attribute, $n = 1$.

(3) Sort the training samples according to the condition attribute values of nth dimension from the smallest to the largest.

(4) Classify the condition attributes sorted of the nth dimension.

(5) Take boundary value of the condition attributes of adjacent two classes as the quantization parameter. Thus, for each continuous-valued attribute, M-1 evaluations will take place. (Analysis shows that the potential quantization parameter for partitioning the continuous-valued attribute will not lie between a pair of training samples with the same class in the sorted sequence. In other words, the best quantization parameter always lies between a pair of training samples belonging to different classes in the sorted sequence. This means that the evaluation of a potential quantization parameter will only be carried out when a pair of training samples with different classes is spotted.)

(6) Quantitate the continuous-valued condition attributes using the quantization parameter of these two condition attributes and then calculate the covariance, $\sigma_{m,n}$, between them for mth class.

(7) Partition the set of training samples into two subsets at each candidate quantization parameter; compute the consistency of classification of the resulting partition. The consistency, denoted by $Q_U(X)$, can be formalized as follows:

$$Q_U(X) = \frac{\sum_{X\in(U/C)} \eta(R_*(X))}{\eta(U)}, \quad (4)$$

where $R_*(X)$, the lower approximation of X, is the set of all elements of U which can be certainly classified as elements of X using the set of attributes $C{\cdot}\eta(R_*(X))$ and $\eta(U)$ denote the cardinality of the sets $R_*(X)$ and U, respectively. The criterion of the consistency of classification (4) gives the ratio of all correctly classified objects to all objects in the system. At first, the consistency of classification for continuous-valued attributes should be computed. Then that of the discrete-valued attributes can be searched for, ensuring the same quality.

(8) Let $n = n + 1$, then return to Step 3 until the final dimension condition attributes.

(9) When $m \leq M_1$, a given positive integer value, check whether the parameter table is comparative or not. If

TABLE 6: Quantization parameters of the continuous-valued condition attributes.

Condition attributes C	Condition features F	Quantization parameters/Codes			
c_1	f_1	<0.08	1	≥0.08	2
c_2	f_2	<0.3	1	≥0.3	2
c_3	f_3	<0.8	1	≥0.8	2
c_4	f_4	<1.4	1	≥1.4	2
c_5	f_5	<0.45	1	≥0.45	2
c_6	f_6	<1.2	1	≥1.2	2
c_7	f_7	<1.0	1	≥1.0	2
c_8	f_8	<0.2	1	≥0.2	2

it is not compatible, let $m = m+1$ and return to Step 2. When $m > M_1$, let $m = m + 1$, and return to Step 2, until $m = M_2$, maximum number of the technical classes, and $M_1 < M_2$.

(10) According to the relationship of m and $\sigma_{m,n}$, $\sigma_{m,n}$ will increase with m increasing for mth dimension condition attributes until $m = m_0$, then it will decrease with m still increasing. So the optimum classification number is at the turning point m_0 on the curve m-$\sigma_{m,n}$.

(11) Again check whether the parameter table it is compatible or not. If it is not, then let $m_0(n) = m_0(n) + 1$, and return to Step 2, until it is compatible.

Let the condition symptoms be training samples, and let the eight features of the frequency spectrum, of the vibration signal be the condition attributes, that is $f_i = c_i$ $(i = 1, 2, \ldots, 8)$, in fault models, respectively. The algorithm mentioned above is used first to determine the quantization parameters of each continuous-valued condition attributes presented in Table 5, with the optimum quantization parameters being shown in Table 6. Then the original values of the features can be translated into subintervals coded by numbers 1, 2,... with the determined quantization parameters. As a result of this translation the so-called coded information table, Table 7, is obtained which is further analysed. The rows in Table 5 describe the training samples of fault classes, and the columns represent the attributes of the samples. Tables 6 and 7 have the same format.

3.3. Derivation of the Diagnosis Rules. In Table 7, there are 30 objects in the universe of discourse or condition discrimination domain U. This means that the information table for the fault diagnosis of the pump is composed of 30 implied diagnosis rules. These rules, however, are not completely meaningful in the practical diagnosis work. According to rough sets theory, the diagnosis rules can be directly extracted by calculating and reducing the data in the information table. Any set of limited decision, expressed by decision logical language ($C \Rightarrow D$), is called decision algorithm [15, 17, 18] or abbreviated as algorithm (C, D). All sets which cannot be omitted in the algorithm (C, D) compose the core of algorithm (C, D), expressed by *core* (C, D). The core values of

TABLE 7: Coded information table for fault diagnosis.

U	c_1	c_2	c_3	c_4	c_5	c_6	c_7	c_8	D
x_1	2	2	1	1	2	1	1	1	1
x_2	2	1	2	1	2	2	1	1	1
x_3	2	1	2	2	2	1	1	1	2
x_4	1	1	2	2	2	1	1	1	2
x_5	2	1	2	2	1	1	1	1	2
x_6	2	2	1	1	2	2	2	2	3
x_7	2	2	1	2	2	2	2	1	3
x_8	2	2	1	1	1	2	2	2	3
x_9	1	1	2	2	1	1	1	2	4
x_{10}	1	1	2	1	1	1	2	2	4
x_{11}	2	1	2	2	1	1	1	2	4
x_{12}	2	2	1	2	1	2	2	2	5
x_{13}	2	2	1	2	1	2	2	2	5
x_{14}	2	2	1	2	1	2	2	2	5

the algorithm of the condition attributes in the information table can be computed by the following equation:

$$\text{Core}(C, D) = \cap\text{Red}(C, D), \quad (5)$$

where Red(C, D) is all reduction family of the condition attributes. Through computation, it is found that the condition attributes c_1, c_2, c_3, c_5, c_6, and c_7 are redundant. This means that the information table will be compatible when removing the condition attributes c_1, c_2, c_3, c_5, c_6, and c_7, respectively, thus, the three reductions $\{c_2, c_4, c_8\}$, $\{c_3, c_4, c_8\}$, and $\{c_3, c_6, c_8\}$ can be obtained and the core values are c_4 and c_8. That is

$$\begin{aligned} \text{Red}(C) &= \{c_2, c_4, c_8\} \cup \{c_3, c_4, c_8\} \cup \{c_4, c_6, c_8\}, \\ \text{Core}(C) &= \{c_2, c_4, c_8\} \cap \{c_3, c_4, c_8\} \cap \{c_4, c_6, c_8\}. \end{aligned} \quad (6)$$

The reduction $\{c_2, c_4, c_8\}$ will be discussed in the following paragraph, while the meanings of the rest of the reductions are similar. After removing the superfluous condition attributes, the decision rule in the reduced information table can be deduced, that is, to compute the core of diagnosis rules $C \Rightarrow D$:

$$\text{Core}(C \Longrightarrow D) = \cap\text{Red}(C \Longrightarrow D), \quad (7)$$

where Red$(C \Rightarrow D)$ is the set of all reduction $C \Rightarrow D$. The reduced diagnosis rules through the computation with (5) and (7) are shown in Table 8. The symbol $(*)$ in Table 8 indicates that the influence of the values of the condition attributes on the fault diagnosis can be ignored.

The fault diagnosis rules in Table 8 can be expressed with decision logical language.

Rule 1. $(c_2 = 1)$ and $(c_8 = 1) \Rightarrow D = d_1 = 1$.

Rule 2. $(c_2 = 1)$ and $(c_4 = 2)$ and $(c_8 = 1) \Rightarrow D = d_2 = 2$.

Rule 3. $(c_2 = 2)$ and $(c_4 = 1)$ and $(c_8 = 2) \Rightarrow D = d_3 = 3$.

Rule 4. $(c_2 = 2)$ and $(c_4 = 2)$ and $(c_8 = 1) \Rightarrow D = d_3 = 3$.

TABLE 8: Reduced information system for the pump fault diagnosis.

U	c_2	c_4	c_8	D
x_1, x_2	*	1	1	1
$x_3 \sim x_5$	1	2	1	2
x_6, x_8	2	1	2	3
x_9	2	2	1	3
x_{10}, x_{11}	1	*	2	4
$x_{12} \sim x_{14}$	2	2	2	5

Rule 5. $(c_2 = 1)$ and $(c_8 = 2) \Rightarrow D = d_4 = 4$.

Rule 6. $(c_2 = 2)$ and $(c_4 = 2)$ and $(c_8 = 2) \Rightarrow D = d_5 = 5$.

The above rules can be translated into the logical statements (if...then...). For example, the meaning of the rule, $(c_2 = 1)$ and $(c_4 = 2)$ and $(c_8 = 1) \Rightarrow D = d_2 = 2$, is as follows.

> *If 5 kHz peak energy (c_2) is less than 0.3 and the peak magnitude in frequency range of 37–66 Hz (c_4) is equal to or larger than 1.4 and the peak magnitude in range of high frequency section of 501–800 Hz (c_8) is less than 0.2, then the fault of the pump takes place likely on the second inlet valve.*

Similar analysis can be carried out to derive reasonable rules for fault diagnosis when a pump has been detected. This shows that the proposed approach is able to gather a set of rules, which describes the general pattern of decision making by domain experts. With the rules extracted, a knowledge-based system can be developed to diagnose the pump faults.

3.4. Test Result Analysis. With the rules described above, the technical states of the pump can be diagnosed. Some of the test samples corresponding to normal running and with various fault conditions of the pump are shown in Table 9. Based on the sample data of the pump under different conditions, the coded condition attributes in the form of an information table (Table 10) are established by the discretization algorithm discussed in Section 3.2. The reduction and the core of the pump with different running conditions are then determined. This result in a minimum attribute set for diagnosing the pump faults. Both the minimum coded condition attributes and one decision attribute (D) are depicted in Table 11. Based on the rules extracted, the final diagnosis results of the detected pump can be discussed as follows.

Rule 1. Because the peak magnitudes in frequency range of 37–66 Hz ($f_4 = 1.38$) are less than 1.4 ($c_4 = 2$) and the peak magnitude in range of high frequency section of 501–800 Hz ($f_8 = 0.14$) of sample y_1 is less than 0.2 ($c_8 = 1$), the sample y_1 represents the pump in normal condition ($d_1 = 1$).

Rule 2. Because 5 kHz peak energies ($f_2 = 0.2292$ and $f_2 = 0.1845$) are less than 0.3 ($c_2 = 1$), the peak magnitudes in frequency range of 37–66 Hz ($f_4 = 1.74$ and $f_4 = 1.65$) are larger than 1.4 ($c_4 = 2$), and the peak magnitudes in range of high frequency section of 501–800 Hz ($f_8 = 0.11$ and $f_8 = 0.13$) are less than 0.2 ($c_8 = 1$) of samples y_2 and y_9, the samples y_2 and y_9 are detected as the fault of the pump takes place likely on the second inlet valve ($d_2 = 2$).

Rule 3. Because 5 kHz peak energies ($f_2 = 0.6403$, $f_2 = 0.7211$, and $f_2 = 0.5785$) are larger than 0.3 ($c_2 = 2$), the peak magnitudes in frequency range of 37–66 Hz ($f_4 = 1.35$, $f_4 = 1.37$, and $f_4 = 1.24$) are less than 1.4 ($c_4 = 1$), and the peak magnitudes in range of high frequency section of 501–800 Hz ($f_8 = 0.25$, $f_8 = 0.31$, and $f_8 = 0.26$) of samples y_3, y_7 and y_8 are larger than 0.2 ($c_8 = 2$), the samples y_3, y_7, and y_8 can be diagnosed as the fault on the second plunger assembly of the pump ($d_3 = 3$).

Rule 5. Because 5 kHz peak energies ($f_2 = 0.1604$, $f_2 = 0.1771$ and $f_2 = 0.1784$) are less than 0.3 ($c_2 = 1$) and the peak magnitudes in range of high frequency section of 501–800 Hz ($f_8 = 0.27$, $f_8 = 0.25$, and $f_8 = 0.24$) of samples y_4, y_{10} and y_{11} are larger than 0.2 ($c_8 = 2$), the samples y_4, y_{10}, and y_{11} indicate that the pump runs under abnormal condition with fault on the forth exhaust valve ($d_4 = 4$).

Rule 6. Because 5 kHz peak energies ($f_2 = 0.5009$ and $f_2 = 0.5697$) are larger than 0.3 ($c_2 = 2$), the peak magnitudes in frequency range of 37–66 Hz ($f_4 = 1.66$ and $f_4 = 1.58$) are larger than 1.4 ($c_4 = 2$), and the peak magnitudes in range of high frequency section of 501–800 Hz ($f_8 = 0.31$ and $f_8 = 0.33$) of samples y_5 and y_6 are larger than 0.2 ($c_8 = 2$), the samples y_5 and y_6 show that the faults of the pump take place likely on the second and third inlet valves at the same time ($d_5 = 5$).

According to the rules, we can reach all the diagnosis results. It can be seen from Table 11 that the decision results of the samples diagnosed are identical to the experiment results (Table 9). Generally, the diagnostic results are quite right. We would like to point out that the samples used here are not sufficient but are practically enough to illustrate the strategy with rough sets theory and how the original data set is transformed into a diagnostic information system directly.

4. Conclusion

This work shows that the basic concepts of rough sets theory provide a way to recognize and diagnose the pump faults. A practical case is used to illustrate the effectiveness of the technique. It is found that the preferential information about certain condition of the pump can be obtained easily and an adapted information table can be constructed readily by using the rough set approach. With the adapted information table established, the minimal diagnostic rule set that exits in a training data set can be obtained to classify technical states of the considered object. This rough approach makes a maintenance engineer in an oil field able to carry out fault diagnosis in an efficient manner and to know what kind of fault information is unimportant or redundant, which will not influence the fault diagnosis result, being missed or mistake occurred. The diagnostic investigation is done on data from a five-plunger pump in outdoor conditions on a

TABLE 9: Samples diagnosed of the pump.

Samples	f_1	f_2	f_3	f_4	f_5	f_6	f_7	f_8	Fault types
y_1	0.1189	0.2896	0.75	1.38	0.54	1.16	0.71	0.14	*Normal*
y_2	0.0802	0.2292	0.83	1.74	0.47	0.86	0.97	0.11	*2nd inlet*
y_3	0.2438	0.6403	0.44	1.35	0.54	1.77	1.12	0.25	*2nd plunger*
y_4	0.0837	0.1604	1.24	1.64	0.32	1.47	0.97	0.27	*4th exhaust*
y_5	0.288	0.5009	0.62	1.66	0.28	1.94	1.3	0.31	*2nd 3rd inlet*
y_6	0.2354	0.5697	0.6	1.58	0.3	1.9	1.4	0.33	*2nd 3rd inlet*
y_7	0.2601	0.7211	0.73	1.37	0.6	1.81	0.8	0.31	*2nd plunger*
y_8	0.2106	0.5785	0.55	1.24	0.4	1.7	1.2	0.26	*2nd plunger*
y_9	0.0755	0.1845	0.8	1.65	0.5	0.91	0.7	0.13	*2nd inlet*
y_{10}	0.0782	0.1771	1.05	1.9	0.45	1.25	0.84	0.25	*4th exhaust*
y_{11}	0.0775	0.1784	1.06	2.1	0.35	1.08	0.6	0.24	*4th exhaust*

TABLE 10: Discretization attributes of the samples in Table 9.

Samples	c_1	c_2	c_3	c_4	c_5	c_6	c_7	c_8
y_1	2	1	1	1	2	1	1	1
y_2	2	1	2	2	2	1	1	1
y_3	2	2	1	1	2	2	2	2
y_4	2	1	2	2	1	2	1	2
y_5	2	2	1	2	1	2	2	2
y_6	2	2	1	2	1	2	2	2
y_7	2	2	1	1	2	2	1	1
y_8	2	2	1	1	1	2	2	2
y_9	1	1	2	2	2	1	1	1
y_{10}	1	1	2	2	2	2	1	2
y_{11}	1	1	2	2	1	1	1	2

TABLE 11: Diagnosis results for different fault types.

Samples	c_2	c_4	c_8	D	Diagnostic results
y_1	*	1	1	$d_1 = 1$	*Normal*
y_2	1	2	1	$d_2 = 1$	*2nd inlet*
y_3	2	1	2	$d_3 = 3$	*2nd plunger*
y_4	1	*	2	$d_4 = 4$	*4th exhaust*
y_5	2	2	2	$d_5 = 5$	*2nd 3rd inlet*
y_6	2	2	2	$d_5 = 5$	*2nd 3rd inlet*
y_7	2	1	2	$d_3 = 3$	2nd plunger
y_8	2	1	2	$d_3 = 3$	*2nd plunger*
y_9	1	2	1	$d_2 = 2$	*2nd inlet*
y_{10}	1	*	2	$d_4 = 4$	*4th exhaust*
y_{11}	1	*	2	$d_4 = 4$	*4th exhaust*

real industrial object. Results show that the new approach can effectively identify different operating states of the pump, which will supply as a basis for the detection and diagnosis of the pump faults.

Acknowledgments

This research was supported by science research plan in Education Department of Shaanxi Province of China (Program no. 07JK365). The author is grateful to Tuha Petroleum Company of China for their kind provision of access to the five-plunger pump and the test-bed. The author would also like to express her appreciation for the useful suggestions put forward by Professors Wei Wang and Xiaohong Zhao of Xi'an Shiyou University.

References

[1] F. Kong and R. Chen, "A combined method for triplex pump fault diagnosis based on wavelet transform, fuzzy logic and neuro-networks," *Mechanical Systems and Signal Processing*, vol. 18, no. 1, pp. 161–168, 2004.

[2] L. J. de Miguel and L. F. Blázquez, "Fuzzy logic-based decision-making for fault diagnosis in a DC motor," *Engineering Applications of Artificial Intelligence*, vol. 18, no. 4, pp. 423–450, 2005.

[3] L. Tong, Y. Chen, Z. Wang, and Q. Qing, "Fusion diagnosis of diesel engine based on a D-S evidence theory," *Science Technology and Engineering*, vol. 10, no. 15, pp. 3750–3753, 2010.

[4] C. R. Parikh, M. J. Pont, and N. Barrie Jones, "Application of Dempster-Shafer theory in condition monitoring applications: a case study," *Pattern Recognition Letters*, vol. 22, no. 6-7, pp. 777–785, 2001.

[5] Z. Pawlak, "Rough sets," *International Journal of Computer & Information Sciences*, vol. 11, no. 5, pp. 341–356, 1982.

[6] F. Kong and R. Chen, "Vibration characteristic analysis on the fluid-end of triplex drilling pump," *Oil Field Equipment*, vol. 25, no. 1, pp. 41–45, 1996.

[7] F. Kong and R. Chen, "Vibration monitoring and fault diagnosis on dynamic-end of triplex drilling pump," *Oil Field Equipment*, vol. 25, no. 2, pp. 19–25, 1996.

[8] M. H. El-Ghamry, R. L. Reuben, and J. A. Steel, "The development of automated pattern recognition and statistical feature isolation techniques for the diagnosis of reciprocating machinery faults using acoustic emission," *Mechanical Systems and Signal Processing*, vol. 17, no. 4, pp. 805–823, 2003.

[9] F. Bie, Y. Liu, G. Zhou, and F. Lü, "Research on fault prediction approach for reciprocating machinery based on local wave method and SVM prediction model," *China Mechanical Engineering*, vol. 22, no. 6, pp. 687–691, 2011.

[10] J. D. Wu, P. H. Chiang, Y. W. Chang, and Y. J. Shiao, "An expert system for fault diagnosis in internal combustion engines using probability neural network," *Expert Systems with Applications*, vol. 34, no. 4, pp. 2704–2713, 2008.

[11] J. Pei, S. Zhang, M. Qi, and G. Wan, "A new method for fault diagnosis of fluid end in drilling pump," *Acta Petrolei Sinica*, vol. 30, no. 4, pp. 617–621, 2009.

[12] C. Wang, W. Li, and Z. Wu, "Fault diagnosis approach based on RBF neural network for valves of reciprocating pumps," *Oil Field Equipment*, vol. 40, no. 1, pp. 24–27, 2011.

[13] S.-Q. Wu and W. Jiang, "Multi source and multi feature fused fault diagnosis on piston pumps based on evidence theory," *Chinese Journal of Construction Machinery*, vol. 9, no. 1, pp. 98–103, 2011.

[14] Z. Pawlak, *Rough Sets, Theoretical Aspects of Reasoning about Data*, Springer, Warsaw, Poland, 1990.

[15] Z. Pawlak, "Rough classification," *International Journal of Human Computer Studies*, vol. 51, no. 2, pp. 369–383, 1999.

[16] Z. Geng and Q. Zhu, "Rough set-based heuristic hybrid recognizer and its application in fault diagnosis," *Expert Systems with Applications*, vol. 36, no. 2, pp. 2711–2718, 2009.

[17] Q. Y. Sun, H. G. Zhang, and J. Dai, "On-line fault diagnose of distribution system based on modified rough sets reduction algorithm," *Proceedings of the Chinese Society of Electrical Engineering*, vol. 27, no. 7, pp. 58–64, 2007.

[18] Q. Wang and R. Cai, "Application of probabilistic rough set models to mechanical fault diagnosis," *Computer Engineering and Applications*, vol. 45, no. 28, pp. 222–224, 2009.

20

Measurement and Model Validation of Nanofluid Specific Heat Capacity with Differential Scanning Calorimetry

Harry O'Hanley, Jacopo Buongiorno, Thomas McKrell, and Lin-wen Hu

Massachusetts Institute of Technology, Cambridge, MA 02139, USA

Correspondence should be addressed to Harry O'Hanley, hohanley@mit.edu

Academic Editor: Kambiz Vafai

Nanofluids are being considered for heat transfer applications; therefore it is important to know their thermophysical properties accurately. In this paper we focused on nanofluid specific heat capacity. Currently, there exist two models to predict a nanofluid specific heat capacity as a function of nanoparticle concentration and material. Model I is a straight volume-weighted average; Model II is based on the assumption of thermal equilibrium between the particles and the surrounding fluid. These two models give significantly different predictions for a given system. Using differential scanning calorimetry (DSC), a robust experimental methodology for measuring the heat capacity of fluids, the specific heat capacities of water-based silica, alumina, and copper oxide nanofluids were measured. Nanoparticle concentrations were varied between 5 wt% and 50 wt%. Test results were found to be in excellent agreement with Model II, while the predictions of Model I deviated very significantly from the data. Therefore, Model II is recommended for nanofluids.

1. Introduction

Nanofluids are unique fluids consisting of nanoparticles suspended in a base fluid. They are currently being considered for use by a wide range of industries from energy to manufacturing to medicine.

The ability to customize the characteristics of the nanofluids—through particle material selection—makes them attractive candidates for heat transfer applications. For example, nanofluids can be used as a medium to more effectively transfer energy captured from solar arrays or cool nuclear reactors.

Recent research has indicated that dispersions of nanoparticles in a base fluid, known as nanofluids, can increase the boiling critical heat flux and overall performance of thermal systems. Typical nanoparticle concentrations may range from 0.01 wt% to 50 wt%, and common particle materials include silica, alumina, copper oxide, zirconia, and carbon nanotubes [1–3]. Water often serves as the base fluid though other liquids such as ethylene glycol have been used [1].

As nanofluids are being considered for thermal applications, it is necessary to be able to predict their thermophysical properties accurately. Because nanofluids were initially considered for thermal conductivity enhancement, this property has been extensively studied [2–4]. However, there have been fewer examinations of nanofluid specific heat capacity [5, 6]. It is the objective of this paper to complement existing research by (i) measuring the specific heat capacity of water-based silica, alumina, and copper oxide nanofluids and (ii) comparing the predictions of two popular nanofluid specific heat capacity models to data.

2. Specific Heat Models

There are two specific heat models widely used in the nanofluid literature. Model I is similar to mixing theory for ideal gas mixtures [3]. It is a straight average relating nanofluid specific heat, $c_{p,nf}$, to basefluid specific heat, $c_{p,f}$, nanoparticle specific heat, $c_{p,n}$, and volume fraction, φ. Using these parameters, Model I calculates the nanofluid specific heat as

$$c_{p,nf} = \varphi c_{p,n} + (1 - \varphi) c_{p,f}. \quad (1)$$

While it is simple and thus commonly used, Model I has little theoretical justification in the context of nanofluids.

Model II [3, 7] is based on the assumption of thermal equilibrium between the particles and the surrounding fluid. Starting from an arbitrary mass of nanofluid m_{nf}, of volume V_{nf} and with a nanoparticle volumetric fraction φ, the nanofluid density is obviously $\rho_{nf} = m_{nf}/V_{nf} = \varphi\rho_n + (1 - \varphi)\rho_f$, where the particle and fluid densities are ρ_n and ρ_f, respectively. The energy required to elevate the nanofluid mass is $\varphi V_{nf}(\rho c_p)_n + (1 - \varphi)V_f(\rho c_p)_f$. Therefore, the nanofluid specific heat capacity per unit mass of nanofluid, that is, the nanofluid specific heat, is

$$c_{p,nf} = \frac{\varphi(\rho c_p)_n + (1 - \varphi)(\rho c_p)_f}{\varphi\rho_n + (1 - \varphi)\rho_f}. \quad (2)$$

A rigorous derivation of (2) is presented also in [8].

Predictions of nanofluid specific heat capacity were made using both models and compared to experimental measurements. Water was the base fluid of all nanofluids used in this investigation. Therefore, handbook values of temperature-dependent water specific heat and density were used in calculating theoretical nanofluid specific heat [9]. Additionally, the specific heat and density of the nanoparticles were assumed to be equal to the respective thermophysical properties of particle material in bulk form.

3. Nanofluids

The specific heat capacities of three nanofluids were analyzed: alumina-water (Nyacol AL20DW), silica-water (Ludox TMA 420859), and copper-oxide-water (Alfa Aesar 45407). Selected nanofluid properties are presented in Table 1.

The stock nanofluids were obtained from commercial vendors and diluted with deionized water to vary their concentrations. Prior to mixing, the nanofluids were manually agitated to ensure uniform dispersion. Dilution was performed by weight percent using a Mettler Toledo XS105 balance. Four unique concentrations were prepared for each nanofluid and are listed in Table 2. For each concentration, two identical samples were prepared and tested.

While nanofluids were diluted and prepared according to their weight fraction, calculations were performed using volume fraction. Using the nanoparticle volume, V_n, and the water volume, V_{H_2O}, the volume fraction can be calculated as

$$\varphi = \frac{V_n}{V_n + V_{H_2O}}. \quad (3)$$

Substituting in nanoparticle mass, m_n, and density ρ_n, and water mass, m_{H_2O}, and density, ρ_{H_2O} (3) can be rewritten as

$$\varphi = \frac{m_n/\rho_n}{m_n/\rho_n + m_{H_2O}/\rho_{H_2O}}. \quad (4)$$

TABLE 1: Nanofluid.

Nanofluid	Particle size (nm)	pH	Specific gravity
NYACOL AL20DW	50 nm	4.0	1.19
Ludox TMA 420859	32 nm	4.0–7.0	1.22–1.24
Alfa Aesar 45407	30 nm	4.6	1.50

TABLE 2: Nanofluid sample concentrations.

Nanofluid	Alumina-water	Silica-water	Copper-oxide-water
Conc. 1 (stock)	20 wt% (6.4 vol%)	34 wt% (19.0 vol%)	50 wt% (13.7 vol%)
Conc. 2	15 wt% (4.6 vol%)	25.5 wt% (13.5 vol%)	37.5 wt% (8.7 vol%)
Conc. 3	10 wt% (2.9 vol%)	15 wt% (8.5 vol%)	25 wt% (5.0 vol%)
Conc. 4	5 wt% (1.4 vol%)	8.5 wt% (4.1 vol%)	12.5 wt% (2.2 vol%)

Equation (4) can be used to determine the nanoparticle volume fraction of nanofluid concentrations created by dilution with deionized water.

4. Measurement Method

A heat-flux-type differential scanning calorimeter (TA Instruments Q2000) was used to measure the nanofluid specific heat capacities. The differential scanning calorimeter (DSC) measures the heat flux into a sample as a function of temperature during a user-prescribed heating regime. It accomplishes this by comparing the heat flux into a pan containing the sample with the heat flux into an empty pan. Hermetically sealed aluminum pans (TA Instruments) were used in the tests presented here.

The classical three-step DSC procedure was followed to measure specific heat capacity [10, 11]. Additionally, testing procedures adhered to protocols set forth in the ASTM Standard Test Method for Determining Specific Heat Capacity by Differential Scanning Calorimetry (E 1269-05).

The three-step DSC procedure begins with designing a heating regime, which should contain the temperature range of interest. Next, a measurement is taken with two empty sample pans loaded into the DSC. During this measurement, the baseline heat flux, Q_0, is obtained. The results of this measurement indicate the bias in the machine, allowing for it to be accounted for during data reduction.

The second measurement is of a reference sample, with a known specific heat, $c_{p,\mathrm{ref}}$. A pan containing the reference sample and an empty pan are loaded into the DSC. The heat flux into the reference sample, Q_{ref}, is recorded throughout the identical heating regime.

The third measurement is made on the actual sample of interest. A pan containing the sample and an empty pan are loaded into the DSC. The heat flux into the sample, Q_{sample}, is recorded during an identical heating regime as the previous two measurements. The heat flux curves from the three measurements are used to comparatively determine the specific heat of the sample, $c_{p,\text{sample}}$, where

$$c_{p,\text{sample}} = \frac{Q_{\text{sample}} - Q_0}{Q_{\text{ref}} - Q_0} \frac{m_{\text{ref}}}{m_{\text{sample}}} c_{p,\text{ref}} \quad (5)$$

and m_{ref} and m_{sample} represent the masses of the reference and sample, respectively. Sample masses were measured using a Perkin Elmer AD6 autobalance.

In these tests, deionized water was used as the reference sample, with specific heat values obtained from Perry's Chemical Engineers Handbook [9]. The DSC heating procedure consisted of three steps:

(1) equilibrate and remain isothermal at 25°C for one minute,

(2) ramp to 75°C at 10°C/min,

(3) remain isothermal at 75°C for one minute.

Heat flux measurement was continuous from 25°C to 75°C. However, for analysis, specific heat capacities were calculated at 35°C, 45°C, and 55°C. For each sample of nanofluid concentration, three measurements were taken. These values were then averaged to yield the data points and related standard deviations presented in Figures 1, 2, 3, 4, 5, 6, 7, 8, and 9.

Prior to nanofluid measurement, this DSC methodology was validated by analyzing two pure fluids, ethylene glycol and glycerin. The results obtained from these measurements were compared against handbook values of specific heat for these liquids [12]. All experimental data are reported in Section 6.

5. Uncertainty Analysis

The measurement uncertainty for the specific heat was calculated by propagating the precision uncertainties of all individual measurements required to determine the specific heat. Equation (5) suggests there is a precision error associated with the heat flux measurements as well as the mass measurements. If all variables are assumed to be distributed normally, the overall measurement uncertainty, μ_{total}, is

$$\mu_{\text{total}} = \sqrt{\sum_i^n \left(\frac{\partial c_{p,\text{sample}}}{\partial x_i} u_i \right)^2}. \quad (6)$$

Here, x_i is the independent variable to be estimated and μ_i is the manufacturer reported precision of the measurement method. Here, (5) is substituted into (6) and the appropriate derivatives are taken to capture all measurement uncertainties. In this instance, each specific heat capacity measurement requires three heat flux measurements and two mass measurements. The associated precision uncertainties are reported in Table 3.

From this analysis, the precision uncertainty in the measurement of specific heat capacity for Model II was found to be about 0.07 (J/g-K).

Table 3: Reported measurement uncertainties.

x_i	μ_i (±)	Equipment type
Q_s	1.2 mW	TA Instruments Q2000
Q_{ref}	3.3 mW	TA Instruments Q2000
Q_0	10 μW	TA Instruments Q2000
m_{sample}	0.2 μg	Perkin Elmer AD6
m_{sample}	0.2 μg	Perkin Elmer AD6

Table 4: Propylene glycol.

Temperature (°C)	Theoretical c_p (J/g-K)	Measured c_p (J/g-K)
35	2.56	2.54 ± 0.191
45	2.62	2.64 ± 0.185
55	2.65	2.65 ± 0.185

Table 5: Glycerin.

Temperature (°C)	Theoretical c_p (J/g-K)	Measured c_p (J/g-K)
35	2.39	2.41 ± 0.008
45	2.41	2.42 ± 0.002
55	2.42	2.44 ± 0.001

6. Results and Discussion

DSC measurements of pure ethylene glycol and glycerin were in good agreement with literature values of specific heat capacity. The results of these tests, presented in Tables 4 and 5, validate the DSC methodology and machine calibration.

Figures 1, 2, 3, 4, 5, 6, 7, 8, and 9 show the nanofluids data and the curves predicted by Models I and II. As expected, as nanoparticle concentration increases, the specific heat capacity decreases. However, Model I largely underestimates the decrease, while Model II offers a much more accurate prediction of nanofluid specific heat capacity. These conclusions are consistent with the alumina nanofluid results reported in [3] and expand the validity of Model II to silica and copper oxide nanofluids. In [3], the researchers also found there to be a sharp departure from Model I with increasing nanoparticle volume fraction.

Even for Model II, there appear to be small discrepancies between the data and predictions. These could come from errors in the listed stock nanofluid concentrations, experimental uncertainties in dilution, or inconsistencies in using the bulk material properties in the model, instead of the actual nanoparticle properties. Recent research suggests that these properties may differ if the material is in nanoparticle form versus bulk form [8]. Investigation of these effects is left for future work.

7. Conclusions

Using a heat flux differential scanning calorimeter (DSC), the specific heat capacities of water-based silica, alumina, and copper oxide nanofluids at various nanoparticle concentrations were measured. The DSC procedure was validated by measuring the specific heat capacities of pure ethylene

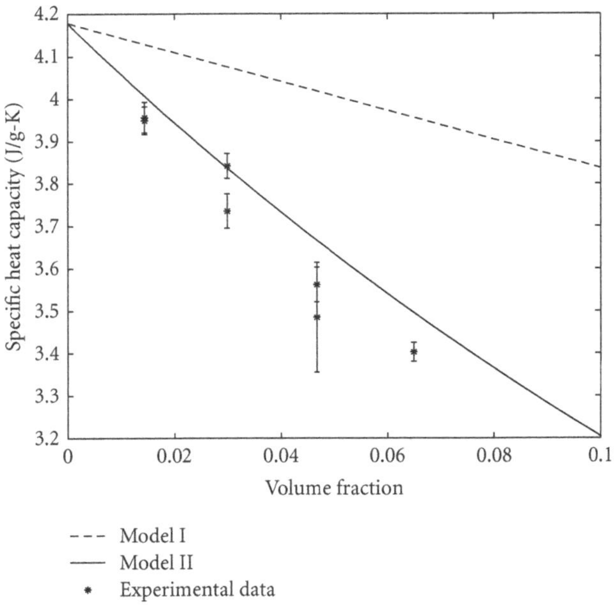

FIGURE 1: Alumina-water at 35°C.

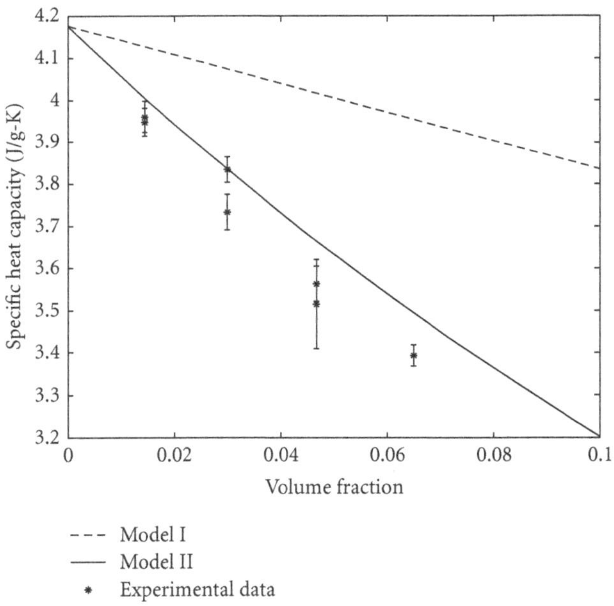

FIGURE 2: Alumina-water at 45°C.

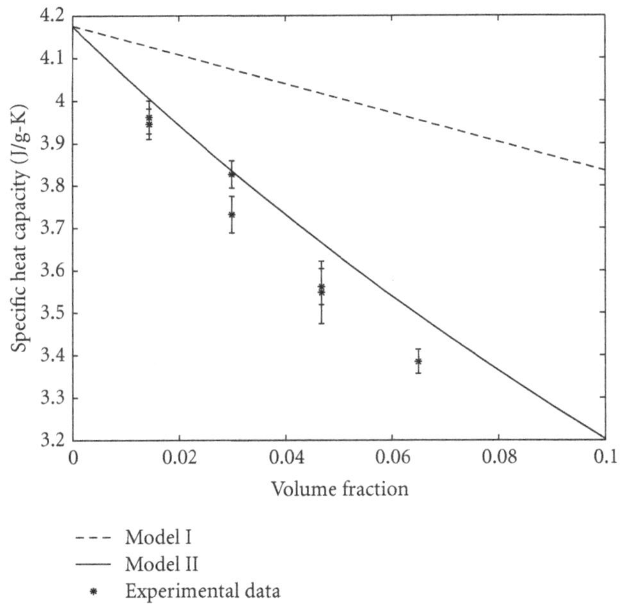

FIGURE 3: Alumina-water at 55°C.

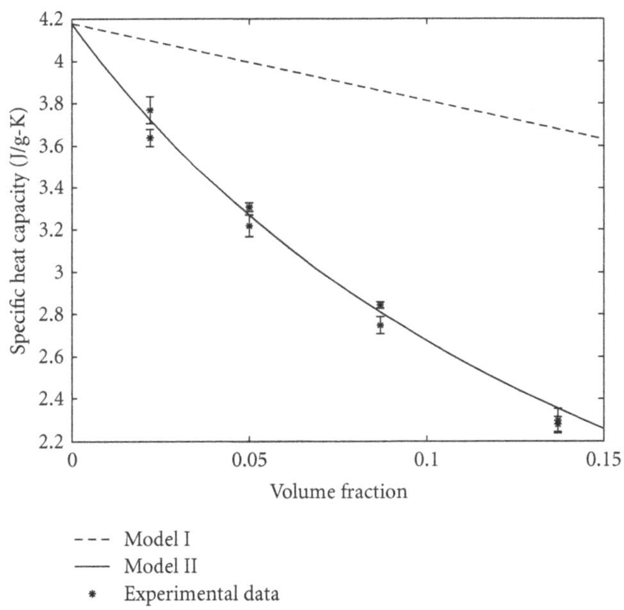

FIGURE 4: Copper-oxide-water at 35°C.

glycol and glycerin, which were confirmed against handbook values. The nanofluid data were used to test the predictions of two popular mixture models for specific heat. The results clearly suggest that the model based on particle/fluid thermal equilibrium (Model II) yields more accurate predictions than the model based on a straight volume-weighted average of the particle and fluid specific heats (Model I). Given its sound theoretical basis, we believeModel II is generally applicable, while Model I should be abandoned. To further improve the accuracy of Model II, future investigations could focus on measuring the actual density and specific heat of the nanoparticles in dispersion and compare them to those of the bulk materials. Additionally, future research to correlate specific heat capacity with agglomeration and sedimentation would be particularly beneficial, as such phenomena are typically unavoidable in certain applications.

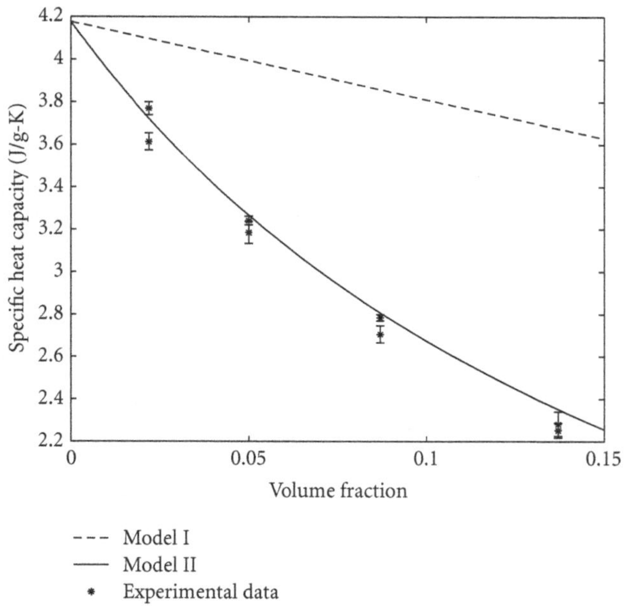

FIGURE 5: Copper-oxide-water at 45°C.

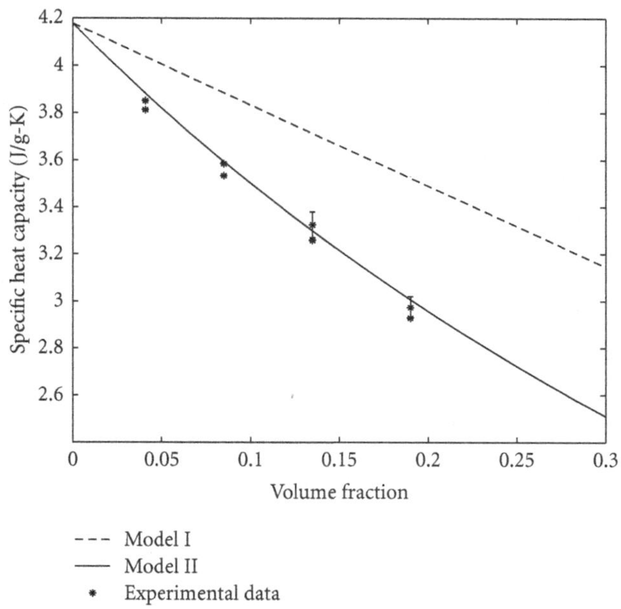

FIGURE 7: Silica-water at 35°C.

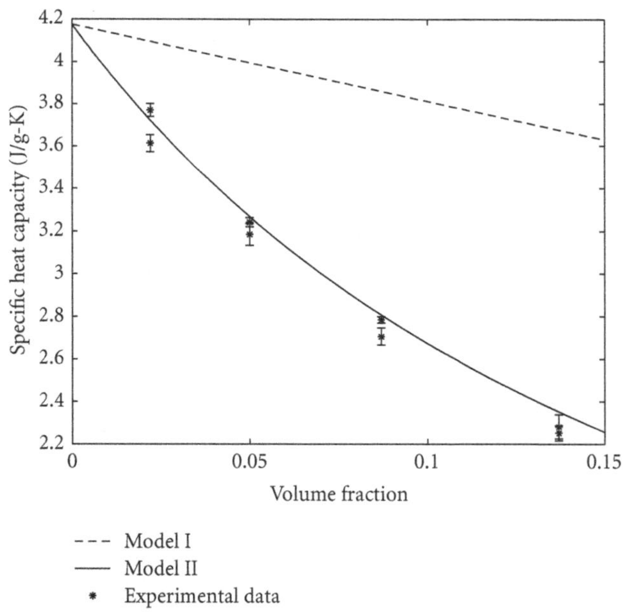

FIGURE 6: Copper-oxide-water at 55°C.

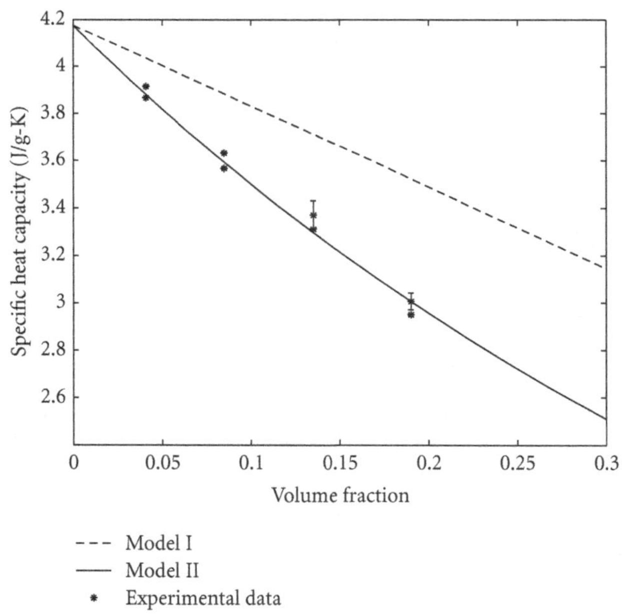

FIGURE 8: Silica-water at 45°C.

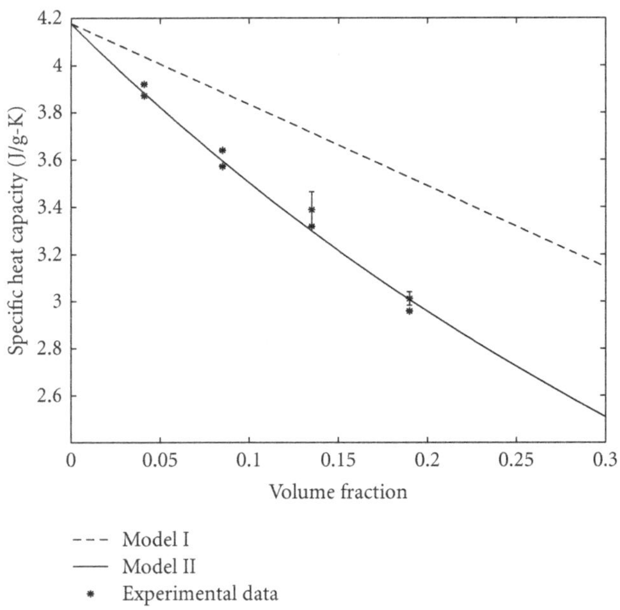

FIGURE 9: Silica-water at 55°C.

Nomenclature

$c_{p,f}$: Specific heat capacity of base fluid (J/g-K)
$c_{p,nf}$: Specific heat capacity of nanofluid (J/g-K)
$c_{p,n}$: Specific heat capacity of nanoparticle (J/g-K)
$c_{p,\mathrm{ref}}$: Specific heat capacity of reference (J/g-K)
$c_{p,\mathrm{sample}}$: Specific heat capacity of sample (J/g-K)
m_{nf}: Mass of nanofluid (g)
m_n: Mass of nanoparticles (g)
$m_{\mathrm{H_2O}}$: Mass of water (g)
m_{ref}: Mass of reference (g)
m_{sample}: Mass of sample (g)
Q_{ref}: Heat flux into reference (Watts)
Q_{sample}: Heat flux into sample (Watts)
Q_0: Heat flux baseline (Watts)
Φ: Volume fraction (unitless)
ρ_f: Density of basefluid (g/cm^3)
ρ_n: Density of nanoparticles (g/cm^3)
$\rho_{\mathrm{H_2O}}$: Density of water (g/cm^3)
V_{nf}: Volume of nanofluid (cm^3)
V_n: Volume of nanoparticles (cm^3)
$V_{\mathrm{H_2O}}$: Volume of water (cm^3).

Acknowledgments

The Materials Research Science and Engineering Center at Harvard University offered their DSC for use in this paper. Within this facility, Dr. Philseok Kim, Dr. Kosta Ladavac, and Roxanne Kimo offered invaluable instruction and advice on the machine operation.

References

[1] K. Kwak and C. Kim, "Viscosity and thermal conductivity of copper oxide nanofluid dispersed in ethylene glycol," *Korea Australia Rheology Journal*, vol. 17, no. 2, pp. 35–40, 2005.

[2] J. Buongiorno, D. C. Venerus, N. Prabhat et al., "A benchmark study on the thermal conductivity of nanofluids," *Journal of Applied Physics*, vol. 106, no. 9, Article ID 094312, 2009.

[3] S. Q. Zhou and R. Ni, "Measurement of the specific heat capacity of water-based Al_2O_3 nanofluid," *Applied Physics Letters*, vol. 92, no. 9, Article ID 093123, 2008.

[4] S. Sinha, S. Barjami, G. Iannacchione, and S. Sinha, "Thermal properties of carbon nanotube based fluids," in *Proceedings of the Memphis-Area Engineering and Sciences Conference*, 2004.

[5] Q. He, W. Tong, and Y. Liu, "Measurement of specific heat of TiO_2-$BaCl_2$-H_2O nano-fluids with DSC," *Air Conditioning & Refrigeration*, vol. 7, no. 5, pp. 19–22, 2007 (Chinese).

[6] X. Peng, X. Yu, and F. Yu, "Experiment study on the specific heat of nanofluids," *Journal of Materials Science & Engineering*, vol. 25, no. 5, pp. 719–722, 2007 (Chinese).

[7] J. Buongiorno, "Convective transport in nanofluids," *Journal of Heat Transfer*, vol. 128, no. 3, pp. 240–250, 2006.

[8] L. P. Zhou, B. X. Wang, X. F. Peng, X. Z. Du, and Y. P. Yang, "On the specific heat capacity of CuO nanofluid," *Advances in Mechanical Engineering*, vol. 2010, Article ID 172085, 4 pages, 2010.

[9] D. Green and R. Perry, *Perry's Chemical Engineers' Handbook*, McGraw-Hill, New York, NY, USA, 8th edition, 2008.

[10] G. Hohne, *Differential Scanning Calorimetry*, Springer, New York, NY, USA, 2003.

[11] U.S. Department of Defense, *Military Handbook—MIL-HDBK-17-3F: Composite Materials Handbook, Volume 3—Polymer Matrix Composites Materials Usage, Design, and Analysis* , U.S. Department of Defense.

[12] C. Yaws, *Chemical Properties Handbook*, McGraw-Hill, New York, NY, USA, 1999.

CFD Simulation of Flow in an Abrasive Water Suspension Jet: The Effect of Inlet Operating Pressure and Volume Fraction on Skin Friction and Exit Kinetic Energy

D. Deepak, D. Anjaiah, K. Vasudeva Karanth, and N. Yagnesh Sharma

Department of Mechanical and Manufacturing Engineering, Manipal Institute of Technology, Manipal University, Manipal 576104, India

Correspondence should be addressed to N. Yagnesh Sharma, yagnesh.sharma@manipal.edu

Academic Editor: Koshi Adachi

Abrasive particles in the suspension mixture in an abrasive water suspension jet (AWSJ) machining causes acute skin friction effect thereby effectively changing the jet diameter due to wear, which in turn influences jet exit kinetic energy. This results in lowering the life of the jet for effective machining. In consideration of this aspect, the present work examines the effect of inlet pressure on skin friction coefficient and jet exit kinetic energy. It is inferred from the analysis that an increase in inlet pressure causes a significant increase in skin friction coefficient and also results in proportional increase in the exit kinetic energy of the jet. Further, it is revealed from the analysis that an increase volume fraction of abrasive (abrasive concentration) in water results in significant decrease in the skin friction coefficient and jet exit kinetic energy.

1. Introduction

Advances in research and development of new engineering materials are indispensable to parallel development of machining methods to process such materials. Machine components with complex shapes that need to be produced from brittle and heat-sensitive materials as well as composites can now be machined by an advanced manufacturing method called abrasive water jet (AWJ) machining. Through computer numerical control of jet movement on the work part, complex profiles with better surface quality and precision can be achieved. Abrasive water suspension jet (AWSJ) is one of the variants of AWJ machining in which suspended abrasive particles in a liquid medium called slurry are pressurized and expelled through the nozzle. Benefit of AWSJ over AWJ is the generation of stable jet with higher power density, which leads to efficient energy transfer to abrasive particles [1–4].

Nozzle wear is a complex phenomenon, which is not only influenced by the material properties of the nozzle but also by the nozzle geometry and operating parameters of AWSJ [5–8]. A host of articles is available for both experimental and numerical aspects of two-phase flow [9–38]. AWSJ machining operates at relatively high pressure (10–1000 MPa). Slurry is accelerated through a fine orifice (0.20–1 mm diameter) to produce a high-velocity stream, which is capable of machining a range of materials. The abrasive particles moving with the corresponding high velocity of flow cause severe wall shear in the nozzle. This causes erosion of the nozzle, due to which the effective diameter of the nozzle may change significantly resulting in reduced exit kinetic energy of the jet. Study of wear characteristics of the nozzle is critical for the growth of AWJ technology. In consideration of this aspect, the present work examines the effect of volume fraction and inlet pressure of suspension mixture on the skin friction effect and jet kinetic energy.

2. Theoretical Formulation

2.1. Numerical Model and Assumptions. Computational domain consists of flow field through abrasive suspension nozzle connected to the focus tube as described in Figure 1. The abrasive water suspension mixture is let into the nozzle at the inlet and is carried down through the converging cone to the focus tube and exits as coherent jet at the nozzle exit, in which the focus tube is used for stabilizing the flow. The numerical model adopted closely follows the work of Hu et al. [9] in which liquid-solid two-phase flow is considered, and the following assumptions are valid for the present work.

(i) Flow is taken to be two-phase flow in which the primary liquid phase mixes homogeneously with the particles of equal diameter, constituting the solid phase.

(ii) The primary liquid phase is continuous and incompressible.

(iii) Two-phase flow assumed is steady and characterized by turbulent flow.

2.2. The Computation of Particulate Loading (β) and Stokes Number (S_t). Particulate loading and the Stokes number are important parameters that help to identify the appropriate multiphase model. Particulate loading has a major impact on phase interactions and is defined as the mass density ratio of the dispersed phase (d) to that of the carrier phase (c). The particulate loading for garnet abrasive is

$$\beta = \frac{\alpha_s \rho_s}{\alpha_l \rho_l} = \frac{0.1 \times 2300}{998} = 0.230. \tag{1}$$

The degree of interaction between the phases is intermediate loading, and the coupling is two ways, that is, the fluid carrier influences the particulate phase via drag and turbulence, but the particles in turn influence the carrier fluid via reduction in mean momentum and turbulence. All multiphase models can handle this type of problem, but it is found that the Eulerian multiphase model seems to be the most accurate one [27]. The average distance between the individual particles of the particulate phase can be estimated by equation:

$$\gamma = \frac{\rho_s}{\rho_l} = \frac{2300}{998} = 2.3. \tag{2}$$

From (1) and (2),

$$k = \frac{\beta}{\gamma} = \frac{0.230}{2.30} = 0.100. \tag{3}$$

From (3),

$$\frac{L}{d_p} = \left(\frac{\pi}{6}\frac{1+k}{k}\right)^{1/3} = \left(\frac{\pi}{6}\frac{1+0.10}{0.10}\right)^{1/3} = 1.7925. \tag{4}$$

The average distance between the individual particles of the particulate phase is calculated from (4) for an abrasive particle size of $d_p = 63\,\mu$m:

$$L = 1.7925 \times d_p = 0.1129\,\text{mm}. \tag{5}$$

Estimating the value of the Stokes number helps to select the most appropriate multiphase model. The Stokes number

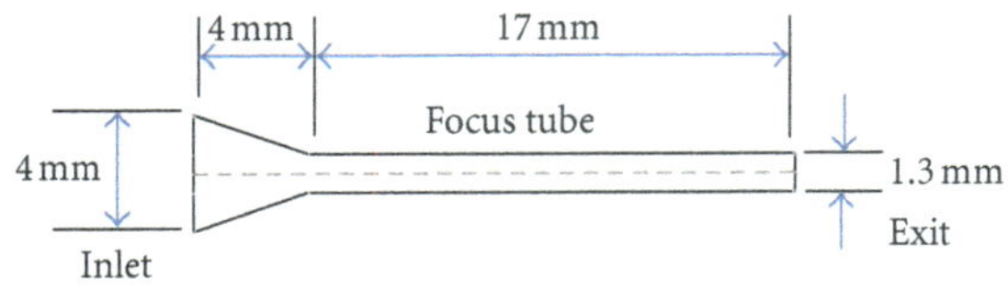

FIGURE 1: Computational model of AWSJ nozzle.

is defined as the ratio of the particle response time (τ_d) to the system response time (t_s). These are obtained from

$$\tau_d = \frac{\rho_d d_d^2}{18\mu_l} = 5.05129 \times 10^{-4},$$
$$t_s = \frac{l}{v} = 1.4218 \times 10^{-3}, \tag{6}$$
$$S_t = \frac{\tau_d}{t_s} = 0.3552.$$

For S_t less than unity, particles will closely follow the fluid flow, and any one of the three multiphase models, namely, Volume of fluid model, mixture model, or eulerian multiphase model, is applicable. Also from the calculation of the effect of particulate loading (β), it is clear that coupling between two phases is intermediate [27]. Hence, present numerical simulation is carried out using Eulerian multiphase model which though is most expensive in computation, it gives most accurate results. Eulerian Multiphase model is embedded in FLUENT software. FLUENT solves a multifluid granular model to describe the flow behavior of fluid solid mixture. The stresses induced in the solid phase are deduced through an analogy between the random particle motion arising from particle to particle collisions and the thermal gradient of molecules in the fluid stream taking into effect the inelasticity of the granular phase. Intensity of the particle velocity fluctuations determines the stresses, viscosity, and pressure of the solid phase [27].

The governing equations for mass and momentum conservation are solved for the steady incompressible flow. The finite-volume method [10] was used for the numerical solution of the system of equations. The algorithm used for the solution of the equations is the interphase slip algorithm (IPSA) [11, 12]. The linkage between pressure and velocity is handled via a more efficient version of the well-known semi-implicit method for pressure-linked equations (SIMPLEs) algorithm developed by Patankar and Spalding [10] which is called SIMPLE shortened (SIMPLEST) [13]. The turbulence is modeled using realizable k-ε turbulence model [27]. The power law scheme was used for obtaining the solution [28]. The governing partial differential equations, for mass and momentum conservations, are detailed below.

Continuity Equation. Consider the following:

$$\frac{1}{\rho_{pq}}\left[\frac{\partial}{\partial t}\left(\alpha_q \rho_q\right) + \nabla \cdot \left(\alpha_q \rho_q v_q\right)\right] = \sum_{p=1}^{N}\left(m_{pq} - m_{qp}\right). \tag{7}$$

Fluid-Solid Momentum Equation. The conservation of momentum equation for the solid phase is as follows:

$$\frac{\partial}{\partial t}(\alpha_s \rho_s v_s) + \nabla \cdot (\alpha_s \rho_s v_s^2) = -\alpha_s \nabla p - \nabla p_s + \nabla \cdot \tau_s + \alpha_s \rho_s g + \sum_{l=1}^{N} [k_{ls}(v_l - v_s) + (m_{ls} v_{ls} - m_{sl} v_{sl})] + (F_s + F_{\text{lift},s} + F_{vm,s}). \quad (8)$$

The conservation of momentum equation for the fluid phase is as follows:

$$\frac{\partial}{\partial t}(\alpha_q \rho_q v_q) + \nabla \cdot (\alpha_q \rho_q v_q^2) = -\alpha_q \nabla p + \nabla \cdot \tau_q + \alpha_q \rho_s g + \sum_{p=1}^{N} \left[k_{pq}(v_p - v_q) + (m_{pq} v_{pq} - m_{qp} v_{qp})\right] + \left(F_q + F_{\text{lift},q} + F_{vm,q}\right). \quad (9)$$

3. Method of Solution

3.1. Numerical Scheme. Conservation equations are solved for each control volume to obtain the velocity and pressure fields. Convergence is effected when all the dependent variable residuals fall below 0.00001 at all grid points. Computational domain is modeled using commercially available preprocessor routine called GAMBIT, and meshing is carried out using appropriate grid cells of suitable size available in the software routine. Wall region in the flow domain is closely meshed using the boundary layer mesh concepts for extracting high-velocity gradients near the boundary walls. According to the structure of nozzle and jet characteristics, computational domain is built as axisymmetric model. Figure 2 shows the computational domain. The grid independence test is performed to check the quality of mesh for solution convergence as shown in Figure 3. It is clear from the grid independence study that there is negligible variation of not more than 1%, in the axial velocity distribution for the three mesh geometries consisting of 21200, 25440, and 29690 control volumes. Hence, considering lower computational time desired, an optimum mesh geometry consisting of 25440 control volumes has been adopted in this work. The runs were performed on computer having Intel Xeon central processing unit with 4 GB RAM and clock speed of 3.10 GHz. A typical run requires 4 hours of CPU time on the above computer.

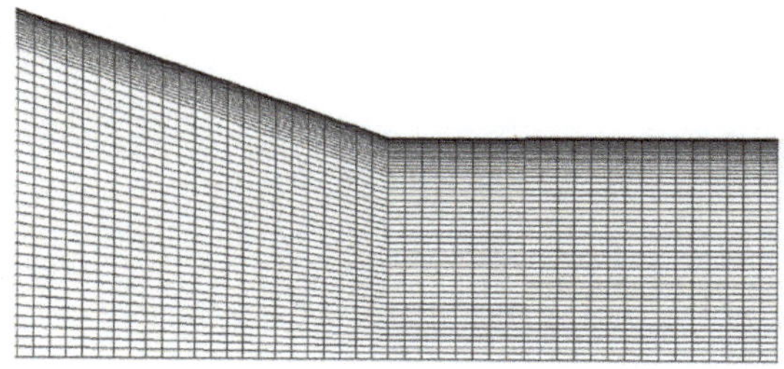

Figure 2: A portion of the meshed domain near the critical section of AWSJ nozzle.

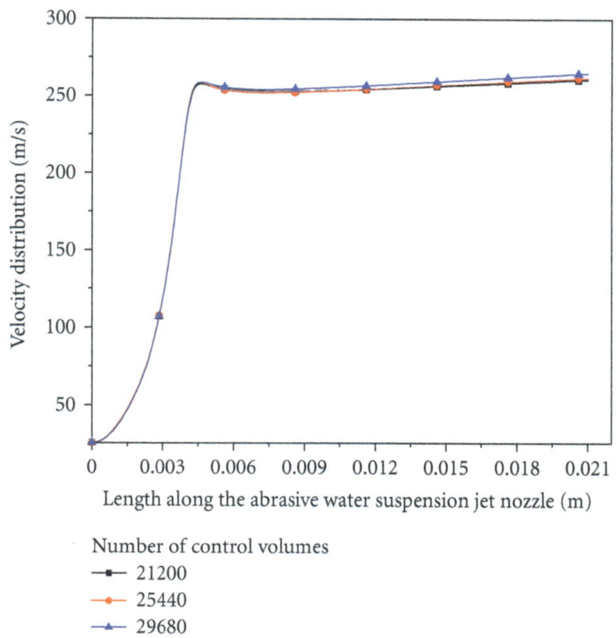

Figure 3: Result of grid independence test for the AWSJ nozzle.

3.2. Boundary Conditions and Operating Parameters. Suitable boundary conditions are imposed on the computational domain, as per the physics of the problem. Inlet boundary condition is specified by the operating pressure entering the nozzle. It is assumed that velocity at inlet is uniform across the cross-section. At the exit, static pressure of effluxing flow is taken to be zero (gauge), so that the computation would proceed by the relative pressure differences across the grid volumes for the entire domain of the flow. Wall boundary conditions are impressed to bound fluid and solid regions. In viscous flow models, as in the present case, velocity components at the wall are set to zero in accordance with the no-slip and impermeability conditions that exist on the wall boundary. The axis of the nozzle is used to reduce computational domain to an axisymmetric problem for which suitable boundary conditions are imposed, that is, the gradients of fluid properties are set to zero across the axis line. In the present numerical simulation, a mixture of water and suspension liquid is treated as phase 1 (suspension liquid) and abrasive as phase II (garnet). The input parameters used in the analysis are as shown in Table 1 taken from [4, 9].

4. Results and Discussion

4.1. Validation of the Numerical Model. The present model is benchmarked against the numerical work cited in [9]. The graph of the velocity distribution of one of the phases (liquid phase) has been calibrated in the present work as shown in

TABLE 1: Input parameters for numerical simulation.

Volume fraction	10%
Density of primary phase	998.2 kg/m^3
Density of secondary phase	2300 kg/m^3
Slip of phases	No slip

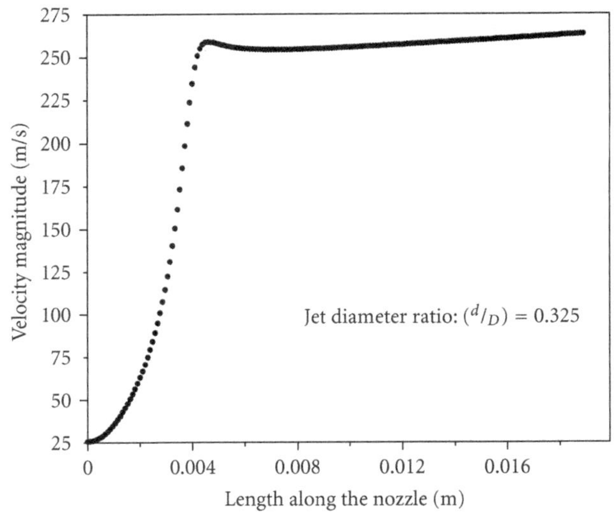

FIGURE 4: The velocity distribution along the length of the nozzle as given in [9].

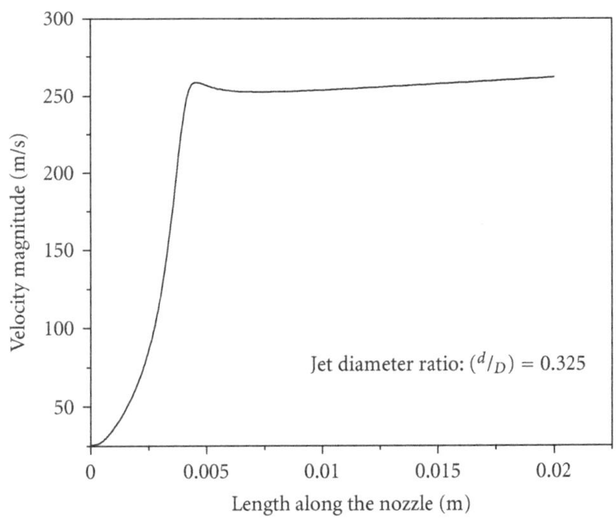

FIGURE 5: The velocity distribution along the length of the nozzle as per the present model.

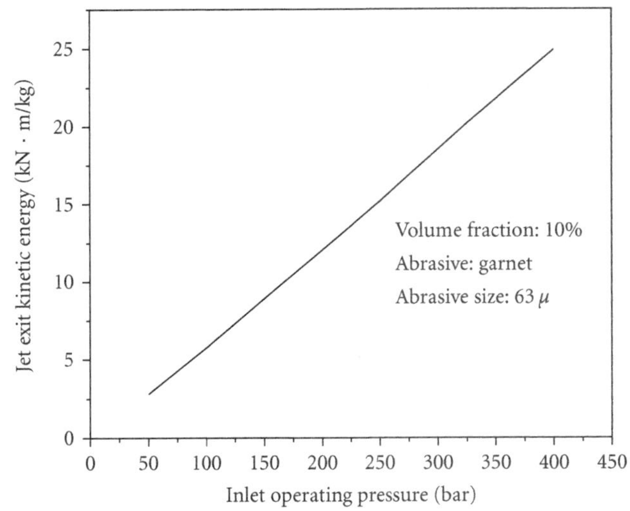

FIGURE 6: Variation of average exit kinetic energy of the jet with inlet operating pressure.

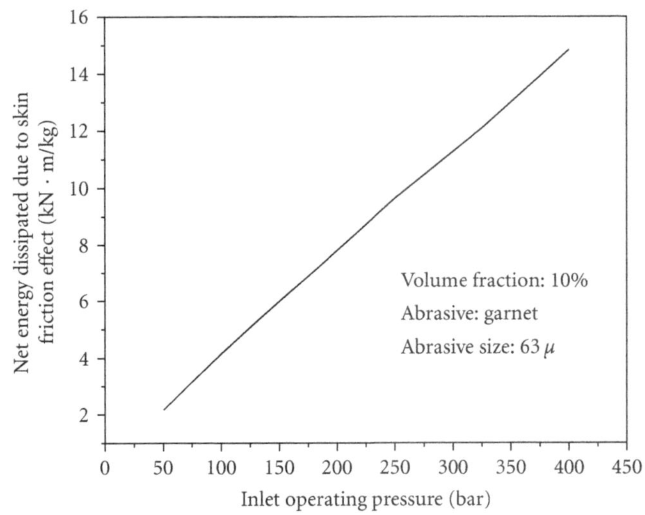

FIGURE 7: Plot of net energy dissipated due to skin friction effect with respect to inlet operating pressure.

Figure 5 with that of the work cited in the literature as shown in Figure 4. It is clear that there is good agreement between the two models as regards to the velocity distribution.

4.2. Effect of Inlet Pressure on Average Exit Kinetic Energy of the Jet. Figure 6 shows the variation of average exit kinetic energy of the jet corresponding to various inlet operating pressures. The average exit kinetic energy is computed from area-weighted average over the control volume at the exit of the nozzle. It is observed from Figure 7 that the jet exit kinetic energy is linearly proportional to corresponding inlet operating pressures. This is due to the fact that the inlet operating pressure energy manifests itself as proportional amount of kinetic energy at the exit of the nozzle with viscous shear stress dissipation representing the loss of energy.

Figure 7 shows net energy dissipated due to skin friction coefficient corresponding to various inlet operating pressure conditions. It is seen from the graph that the viscous shear force dissipation is also proportional to various inlet operating pressure conditions.

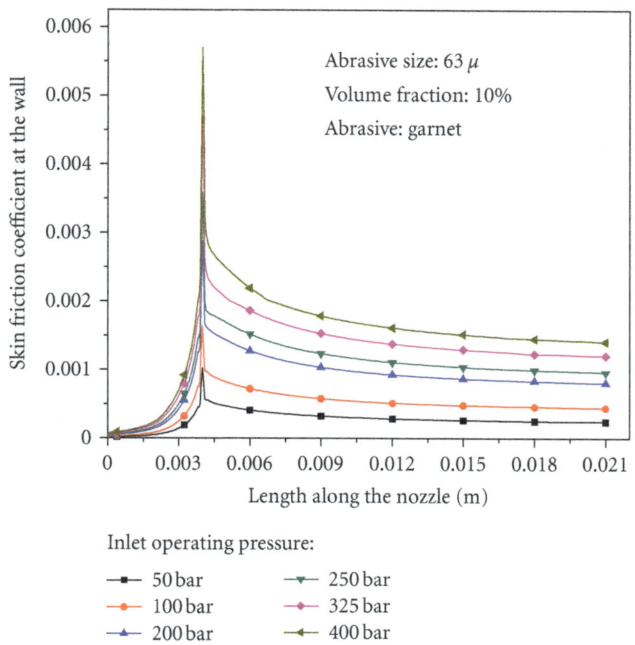

Figure 8: Distribution of skin friction coefficient along the length of the nozzle corresponding to various inlet operating pressure.

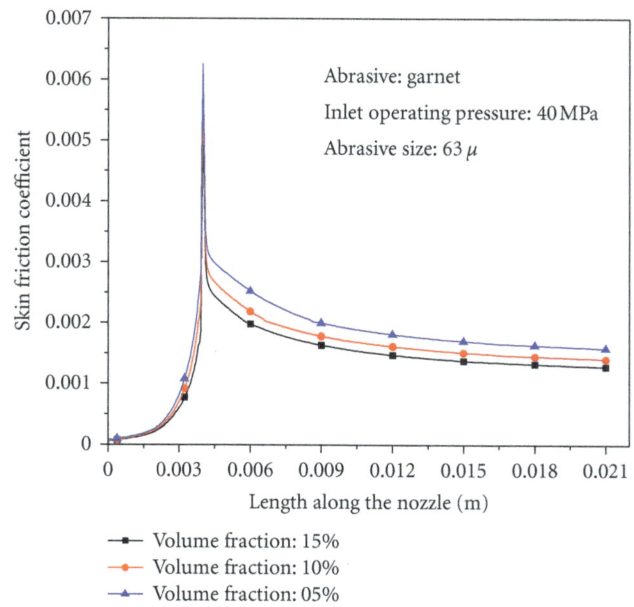

Figure 9: Effect of volume fraction on skin friction coefficient along the length of the nozzle.

The net energy dissipated due to viscous shear is defined as

$$\begin{bmatrix}\text{Net energy dissipated due}\\ \text{to viscous shear}\end{bmatrix} = \begin{bmatrix}\text{Area-weighted average of}\\ \text{inlet operating pressure energy}\end{bmatrix} - \begin{bmatrix}\text{Area-weighted average of}\\ \text{jet exit kinetic energy}\end{bmatrix}. \quad (10)$$

In addition, abrasive particles moving with flow cause severe wall shear that leads to erosion of the inside surface of the nozzle, resulting in decreased jet kinetic energy, and thereby affecting the performance of the nozzle for effective machining.

4.3. Effect of Inlet Operating Pressure on Skin Friction Coefficient. Figure 8 shows the variation of skin friction coefficient developed along the length of the nozzle corresponding to various inlet operating pressure conditions. It is seen that along the length of nozzle corresponding to various inlet operating pressure, the skin friction coefficient shows sharp increase with spike occurring followed by a short drop in its values exactly where there is a change in cross-section from conical to straight. It is well known that local skin friction coefficient is proportional to local Reynolds number of the flow or, in other words, the corresponding velocity of flow [21]. Thus, within conical portion of nozzle, velocity increases rapidly, and hence, skin friction coefficient spikes up as shown in the graph. As the fluid has not fully developed in the short conical tube when flow velocity changed rapidly at the critical section as explained above, it is found from the graph that skin friction coefficient shows sharp fall due to sudden contraction which corresponded to a loss of energy of flow. However, as the fluid moves further downstream into a well-developed turbulent flow profile all along the long focus tube, the skin friction coefficient does not show much variation and remains almost constant as shown in the graph.

But at higher inlet operating pressures and hence correspondingly at higher kinetic energy levels along the flow paths, the velocities are also high. As stated earlier, higher velocities contribute to higher skin friction coefficient at the wall. Hence, there is higher level of skin friction coefficient at higher pressure, and conversely at lower operating pressure, skin friction coefficient will be lower.

4.4. Effect of Volume Fraction of Abrasive Particle on Skin Friction Coefficient. Simulation has been carried out using garnet abrasive for volume fraction of 5%, 10%, and 15%. The effect of volume fraction on skin friction coefficient may be observed from Figures 9 and 10. It is clear from these figures that the skin friction coefficient increases with reduction in volume fraction of abrasive particles along the length of the AWSJ nozzle. Skin friction coefficient attains a peak value at critical section of nozzle where cross-section of the nozzle changes from conical to straight portion and thereafter remains almost constant along the straight focus section of the nozzle. This may be attributed to the fact that the velocity of the jet increases with corresponding decrease in volume fraction of abrasive particles, which results in an increase in the velocity gradient near to the wall of the nozzle. This leads to higher skin friction coefficient. Further, the skin friction coefficient shows a peak near the point corresponding to the critical region where flow changes from converging cone to cylindrical portion. This could be

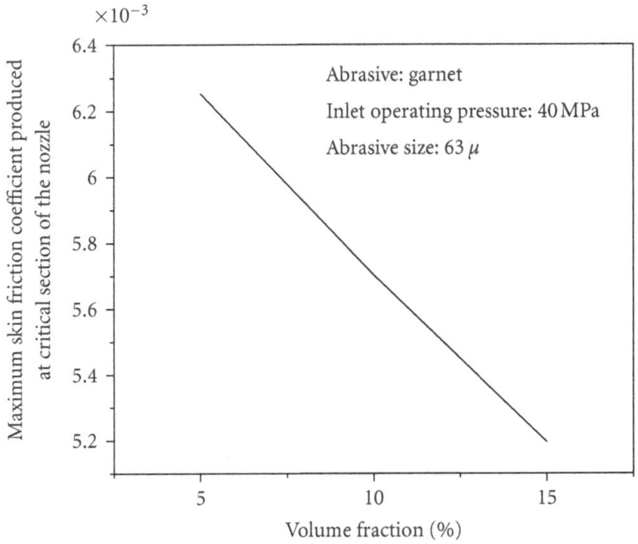

Figure 10: Effect of volume fraction on skin friction coefficient at the critical section of the nozzle.

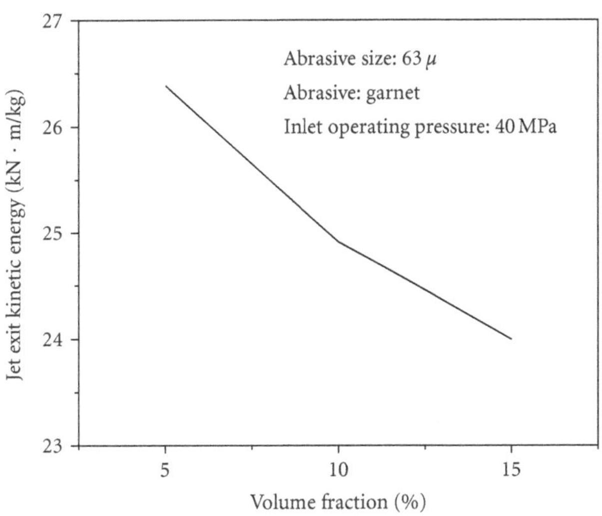

Figure 11: Effect of volume fraction on average exit kinetic energy of the jet.

explained by the fact that a sudden velocity change caused due to a higher-velocity gradient in the near wall region at the critical section produces a peak in the skin friction coefficient at that region. However, the skin friction coefficient seems to attain near constant value along the focus tube as there is no significant velocity change in the constant diameter focus tube till the exit of the nozzle for any given volume fraction.

It is observed from Figure 10 that increase in the volume fraction of abrasive in suspension mixture results in decrease in skin friction coefficient. This is due to the fact that increase in volume fraction of abrasive particles results in increased concentration of abrasive in the suspension mixture. There is increased fluid energy loss in transporting these abrasive particles, leading to a decrease in kinetic energy resulting in decreased jet velocity. This leads to reduce skin friction coefficient along the wall of the nozzle.

4.5. Effect of Volume Fraction of Abrasive Particles on Exit Kinetic Energy of the Jet. Figure 11 shows the variation of average exit kinetic energy of the jet corresponding to various volume fractions of abrasive particles. The average exit kinetic energy is computed from area-weighted average over the control volume at the exit of the nozzle. It is observed from Figure 11 that there is a marginal decrease in jet exit kinetic energy with increase in volume fraction of abrasive particles. As explained previously, the higher the concentration of abrasive particles in the fluid, the higher the fluid inertial resistance while transporting these abrasive particles leading to reduction in exit velocity. Hence, there is a corresponding decrease in jet kinetic energy with increase in volume fraction of abrasive particles.

5. Conclusion

The following conclusions are deduced from the above CFD simulation:

(i) increase in inlet operating pressure results in significant increase in the skin friction coefficient;

(ii) the skin friction coefficient approach peak values correspond to the sudden change in the flow passage geometry at the critical section;

(iii) increase in the inlet operating pressure results in linear increase in the average exit kinetic energy of jet;

(iv) the net energy dissipated due to skin friction coefficient is linearly proportion to the inlet operating pressure;

(v) for the same operating parameters, increase in the abrasive volume fraction results in significant decrease in the skin friction coefficient;

(vi) increase in volume fraction of abrasive particles results in decrease in exit kinetic energy of the jet.

Nomenclatures

d: Focus tube diameter (mm)
d_p: Abrasive particle size (μm)
D: Inlet diameter of nozzle (mm)
F_{lift}: Lift force (N)
F_s: External body force (N)
F_{vm}: Virtual mass force (N)
K: Momentum exchange coefficient
L: Average distance between particles (mm)
$\dot{m}$: Mass flow rate of mixture (m/s)
v: Velocity of phase (m/s)
α: Volume fraction of the phase
β: Particulate loading
ρ: Density of suspension mixture (kg/m)
L_s: Length of flow domain (mm)
V_s: Characteristic velocity of the flow (m/s)

τ_d: Particle response time (s)
t_s: System response time (s)
S_t: Stokes number.

Subscripts

p, q: Phases
l: Liquid phase
s: Solid phase.

References

[1] J. Folkes, "Waterjet-an innovative tool for manufacturing," *Journal of Materials Processing Technology*, vol. 209, no. 20, pp. 6181–6189, 2009.

[2] M. Hashish, "Observations of wear of abrasive-waterjet nozzle materials," *Journal of Tribology*, vol. 116, no. 3, pp. 439–444, 1994.

[3] T. Nguyen, D. K. Shanmugam, and J. Wang, "Effect of liquid properties on the stability of an abrasive waterjet," *International Journal of Machine Tools and Manufacture*, vol. 48, no. 10, pp. 1138–1147, 2008.

[4] D. Anjaiah and A. M. Chincholkar, "Cutting of glass using low pressure abrasive water suspension jet with the addition of zycoprint polymer," in *Proceedings of the 19th International Conference on Water Jetting*, pp. 105–119, BHR Group, October 2008.

[5] U. Anand and J. Katz, "Prevention of nozzle wear Abrasive Water Suspension Jets (AWSJ) using porous lubricated nozzles," *Journal of Tribology*, vol. 125, no. 1, pp. 168–180, 2003.

[6] E. Ness and R. Zibbell, "Abrasion and erosion of hard materials related to wear in the abrasive waterjet," *Wear*, vol. 196, no. 1-2, pp. 120–125, 1996.

[7] M. Nanduri, D. G. Taggart, and T. J. Kim, "The effects of system and geometric parameters on abrasive water jet nozzle wear," *International Journal of Machine Tools and Manufacture*, vol. 42, no. 5, pp. 615–623, 2002.

[8] V. N. Pi and N. Q. Tuan, "A study on nozzle wear modeling in abrasive waterjet cutting," *Advanced Materials Research*, vol. 76, pp. 345–350, 2009.

[9] G. Hu, W. Zhu, Y. Tao, and J. Yuan, "Numerical simulation and experimental study of liquid-solid two-phase flow in nozzle of DIA jet," in *Proceedings of the 6th IEEE International Conference on Industrial Informatics (INDIN '08)*, pp. 1700–1705, July 2008.

[10] S. V. Patankar and D. B. Spalding, "A calculation procedure for heat, mass and momentum transfer in three-dimensional parabolic flows," *International Journal of Heat and Mass Transfer*, vol. 15, no. 10, pp. 1787–1806, 1972.

[11] D. B. Spalding, "A general purpose computer program for multi-dimensional one- and two-phase flow," *Mathematics and Computers in Simulation*, vol. 23, no. 3, pp. 267–276, 1981.

[12] N. C. Markatos, "Modelling of two-phase transient flow and combustion of granular propellants," *International Journal of Multiphase Flow*, vol. 12, no. 6, pp. 913–933, 1986.

[13] M. K. Patel and N. C. Markatos, "Evaluation of eight discretization schemes for two-dimensional convection-diffusion equations," *International Journal for Numerical Methods in Fluids*, vol. 6, no. 3, pp. 129–154, 1986.

[14] N. C. Markatos and M. G. Psaltaki, "Modelling the behaviour of an oil spill in marine environments," in *Proceedings of the 6th International Conference on Engineering Computational Technology*, no. 47, Civil-Comp Press, Stirlingshire, UK, 2008.

[15] N. C. Markatos and D. Kirkcaldy, "Analysis and computation of three-dimensional, transient flow and combustion through granulated propellants," *International Journal of Heat and Mass Transfer*, vol. 26, no. 7, pp. 1037–1053, 1983.

[16] N. C. Markatos, "Computational fluid flow capabilities and software," *Ironmaking and Steelmaking*, vol. 16, no. 4, pp. 266–273, 1989.

[17] H. H. Liakos, M. K. Koukou, M. A. Founti, and N. C. Markatos, "Industrial-scale processing of granite surfaces by natural gas jet flames," *Applied Thermal Engineering*, vol. 22, no. 4, pp. 393–405, 2002.

[18] P. J. Oliveira and R. I. Issa, "Numerical aspects of an algorithm for the Eulerian simulation of two-phase flows," *International Journal for Numerical Methods in Fluids*, vol. 43, no. 10-11, pp. 1177–1198, 2003.

[19] A. K. Singhal and D. B. Spalding, "Predictions of two-dimensional boundary layers with the aid of the k-ε model of turbulence," *Computer Methods in Applied Mechanics and Engineering*, vol. 25, no. 3, pp. 365–383, 1981.

[20] Z. Shangxian, L. Yan, W. Quan, B. Discretization, and G. Space, "Track calculation and numerical simulation on particles in high pressure abrasive water jet nozzle," in *IEEE International Conference on Measuring Technology and Mechatronics Automation*, pp. 1039–1042, 2010.

[21] J. Ling, P. V. Skudarnov, C. X. Lin, and M. A. Ebadian, "Numerical investigations of liquid-solid slurry flows in a fully developed turbulent flow region," *International Journal of Heat and Fluid Flow*, vol. 24, no. 3, pp. 389–398, 2003.

[22] B. K. Gandhi, S. N. Singh, and V. Seshadri, "Study of the parametric dependence of erosion wear for the parallel flow of solid-liquid mixtures," *Tribology International*, vol. 32, no. 5, pp. 275–282, 1999.

[23] J. Chahed, V. Roig, and L. Masbernat, "Eulerian—eulerian two-fluid model for turbulent gas-liquid bubbly flows," *International Journal of Multiphase Flow*, vol. 29, no. 1, pp. 23–49, 2003.

[24] L. J. Graham, D. Lester, and J. Wu, "Slurry erosion in complex flows: experiment and CFD," in *Proceedings of the 7th International Conference on CFD in the Minerals and Process Industries*, pp. 1–6, CSIR, Melbourne, Australia, 2009.

[25] T. Mabrouki, K. Raissi, and A. Cornier, "Numerical simulation and experimental study of the interaction between a pure high-velocity waterjet and targets: contribution to investigate the decoating process," *Wear*, vol. 239, no. 2, pp. 260–273, 2000.

[26] F. J. Moraga, A. E. Larreteguy, D. A. Drew, and R. T. Lahey, "Assessment of turbulent dispersion models for bubbly flows in the low Stokes number limit," *International Journal of Multiphase Flow*, vol. 29, no. 4, pp. 655–673, 2003.

[27] *Fluent User's Guide*, vol. 3, Fluent Incorporated Publishers, Lebanon, 1998.

[28] J. D. Anderson, *Computational Fluid Dynamics: The Basics with Applications*, McGraw-Hill, 6th edition, 1995.

[29] D. Lakehal, "On the modelling of multiphase turbulent flows for environmental and hydrodynamic applications," *International Journal of Multiphase Flow*, vol. 28, no. 5, pp. 823–863, 2002.

[30] J. Aubin, D. F. Fletcher, and C. Xuereb, "Modeling turbulent flow in stirred tanks with CFD: the influence of the modeling approach, turbulence model and numerical scheme," *Experimental Thermal and Fluid Science*, vol. 28, no. 5, pp. 431–445, 2004.

[31] C. P. Chou, J. Y. Chen, J. Janicka, and E. Mastorakos, "Modeling of turbulent opposed-jet mixing flows with *j* model and second-order closure," *International Journal of Heat and Mass Transfer*, vol. 47, no. 5, pp. 1023–1035, 2004.

[32] F. Mashayek and R. V. R. Pandya, "Analytical description of particle/droplet-laden turbulent flows," *Progress in Energy and Combustion Science*, vol. 29, no. 4, pp. 329–378, 2003.

[33] J. A. Moríñigo and J. J. Salvá, "Numerical study of the start-up process in an optimized rocket nozzle," *Aerospace Science and Technology*, vol. 12, no. 6, pp. 485–489, 2008.

[34] K. Sengupta, K. Russell, W. J. Minkowycz, and F. Mashayek, "Numerical simulation data for assessment of particle-laden turbulent flow models," *International Journal of Heat and Mass Transfer*, vol. 48, no. 15, pp. 3035–3046, 2005.

[35] K. A. Hafez, O. A. Elsamni, and K. Y. Zakaria, "Numerical investigation of the fully developed turbulent flow over a moving wavy wall using k-ε turbulence model," *Alexandria Engineering Journal*, vol. 50, no. 2, pp. 145–162, 2011.

[36] M. J. Steinkamp, T. T. Clark, and F. H. Harlow, "Two-point description of two-fluid turbulent mixing-II. Numerical solutions and comparisons with experiments," *International Journal of Multiphase Flow*, vol. 25, no. 4, pp. 639–682, 1999.

[37] B. S. Nie, J. Q. Meng, and Z. F. Ji, "Numerical simulation on flow field of pre-mixed abrasive water jet nozzle," in *Proceedings of the Asia Simulation-7th International Conference on System Simulation and Scientific Computing (ICSC '08)*, pp. 247–251, October 2008.

[38] G. J. Brown, "Erosion prediction in slurry pipeline tee-junctions," *Applied Mathematical Modelling*, vol. 26, no. 2, pp. 155–170, 2002.

Permissions

The contributors of this book come from diverse backgrounds, making this book a truly international effort. This book will bring forth new frontiers with its revolutionizing research information and detailed analysis of the nascent developments around the world.

We would like to thank all the contributing authors for lending their expertise to make the book truly unique. They have played a crucial role in the development of this book. Without their invaluable contributions this book wouldn't have been possible. They have made vital efforts to compile up to date information on the varied aspects of this subject to make this book a valuable addition to the collection of many professionals and students.

This book was conceptualized with the vision of imparting up-to-date information and advanced data in this field. To ensure the same, a matchless editorial board was set up. Every individual on the board went through rigorous rounds of assessment to prove their worth. After which they invested a large part of their time researching and compiling the most relevant data for our readers. Conferences and sessions were held from time to time between the editorial board and the contributing authors to present the data in the most comprehensible form. The editorial team has worked tirelessly to provide valuable and valid information to help people across the globe.

Every chapter published in this book has been scrutinized by our experts. Their significance has been extensively debated. The topics covered herein carry significant findings which will fuel the growth of the discipline. They may even be implemented as practical applications or may be referred to as a beginning point for another development. Chapters in this book were first published by Hindawi Publishing Corporation; hereby published with permission under the Creative Commons Attribution License or equivalent.

The editorial board has been involved in producing this book since its inception. They have spent rigorous hours researching and exploring the diverse topics which have resulted in the successful publishing of this book. They have passed on their knowledge of decades through this book. To expedite this challenging task, the publisher supported the team at every step. A small team of assistant editors was also appointed to further simplify the editing procedure and attain best results for the readers.

Our editorial team has been hand-picked from every corner of the world. Their multi-ethnicity adds dynamic inputs to the discussions which result in innovative outcomes. These outcomes are then further discussed with the researchers and contributors who give their valuable feedback and opinion regarding the same. The feedback is then collaborated with the researches and they are edited in a comprehensive manner to aid the understanding of the subject.

Apart from the editorial board, the designing team has also invested a significant amount of their time in understanding the subject and creating the most relevant covers. They scrutinized every image to scout for the most suitable representation of the subject and create an appropriate cover for the book.

The publishing team has been involved in this book since its early stages. They were actively engaged in every process, be it collecting the data, connecting with the contributors or procuring relevant information. The team has been an ardent support to the editorial, designing and production team. Their endless efforts to recruit the best for this project, has resulted in the accomplishment of this book. They are a veteran in the field of academics and their pool of knowledge is as vast as their experience in printing. Their expertise and guidance has proved useful at every step. Their uncompromising quality standards have made this book an exceptional effort. Their encouragement from time to time has been an inspiration for everyone.

The publisher and the editorial board hope that this book will prove to be a valuable piece of knowledge for researchers, students, practitioners and scholars across the globe.

List of Contributors

Hiroshi Tanaka
Department of Mechanical Engineering, Kurume National College of Technology, Komorino, Kurume, Fukuoka 830-8555, Japan

Jun Huang, Zude Zhou and Juntao Chen
School of Mechanical and Electronic Engineering, Wuhan University of Technology, Wuhan, Hubei 430070, China

Litong Li, Xiwang Deng and Dongsheng Zhang
National Engineering Laboratory for Optical Fiber Sensors, Wuhan University of Technology, Wuhan, Hubei 430070, China

Essam L. Esmail
College of Engineering, University of Qadisiyah, Al Qadisiyah, Iraq

Yue Liu and Christopher W. Wilson
Department of Mechanical Engineering, University of Sheffield, Mappin Street, Sheffield S1 3JD, UK

M. McGarry and James G. Kohl
University of San Diego, Department of Engineering, 5998 Alcala Park, Loma Hall, San Diego, CA 92110, USA

Semin Chun, Young-Tark Kim and Tae-Hyoung Kim
School of Mechanical Engineering, College of Engineering, Chung-Ang University, 221 Heukseok-dong, Dongjak-gu, Seoul 156-756, Republic of Korea

Oronzio Manca, Sergio Nardini, Daniele Ricci and Salvatore Tamburrino
Dipartimento di Ingegneria Aerospaziale e Meccanica, Seconda Universit`a degli Studi di Napoli, Real Casa dell'Annunziata, Via Roma 29, 81031 Aversa (CE), Italy

Bo Yu
National Engineering Laboratory for Pipeline Safety, Beijing Key Laboratory of Urban Oil and Gas Distribution Technology, China University of Petroleum, Beijing 102249, China

Yi Wang
National Engineering Laboratory for Pipeline Safety, Beijing Key Laboratory of Urban Oil and Gas Distribution Technology, China University of Petroleum, Beijing 102249, China
Computational Transport Phenomena Laboratory, Division of Physical Science and Engineering, King Abdullah University of Science and Technology, Thuwal 23955-6900, Saudi Arabia

Shuyu Sun
Computational Transport Phenomena Laboratory, Division of Physical Science and Engineering, King Abdullah University of Science and Technology, Thuwal 23955-6900, Saudi Arabia

Sandhyarani Biswas
Department of Mechanical Engineering, NIT, Rourkela 769008, India

Amar Patnaik and Ritesh Kaundal
Department of Mechanical Engineering, NIT, Hamirpur 177005, India

S. M. Abdul Khader and B. Raghuvir Pai
Department of Mechanical Engineering, MIT, Manipal 576104, India

V. R. K. Rao and Anurag Ayachit
Department of Radio diagnosis & Imaging, KMC, Manipal 576104, India

S. Ganesh Kamath
Department of Cardiovascular & Thoracic Surgery, KMC, Manipal 576104, India

M. Reggio
Ecole Polytechnique, Universit´e de Montr´eal, C.P. 6079, Succ. "Centre Ville", Montr´eal, QC, Canada H3C 3A7
D´epartement de G´enie M´ecanique, ´Ecole Polytechnique, Universit´e de Montr´eal, C.P. 6079, Succ. "Centre Ville", Montr´eal, QC, Canada H3C 3A7

P. Vasseur
Ecole Polytechnique, Universit´e de Montr´eal, C.P. 6079, Succ. "Centre Ville", Montr´eal, QC, Canada
Laboratoire des Technologies Innovantes, Universit´e de Picardie Jules Verne d'Amiens, rue des facult´es, le Bailly, 80025 Amiens Cedex, France

J. Guiet
Ecole Polytechnique, Universit´e de Montr´eal, C.P. 6079, Succ. "Centre Ville", Montr´eal, QC, Canada

W. Limtrakarn and P. Boonmongkol
Department of Mechanical Engineering, Thammasat University, Pathumthani 12120, Thailand

A. Chompupoung
Royal Project Foundation, Chiang Mai 50200, Thailand

K. Rungprateepthaworn
National Electronics and Computer Technology Center, Klong Luang, Pathumthani 12120, Thailand

J. Kruenate
National Metal and Materials Technology Center, Klong Luang, Pathumthani 12120, Thailand

P. Dechaumphai
Department of Mechanical Engineering, Chulalongkorn University, Bangkok 10330, Thailand

C. H. Li, Z. L. Han, Q. Zhang and S. Wang
School of Mechanical Engineering, Qingdao Technological University, Qingdao 266033, China

B. Burak Özhan and Mehmet Pakdemirli
Department of Mechanical Engineering, Celal Bayar University, Muradiye, 45140 Manisa, Turkey
Applied Mathematics and Computation Center, Celal Bayar University, Muradiye, 45140 Manisa, Turkey

S. Ramana Babu
Department of Mechanical Engineering, G.V.P. College of Engineering, Visakhapatnam 530048, India

V. Ramachandra Raju
Department of Mechanical Engineering, JNTU College of Engineering, Vizianagaram 535003, India

K. Ramji
Department of Mechanical Engineering, AU College of Engineering, Visakhapatnam 530003, India

N. M. L. Huq
Department of Mechanical Engineering, Dhaka University of Engineering and Technology (DUET), Gazipur 1700, Bangladesh

A. M. Afsar
Department of Mechanical Engineering, Bangladesh University of Engineering and Technology (BUET), Dhaka 1000, Bangladesh

Peter Wachtel, Benn Gleason, J. David Musgraves and Kathleen Richardson
School of Materials Science & Engineering, COMSET, Clemson, SC 29634, USA

Peiman Mosaddegh
Department of Mechanical Engineering, Clemson University, Clemson, SC 29634, USA

Jiangping Wang
School of Mechanical Engineering, Xi'an Shiyou University, Xi'an, Shaanxi 710065, China

Harry O'Hanley, Jacopo Buongiorno, Thomas McKrell and Lin-wen Hu
Massachusetts Institute of Technology, Cambridge, MA 02139, USA

D. Deepak, D. Anjaiah, K. Vasudeva Karanth and N. Yagnesh Sharma
Department of Mechanical and Manufacturing Engineering, Manipal Institute of Technology, Manipal University, Manipal 576104, India